Molecular Biology of Disease Vectors

Molecular Biology of Disease Vectors

Editor

Michail Kotsyfakis

MDPI • Basel • Beijing • Wuhan • Barcelona • Belgrade • Manchester • Tokyo • Cluj • Tianjin

Editor
Michail Kotsyfakis
Institute of Parasitology
Biology Center of the Czech
Academy of Sciences
Budweis
Czech Republic

Editorial Office
MDPI
St. Alban-Anlage 66
4052 Basel, Switzerland

This is a reprint of articles from the Special Issue published online in the open access journal *International Journal of Molecular Sciences* (ISSN 1422-0067) (available at: www.mdpi.com/journal/ijms/special_issues/Disease_Vectors).

For citation purposes, cite each article independently as indicated on the article page online and as indicated below:

LastName, A.A.; LastName, B.B.; LastName, C.C. Article Title. *Journal Name* **Year**, *Volume Number*, Page Range.

ISBN 978-3-0365-6697-9 (Hbk)
ISBN 978-3-0365-6696-2 (PDF)

Contents

About the Editor

Michail Kotsyfakis

Michail Kotsyfakis obtained his Ph.D. in the topic of malaria/mosquitoes interaction in November 2004 from the Biology Department, University of Crete, Greece, under the supervision of Professor Kitsos Louis at the Institute of Molecular Biology and Biotechnology-FORTH, Greece. In 2005, he was hired at the NIAID/NIH, USA as a post-doctoral visiting fellow (Rockville, Maryland, USA) where he specialized in tick–vertebrate host interactions under the mentorship of Professor Jose'MC Ribeiro. In 2009, he established his laboratory (Laboratory of Genomics and Proteomics of Disease Vectors) in the Biology Center of the Czech Academy of Sciences in Budweis, Czechia as a recipient of the prestigious Jan Evangelista Purkyně fellowship of the Czech Academy of Sciences. In July 2016, he was appointed as a Full research Professor in the same Academic Institution.

His group has received support from EU, EMBO Short Term fellowships, Alexander von Humboldt Foundation and the Czech Science Foundation, among others. He has published many papers in prestigious journals, and he has delivered a number of invited talks worldwide. He is an Editorial Board member in five international journals. His current research focuses on using ticks as model organisms to address two fundamental research questions: how to improve our understanding of the basic molecular and biochemical mechanisms which mediate the disease transmission lifecycle of the specific disease vectors and how the mechanisms adopted by ticks to modulate vertebrate host physiological mechanisms may be translated into treatments for human diseases (e.g., related to vertebrate immunity and hemostasis dysfunctions).

International Journal of
Molecular Sciences

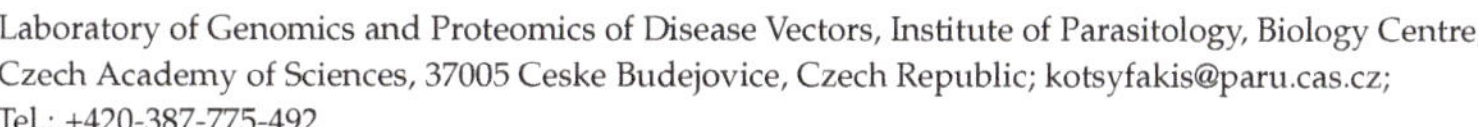

Editorial

Editorial: Special Issue on the "Molecular Biology of Disease Vectors"

Michail Kotsyfakis

Laboratory of Genomics and Proteomics of Disease Vectors, Institute of Parasitology, Biology Centre, Czech Academy of Sciences, 37005 Ceske Budejovice, Czech Republic; kotsyfakis@paru.cas.cz; Tel.: +420-387-775-492

Citation: Kotsyfakis, M. Editorial: Special Issue on the "Molecular Biology of Disease Vectors". *Int. J. Mol. Sci.* **2023**, *24*, 2881. https://doi.org/10.3390/ijms24032881

Received: 28 January 2023
Revised: 30 January 2023
Accepted: 31 January 2023
Published: 2 February 2023

Arthropod disease vectors not only transmit malaria but many other serious diseases, many of which are, to a greater or lesser degree, neglected. There is therefore a need for concerted efforts to develop new means with which to prevent disease transmission. In most cases, disease transmission involves a tripartite interaction between the arthropod disease vector, the vertebrate host, and the vector-borne pathogen. This Special Issue provides a compilation of the latest research in this area, together with up-to-date information on the molecular and biochemical events that mediate this tripartite interaction.

Two papers report on the application of systems biology approaches to hard ticks, which serve as important disease vectors in the Western world. *Ixodes ricinus* ticks are distributed across Europe and are important vectors of tick-borne encephalitis as well as Lyme disease. The last decade has seen intensive efforts in characterizing and understanding the roles of long non-coding RNAs (lncRNA) in health and disease, including in vector biology [1,2]. Here, Medina et al. present an exhaustive analysis of *I. ricinus* lncRNAs based on 131 RNA-seq datasets from three different BioProjects [3]. Their data analysis suggests that lncRNAs may act as sponges (scavengers/binders) of host miRNAs and thus exert diverse biological roles related to tick–host interactions in different tick tissues. Similarly, microRNAs (miRNAs) are a class of small non-coding RNAs involved in many biological processes, including in the immune pathways that control bacterial, parasitic, and viral infections. There are little data on differentially expressed miRNAs in the black-legged tick *Ixodes scapularis* after infection with *Borrelia burgdorferi*, the causative agent of Lyme disease in the United States. Kumar et al. used small RNA sequencing and qRT-PCR analyses to identify and validate differentially expressed *I. scapularis* salivary miRNAs [4], and in doing so provided new insights into the miRNAs expressed in *I. scapularis* salivary glands in addition to paving the way for their functional manipulation to prevent or treat *B. burgdorferi* infection.

Hard ticks feed for several days or weeks on their hosts, and their saliva contains thousands of polypeptides belonging to dozens of families, as identified by salivary transcriptomic analyses [5]. Mapping coding sequences to protein databases helps to identify putative secreted proteins and their potential functions at the tick–host interface, where pathogen transmission takes place. Mans et al. analyzed the classification of tick salivary proteins given recent developments in the Alphafold2/Dali programs, and in doing so detected novel protein families and revealed new insights that connected the structures and functions of tick salivary proteins [6]. Tick saliva is a rich source of antihemostatic, anti-inflammatory, and immunomodulatory molecules that actively help ticks to finish their blood meal [7,8]. Kotál et al. presented the functional and structural characterization of Iripin-8, a salivary serpin from *I. ricinus* [9]. The first crystal structure of a tick serpin in the native state demonstrated that Iripin-8 is a tick serpin with a conserved reactive center loop that possesses antihemostatic activity that may mediate interference with a host's innate immunity.

Host blood protein digestion, essential for tick development and reproduction, occurs in a tick's midgut digestive cells, driven by cathepsin proteases. Little is known about the regulation of the digestive proteolytic machinery in *I. ricinus*. In another paper from Kotál et al. the team present the functional and structural characterization of a novel cystatin-type protease inhibitor, mialostatin, from the *I. ricinus* midgut, which is likely to be involved in the regulation of gut-associated proteolytic pathways, making midgut cystatins promising targets for tick control strategies [10].

Arthropod-borne viruses, referred to collectively as arboviruses, infect millions of people worldwide each year and have the potential to cause severe disease. They are predominately transmitted to humans through the blood-feeding behavior of three main groups of biting arthropods: ticks, mosquitoes, and sandflies. The pathogens harbored by these blood-feeding arthropods are transferred to animal hosts through the deposition of virus-rich saliva into the skin. These infections sometimes become systemic and can lead to neuroinvasion as well as life-threatening viral encephalitis. Schneider et al. review the ways in which arthropod vectors influence viral pathogenesis [11]. They particularly emphasize how saliva and salivary gland extracts from the three dominant arbovirus vectors impact the trajectory of the cellular immune response to arbovirus infection in the skin.

The increase in the global disease burden and distribution of arboviruses is driven primarily by the spread of the two key invasive disease vectors, *Aedes aegypti* and *Ae. albopictus*, and by the spread of new and re-emerging viruses through international travel. Ahmed et al. present data supporting the emergence of *Ae. albopictus* in Sudan. This is a serious public health concern and argues for urgent improvements in vector surveillance as well as control through the implementation of integrated molecular xenosurveillance [12]. The threat of major arboviral diseases in the region underlines the need for the institutionalization of the One Health strategy for the prevention and control of future pandemics.

Cysteine-rich trypsin-inhibitor-like domain (TIL)-harboring proteins are broadly distributed in nature but remain understudied in vector mosquitoes. Tikhe et al. provide new insights into the role of a TIL-domain-containing protein of the arbovirus vector *Ae. Aegypti*, called cysteine-rich venom protein 379 (CRVP379) [13]. CRVP379 was previously shown to be essential for dengue virus infection in *Ae. aegypti* mosquitoes. Here, the importance of CRVP379 is demonstrated in *Ae. aegypti* reproductive biology, which makes this molecule an interesting candidate for the development of *Ae. aegypti* population control methods.

The PIWI-interacting RNA (piRNA) pathway, first characterized in *Drosophila*, provides an RNA interference (RNAi) mechanism with which to maintain the integrity of the germline genome by silencing transposable elements. *Ae. aegypti* mosquitoes exhibit an expanded repertoire of PIWI proteins involved in the piRNA pathway, suggesting their functional divergence. Williams et al. investigated the RNA-binding dynamics and subcellular localization of *Ae. aegypti* Piwi4 (AePiwi4), a PIWI protein involved in antiviral immunity and embryonic development [14]. Their experiments provide insights into the dynamic role played by AePiwi4 in RNAi and pave the way for future studies in order to understand PIWI interactions with diverse RNA populations.

The sole currently approved malaria vaccine targets the circumsporozoite protein that densely coats the surface of sporozoites, the parasite stage deposited into the skin of the mammalian host by infected mosquitoes; however, this vaccine only confers moderate protection against clinical disease in children, driving the search for novel candidates. Sá et al. demonstrate the importance of the membrane-associated erythrocyte binding-like protein (MAEBL) for infection by *Plasmodium* sporozoites [15]. Their data provide further insights into the role of MAEBL in sporozoite infectivity and may contribute to the design of future immune interventions.

Climate change is probably the foremost threat to human health in the 21st century. Climate directly impacts health through climatic extremes, air quality, rises in sea level, and multifaceted influences on food production systems as well as water resources. Climate also affects infectious diseases, which have played a significant role in human history—not least recently with the COVID-19 pandemic—impacting the rise and fall of civilizations in addition to facilitating the conquest of new territories [16]. Research into neglected vector-borne diseases must be a priority as the effects of climate change become ever more apparent. Together, the articles in this Special Issue highlight significant aspects of the physiology of different disease vectors, shed light on the molecular biology of a vector-borne pathogen, provide data on the tripartite interactions between vector-borne pathogens, disease vectors, and vertebrate hosts, and present evidence about the emergence of disease vectors in new geographical territories.

Funding: Michail Kotsyfakis received funding from the Grant Agency of the Czech Republic (grant 19-382 07247S) and ERD Funds, project CePaVip OPVVV (no. 384 CZ.02.1.01/0.0/0.0/16_019/0000759).

Conflicts of Interest: The author declares no conflict of interest.

References

1. Ahmad, P.; Bensaoud, C.; Mekki, I.; Rehman, M.U.; Kotsyfakis, M. Long Non-Coding RNAs and Their Potential Roles in the Vector-Host-Pathogen Triad. *Life* **2021**, *11*, 56. [CrossRef] [PubMed]
2. Bensaoud, C.; Martins, L.A.; Aounallah, H.; Hackenberg, M.; Kotsyfakis, M. Emerging Roles of Non-Coding RNAs in Vector-Borne Infections. *J. Cell Sci.* **2020**, *134*, jcs246744. [CrossRef] [PubMed]
3. Medina, J.M.; Abbas, M.N.; Bensaoud, C.; Hackenberg, M.; Kotsyfakis, M. Bioinformatic Analysis of *Ixodes ricinus* Long Non-Coding RNAs Predicts Their Binding Ability of Host miRNAs. *Int. J. Mol. Sci.* **2022**, *23*, 9761. [CrossRef] [PubMed]
4. Kumar, D.; Downs, L.P.; Embers, M.; Flynt, A.S.; Karim, S. Identification of microRNAs in the Lyme Disease Vector *Ixodes scapularis*. *Int. J. Mol. Sci.* **2022**, *23*, 5565. [CrossRef] [PubMed]
5. Martins, L.A.; Bensaoud, C.; Kotal, J.; Chmelar, J.; Kotsyfakis, M. Tick Salivary Gland Transcriptomics and Proteomics. *Parasite Immunol.* **2021**, *43*, e12807. [CrossRef] [PubMed]
6. Mans, B.J.; Andersen, J.F.; Ribeiro, J.M.C. A Deeper Insight into the Tick Salivary Protein Families under the Light of Alphafold2 and Dali: Introducing the TickSialoFam 2.0 Database. *Int. J. Mol. Sci.* **2022**, *23*, 5613. [CrossRef] [PubMed]
7. Jmel, M.A.; Voet, H.; Araujo, R.N.; Tirloni, L.; Sa-Nunes, A.; Kotsyfakis, M. Tick Salivary Kunitz-Type Inhibitors: Targeting Host Hemostasis and Immunity to Mediate Successful Blood Feeding. *Int. J. Mol. Sci.* **2023**, *24*, 1556. [CrossRef] [PubMed]
8. Jmel, M.A.; Aounallah, H.; Bensaoud, C.; Mekki, I.; Chmelar, J.; Faria, F.; M'Ghirbi, Y.; Kotsyfakis, M. Insights into the Role of Tick Salivary Protease Inhibitors during Ectoparasite-Host Crosstalk. *Int. J. Mol. Sci.* **2021**, *22*, 892. [CrossRef] [PubMed]
9. Kotal, J.; Polderdijk, S.G.I.; Langhansova, H.; Ederova, M.; Martins, L.A.; Berankova, Z.; Chlastakova, A.; Hajdusek, O.; Kotsyfakis, M.; Huntington, J.A.; et al. *Ixodes ricinus* Salivary Serpin Iripin-8 Inhibits the Intrinsic Pathway of Coagulation and Complement. *Int. J. Mol. Sci.* **2021**, *22*, 9480. [CrossRef] [PubMed]
10. Kotal, J.; Busa, M.; Urbanova, V.; Rezacova, P.; Chmelar, J.; Langhansova, H.; Sojka, D.; Mares, M.; Kotsyfakis, M. Mialostatin, a Novel Midgut Cystatin from *Ixodes ricinus* Ticks: Crystal Structure and Regulation of Host Blood Digestion. *Int. J. Mol. Sci.* **2021**, *22*, 5371. [CrossRef] [PubMed]
11. Schneider, C.A.; Calvo, E.; Peterson, K.E. Arboviruses: How Saliva Impacts the Journey from Vector to Host. *Int. J. Mol. Sci.* **2021**, *22*, 9173. [CrossRef] [PubMed]
12. Ahmed, A.; Abubakr, M.; Sami, H.; Mahdi, I.; Mohamed, N.S.; Zinsstag, J. The First Molecular Detection of *Aedes albopictus* in Sudan Associates with Increased Outbreaks of Chikungunya and Dengue. *Int. J. Mol. Sci.* **2022**, *23*, 1802. [CrossRef] [PubMed]
13. Tikhe, C.V.; Cardoso-Jaime, V.; Dong, S.; Rutkowski, N.; Dimopoulos, G. Trypsin-like Inhibitor Domain (TIL)-Harboring Protein Is Essential for *Aedes aegypti* Reproduction. *Int. J. Mol. Sci.* **2022**, *23*, 7736. [CrossRef] [PubMed]
14. Williams, A.E.; Shrivastava, G.; Gittis, A.G.; Ganesan, S.; Martin-Martin, I.; Valenzuela Leon, P.C.; Olson, K.E.; Calvo, E. *Aedes aegypti Piwi4* Structural Features Are Necessary for RNA Binding and Nuclear Localization. *Int. J. Mol. Sci.* **2021**, *22*, 2733. [CrossRef] [PubMed]
15. Sa, M.; Costa, D.M.; Teixeira, A.R.; Perez-Cabezas, B.; Formaglio, P.; Golba, S.; Sefiane-Djemaoune, H.; Amino, R.; Tavares, J. *MAEBL* Contributes to Plasmodium Sporozoite Adhesiveness. *Int. J. Mol. Sci.* **2022**, *23*, 5711. [CrossRef] [PubMed]
16. Caminade, C.; McIntyre, K.M.; Jones, A.E. Impact of Recent and Future Climate Change on Vector-Borne diseases. *Ann. N. Y. Acad. Sci.* **2019**, *1436*, 157–173. [CrossRef] [PubMed]

International Journal of
Molecular Sciences

MDPI

Article

Bioinformatic Analysis of *Ixodes ricinus* Long Non-Coding RNAs Predicts Their Binding Ability of Host miRNAs

José María Medina [1,2], Muhammad Nadeem Abbas [3], Chaima Bensaoud [4], Michael Hackenberg [1,2,*,†] and Michail Kotsyfakis [4,*,†]

1 Departamentode Genética, Facultad de Ciencias, Universidad de Granada, Campus de Fuentenueva s/n, 18071 Granada, Spain
2 Laboratorio de Bioinformática, Centro de Investigación Biomédica, PTS, Instituto de Biotecnología, Avda. del Conocimiento s/n, 18016 Granada, Spain
3 State Key Laboratory of Silkworm Genome Biology, Key Laboratory of Sericultural Biology and Genetic Breeding, Ministry of Agriculture, Southwest University, Chongqing 400715, China
4 Institute of Parasitology, Biology Centre, Czech Academy of Sciences, 37005 Ceske Budejovice, Czech Republic
* Correspondence: hackenberg@go.ugr.es (M.H.); mich_kotsyfakis@yahoo.com (M.K.)
† These authors contributed equally to this work.

Abstract: *Ixodes ricinus* ticks are distributed across Europe and are a vector of tick-borne diseases. Although *I. ricinus* transcriptome studies have focused exclusively on protein coding genes, the last decade witnessed a strong increase in long non-coding RNA (lncRNA) research and characterization. Here, we report for the first time an exhaustive analysis of these non-coding molecules in *I. ricinus* based on 131 RNA-seq datasets from three different BioProjects. Using this data, we obtained a consensus set of lncRNAs and showed that lncRNA expression is stable among different studies. While the length distribution of lncRNAs from the individual data sets is biased toward short length values, implying the existence of technical artefacts, the consensus lncRNAs show a more homogeneous distribution emphasizing the importance to incorporate data from different sources to generate a solid reference set of lncRNAs. KEGG enrichment analysis of host miRNAs putatively targeting lncRNAs upregulated upon feeding showed that these miRNAs are involved in several relevant functions for the tick-host interaction. The possibility that at least some tick lncRNAs act as host miRNA sponges was further explored by identifying lncRNAs with many target regions for a given host miRNA or sets of host miRNAs that consistently target lncRNAs together. Overall, our findings suggest that lncRNAs that may act as sponges have diverse biological roles related to the tick–host interaction in different tissues.

Keywords: *Ixodes ricinus*; ectoparasite-host interactions; host immunity; RNA-sequencing; lncRNA

Citation: Medina, J.M.; Abbas, M.N.; Bensaoud, C.; Hackenberg, M.; Kotsyfakis, M. Bioinformatic Analysis of *Ixodes ricinus* Long Non-Coding RNAs Predicts Their Binding Ability of Host miRNAs. *Int. J. Mol. Sci.* **2022**, *23*, 9761. https://doi.org/10.3390/ijms23179761

Academic Editor: Tamir Tuller

Received: 24 July 2022
Accepted: 24 August 2022
Published: 28 August 2022

1. Introduction

Ticks are an economically and medically important group of arthropods that feed on the blood of a wide range of vertebrate hosts. The attachment and insertion of tick mouthparts into the host activate the hemostatic processes of blood coagulation, vaso-constriction, and platelet aggregation to reduce blood loss [1,2]. Host innate immunity is induced in response to extended tick feeding and/or long-term tick exposure, and then adaptive immunity is triggered to further improve host defenses [3]. Ticks have adapted to regulate host immune reactions and hemostasis by secreting a complex cocktail of pharmaco-active substances into the host through their saliva [3,4]. There has been—and continues to be—extensive research to identify and functionally characterize the salivary gland or saliva components in ticks. For example, recent studies have tried to identify protective antigens from ticks through the transcriptomes and proteomes of the tick midgut and salivary glands, and the resulting sialomes and mialomes have been annotated and examined for the selection and characterization of antigenic candidates [5,6].

Recent advances in high-throughput sequencing technologies have provided a deeper understanding of non-coding (nc)RNAs—which range from micro (mi)RNAs to long non-coding (lnc)RNAs—in tick tissues [7]. LncRNAs are RNA strands longer than 200 nucleotides that are structurally identical to mRNAs but are not translated into proteins [8]. LncRNAs can regulate coding genes in various ways. For instance, they can modify gene expression through transcription factor activation or repression by binding and localization, chromatin remodeling, imprinting, and enhancer regulation [8]. They are involved in a number of regulatory processes, including post-transcriptional modulation, mRNA processing, and protein trafficking [8]. Chromatin remodeling is modified by lncRNA-induced histone methylation, which alters chromatin structure to increase or decrease DNA access to the transcriptional machinery [9]. LncRNAs participate in this critical role by acting as a protein scaffold with many binding domains that allow methylation and demethylation and inter-action with the target histone [9,10]. LncRNAs can also influence translation by binding to mRNAs, increasing or decreasing translation, or triggering mRNA degradation. Further-more, cytoplasmic lncRNAs can function as miRNA precursors or as "miRNA sponges": as a result of the competition between lncRNAs and mRNA for miRNA recognition elements, miRNA functions are lost [11]. In eukaryote pathophysiology, these miRNA sponges are widespread regulators of miRNA activity. They are also considered crucial regulators in ectoparasite-host crosstalk, such as during vector–host interactions [12].

LncRNAs have been studied most in humans and animals that are used in medical research [13]. However, there is very little research on lncRNAs in host–parasite interactions in arthropods. In the context of vector-host-pathogen interactions, recent reviews have high-lighted a role for small ncRNAs in trans-kingdom and inter-species communication [14,15]. So far, lncRNAs in arthropods have received little attention, and many of their functions are unclear. They could be expressed by the vector to counteract host defenses. It has been proposed that vector lncRNAs are transported in salivary exosomes to the host, where they act as host miRNA sponges to disrupt natural defense reactions; however, this hypothesis still requires validation [12].

The main objective of the present study was to elucidate the biological roles of lncRNAs in ectoparasite–host interactions, with bioinformatic means and through a consensus strategy, and particularly their potential role in tick–host interactions. Thus, we aimed to offer a new mechanistic understanding to inform the development of novel treatments for various diseases. *Ixodes ricinus* was used as a model ectoparasite, as it is a common vector of human and animal diseases, particularly in Europe.

2. Results

2.1. Summary of Transcriptome Assembly and lncRNA Isolation

By means of transcriptome assembly and lncRNA isolation, a total of 14,079, 677,278, and 358,278 lncRNAs were isolated from the transcriptome from midguts and salivary glands (MG-SG lncRNAs), salivary glands only (SG lncRNAs), and whole body samples (WB lncRNAs), respectively. The relatively low number of lncRNAs obtained from the transcriptome from midguts and salivary glands reflects the additional and more stringent steps taken in its transcriptome assembly process as explained in the Material and methods correspondent section.

To generate a consensus set we clustered the three sets of lncRNAs. The consensus lncRNA was defined as the longest transcript from a given cluster that could be detected in the MG-SG dataset (the most reliable lncRNA set, since the selection was more stringent) and at least in one of the other two transcriptomes. In this way we obtained a final set of 3118 consensus lncRNAs.

With respect to coding RNAs, we isolated 12,158, 55,825 and 47,472 coding sequences from the transcriptome from midguts and salivary glands (MG-SG coding RNAs), salivary glands only (SG coding RNAs), and whole body samples (WB coding RNAs), respectively. We followed the same protocol that we used for lncRNAs to get a final consensus set of 5659 coding RNAs.

2.2. Expression Patterns of lncRNAs in the Midgut and Salivary Glands under Different Feeding Treatments

Initially, four *I. ricinus* sets of lncRNAs were analyzed: consensus lncRNAs, MG-SG lncRNAs, SG lncRNAs, and WB lncRNAs. To assess the quality of these sets, we evaluated and visualized the length distributions (Figure 1). LncRNAs obtained from the individual projects show a strikingly different distribution compared to the consensus sequences. Short lncRNAs are clearly overrepresented (maximum around 250 nt) in the individual sets while the consensus shows a much more homogeneous distribution. We interpret this finding as the successful recovery of longer sequences by means of the consensus approach i.e., lncRNAs that are represented decently only in one of the individual sets.

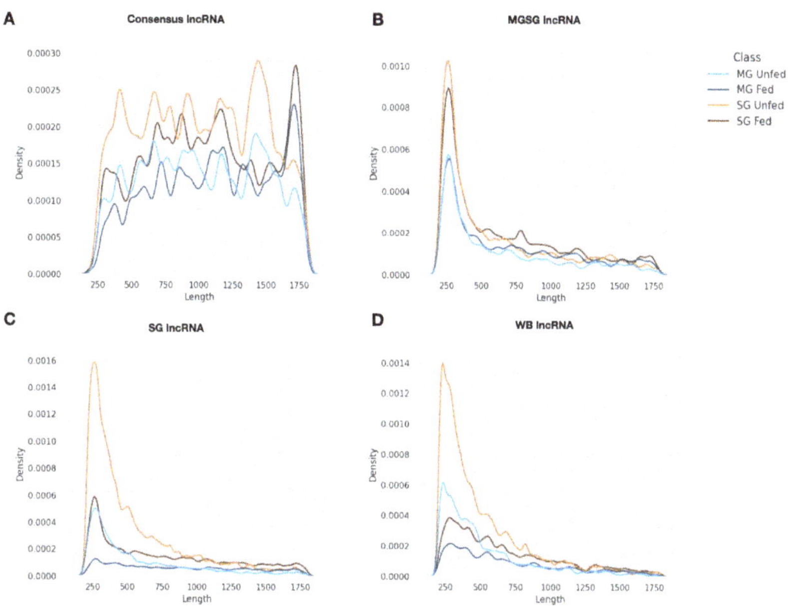

Figure 1. Length distribution of differentially expressed lncRNAs in the midgut and salivary glands of unfed and fed *Ixodes ricinus*. (**A**) Length distribution of significantly differentially expressed consensus lncRNAs (**A**), MG-SG lncRNAs (**B**), SG lncRNAs (**C**), and WB lncRNAs (**D**) upregulated in unfed and fed ticks in the midguts and salivary glands. Different lines represent lncRNAs upregulated or downregulated upon feeding in midgut (MG fed, MG unfed) and salivary glands (SG fed, SG unfed) with respect to midgut and salivary gland dataset (BioProject: PRJNA716261).

Therefore, we will base this study in the consensus transcriptome.

Since this is, to our best knowledge, the first time that lncRNAs are analyzed in *I. ricinus*, we first characterize the overall expression levels of lncRNAs in comparison to coding sequences. Figure 2 shows the mean normalized expression values for lncRNAs and coding sequences as a function of tissue or feeding state. The mean expression value of the consensus lncRNA was consistent among the different conditions representing approximately 30% of the total expression. We found some samples with strongly increased lncRNA expression, interestingly most of which belong to unfed or early feeding samples.

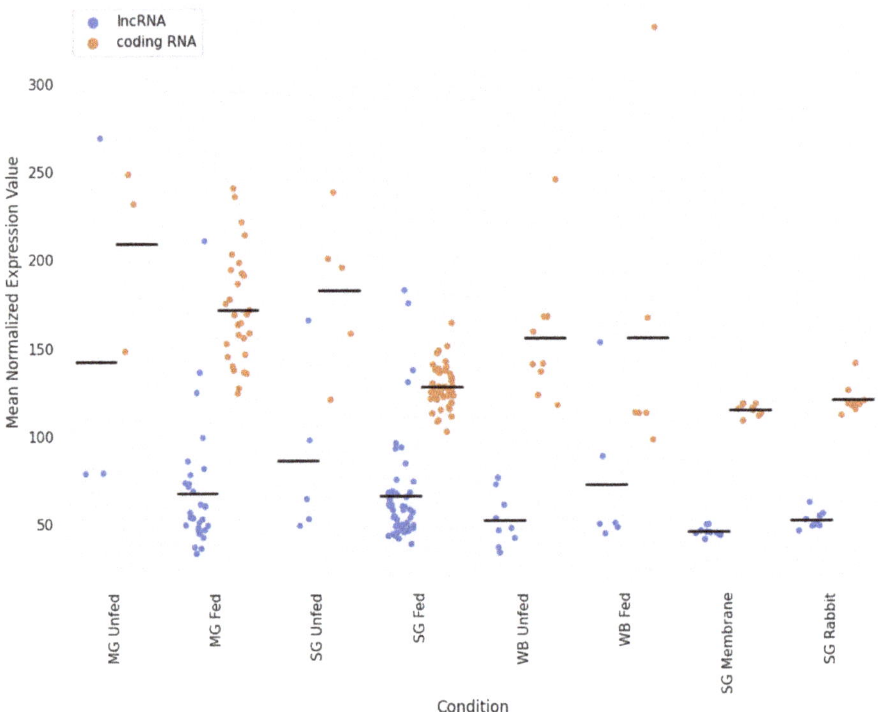

Figure 2. Expression analysis of lncRNAs and coding RNAs for different conditions of the samples retrieved for this study. The expression of lncRNAs and coding RNAs was analyzed for different feeding treatments: MG unfed, MG fed, SG unfed, and SG fed correspond to samples from midgut and salivary glands from ticks not fed or in the slow feeding phase, respectively (BioProject PRJNA716261); WB unfed and WB fed correspond to samples from whole body from ticks not fed or that were in the slow feeding phase, respectively (BioProject: PRJNA395009); lastly, SG Membrane and SG rabbits show the expression value for samples from salivary glands only that were fed on rabbits or from a membrane (BioProject: PRJNA312361). Mean values per condition are marked with a black line.

Additionally, to better understand the impact of lncRNA in ticks that are not feeding from a host and ticks that are actively feeding, we analyzed the differential expression of lncRNAs and coding RNAs between samples from unfed and fed ticks for both midgut and salivary glands (PRJNA716261). Differentially expressed (DE) lncRNAs and coding RNAs in midgut and salivary glands are displayed using volcano plots (Figure S1). For midgut, we found a total of 1110 DE lncRNAs (35.6% of the total of consensus lncRNAs) and 2727 DE coding RNAs (48.2% of the total of consensus coding RNAs). Regarding the salivary glands, 1311 lncRNAs (42% of the consensus lncRNAs) and 2965 coding RNAs (52.4% of the consensus coding RNAs) were differentially expressed between feeding treatments (Figure 3A). The DE lncRNAs and coding RNAs were then classified in two classes: (1) overexpressed when *I. ricinus* is not feeding and (2) overexpressed when *I. ricinus* is actively feeding. Here, we found that, for both tissues, the majority of DE lncRNAs and DE coding RNAs are overexpressed when *I. ricinus* is actively feeding (Figure 3B,C). Specifically, in midgut, 74.6% of the lncRNAs and 63% of coding RNAs were overexpressed in fed ticks whereas 59.9% and 68.4% of the lncRNAs and coding RNAs, respectively, were overexpressed in salivary glands of feeding ticks.

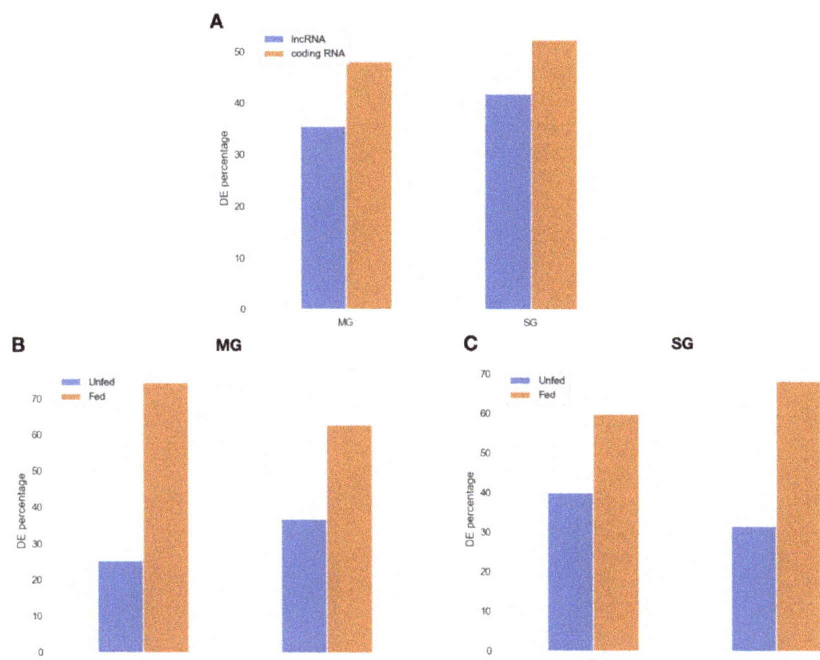

Figure 3. Statistics of differentially expressed lncRNAs and coding RNAs for midgut and salivary glands of *Ixodes ricinus*. (**A**) Percentage of lncRNAs and coding RNAs differentially expressed in midgut and salivary glands. (**B**) Distribution of DE lncRNAs and cRNAs between feeding threads in midgut. (**C**) Distribution of DE lncRNAs and cRNAs between feeding threads in salivary glands.

2.3. Target Prediction and KEGG Analysis of the Consensus Differentially Expressed lncRNAs

In order to explore putative roles of lncRNAs in tick feeding, we analyzed the putative functions of host miRNA targets. Applying miRNAconsTarget from sRNAtoolbox [16], we carried out a target prediction using human microRNAs from MirGeneDB v2.1 [17] and DE lncRNA sequences. We selected and functionally analyzed those miRNAs with the overall highest ratio of targets. We found that hsa-mir-5683, hsa-mir-29b-3p, and hsa-mir-154-3p are the miRNAs with the highest ratio of targets in feeding regulated lncRNAs. According to MirPath [18], these three miRNAs target genes are involved in focal adhesion and PI3K-Akt signaling pathway. They also target genes with a role in ECM-receptor interaction, fatty acids metabolism and biosynthesis, among other functions. Regarding the salivary glands, the miRNAs with the highest ratio of targets for lncRNAs upregulated in fed ticks were hsa-miR-4664-3p, hsa-miR-431-5p, and has-miR-28-3p. Among the KEGG pathways enriched for the genes they target, we identified Hippo signaling pathway, adherens junction, cell cycle, choline metabolism in cancer, and mucin type O-Glycan biosynthesis. All KEGG pathways enriched for miRNAs in both tissues, *p*-value, number of genes involved and number of miRNAs targeting genes involved in the KEGG pathway may be found in File S1.

2.4. Analysis of the Reproducibility of the Differential Expression of lncRNAs

Reproducibility is a key concept to increase the reliability of scientific results. We took advantage of having two different studies with samples from salivary glands of ticks fed on rabbits till 24 and 72 h obtaining differentially expressed lncRNAs and coding sequences for both studies (PRJNA716261, PRJNA312361). The distribution of differential expression in form of Volcano plots can be found in Figure S2.

We show that the percentage of differentially expressed sequences that coincide between the two studies were similar for lncRNAs and coding RNAs (23.97% and 21.16%, Figure 4A,B). Figure 4C–F shows a breakdown of these numbers into up and downregulated at 72 h compared to 24 h. Interestingly, the overlap between both studies is much higher for upregulated sequences (around 27%, Figure 4E,F) than for downregulated ones (8% for lncRNAs, Figure 4C,D).

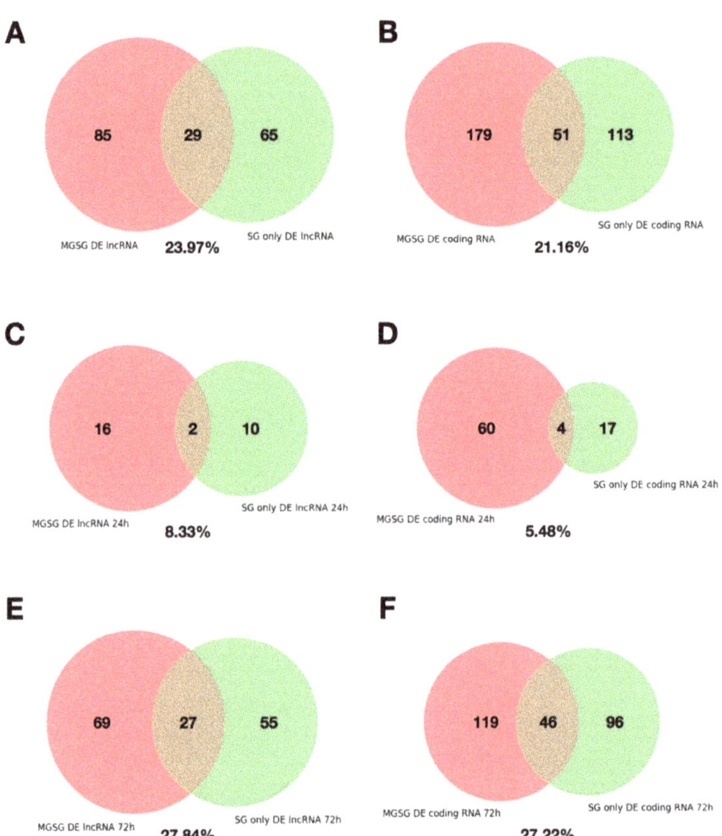

Figure 4. Analysis of the reproducibility of the differential expression of consensus lncRNAs and coding RNAs between the studies MGSG (PRJNA716261) and SG only (PRJNA312361) in salivary glands. The Venn diagrams shows the overlap between (**A**) DE lncRNAs in MGSG samples and DE lncRNAs in SG only samples, (**B**) DE coding RNAs in MGSG samples and DE coding RNAs in SG only samples, (**C**) DE lncRNAs downregulated at 72 h of feeding in MGSG and SG only samples (**D**) DE coding RNAs downregulated at 72 h of feeding in MGSG and SG only samples, (**E**) upregulated lncRNAs at 72 h of feeding in MGSG and SG only samples and (**F**) upregulated coding RNAs at 72 h of feeding in MGSG and SG only.

2.5. Prediction of Sponge Candidates and Functional Analysis

Cytoplasmic lncRNAs can serve as "miRNA sponges" competing with mRNA molecules for miRNA binding which can lead to the loss of miRNA function. To explore the possibility that tick lncRNAs play a role in the feeding process we explore two different sponge models: (i) high frequency targets and (ii) recurrent host miRNA combinations. We first performed target prediction analysis for all consensus lncRNAs as mentioned before. The results of target prediction can be found in File S2. As a negative control we carry out target prediction for a randomized set of consensus lncRNAs using shuffleseq (File S3). Target prediction for the consensus and randomized lncRNAs is shown for the top 200 miRNA-lncRNA combinations with the highest number of targets (Files S4 and S5). Most lncRNAs of this set contained more targets than its randomized version for the same miRNA (Files S6 and S7 for normalized target counts).

Next, we tried to extract the subset of lncRNAs with the highest possibility of being sponge candidates applying three different approaches: (i) highest number of targets for a given miRNA (Files S4 and S5), (ii) highest density of targets (Files S6 and S7), (iii) combination of microRNA target sites. For all, a negative control was performed using shuffled sequences. Significant microRNA combinations were detected using the Apriori algorithm. Thus, we identified sets of miRNAs that consistently target lncRNAs (File S8). To further understand the function of lncRNAs that may act as sponges, we selected the best sponge candidates from each analysis defined as the consensus lncRNA with the greatest number of targets for a miRNA, the consensus lncRNAs with the greatest number of targets per kb for a given miRNA, and three groups of consensus lncRNAs, with no common lncRNAs between them, that are significantly and consistently targeted by a group of host miRNAs.

A heat map was constructed for potential sponge candidates to analyze variability in lncRNA expression in unfed and fed midguts and salivary glands. The expression of many DE lncRNAs was strongly increased by feeding in the salivary glands compared with the midgut (Figure 5). To further understand the function of the sponge candidates, we used miTALOS v2.0 [19] to characterize human miRNAs or miRNA combinations which may be target by tick lncRNAs. These individual host miRNAs or miRNA combinations were found to control hemostasis-associated (e.g., glycosaminoglycan biosynthesis–heparan sulfate/heparin) or immune response (e.g., TGF-β signaling pathway, BMP signaling, TCF-dependent signaling in response to WNT, and PDGF pathway)-associated pathways in humans. In addition, miRNAs or miRNA combinations also had targets on a variable number of genes in their corresponding pathways (File S9), suggesting potential functions of lncRNAs in tick feeding and ectoparasite–host interactions.

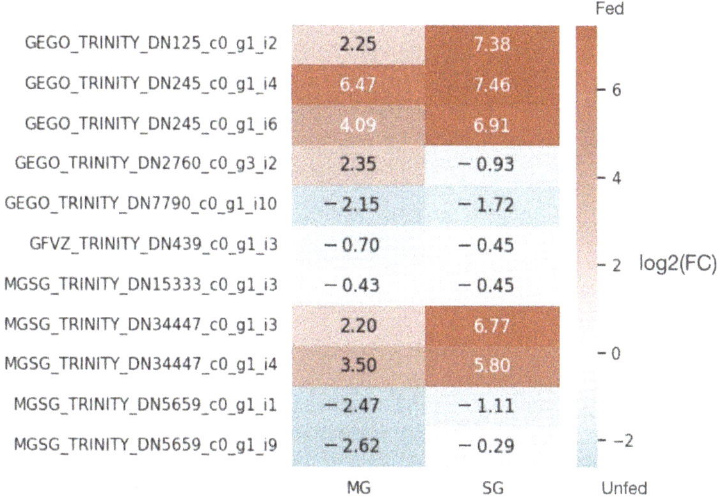

Figure 5. Differential expression of the sponge candidates in the midgut and salivary glands of unfed and fed *Ixodes ricinus*. The heatmap displays the logarithm fold change to base 2 between the unfed and fed *I. ricinus* for each sponge candidate in both the midgut and salivary glands. A logarithm of fold change to base 2 greater than 0 indicates that the sponge candidate is upregulated in fed ticks, while a logarithm of fold change to base 2 less than 0 indicates that the sponge candidate is upregulated in unfed ticks.

3. Discussion

The molecular mechanisms of action and physiological impacts of lncRNAs are an intense focus of applied human and animal research [13]. However, there is currently very little literature on the identification and functional analysis of lncRNAs in tick–host interactions, limiting our understanding of lncRNAs and their involvement in ectoparasite–host interactions. Therefore, to evaluate the potential regulatory role of lncRNAs in *I. ricinus*–host interactions, we investigated the differential expression of lncRNAs in the midguts and salivary glands of unfed and fed *I. ricinus* ticks. Previous studies have shown that lncRNAs are involved in different physiological processes such as mRNA processing, post-transcriptional regulation, and protein trafficking [8,9]. In addition, cytoplasmic lncRNAs can also act as miRNA precursors or as "miRNA sponges", where they compete with mRNAs for miRNA recognition elements, causing the loss of miRNA function [11]. For this reason, we performed transcriptome assembly, lncRNA identification, differential expression analysis, and functional in silico analysis to identify lncRNAs in the midguts and salivary glands of *I. ricinus* ticks under different feeding treatments (unfed and fed) with potential functions related to tick-host interactions. We focused on midgut and salivary gland-related differentially expressed lncRNAs and showed that the production of lncRNAs in the salivary gland may play a role in the ectoparasite–host interaction.

Ixodes ricinus and other Ixodid ticks are unique in their long period of attachment to the host [20,21]). The tick salivary gland mediates various activities that ensure the tick's biological success during feeding. The tick salivary gland produces biologically active molecules that facilitate blood meal acquisition [4]. Therefore, the composition of tick saliva requires careful and comprehensive molecular resolution to understand the complex feeding biology of ticks. Such resolution could underpin the discovery of pharmacologically active molecules of clinical interest. So far, many transcript and protein profiling studies of tick saliva have been conducted at different developmental stages, sexes, and feeding conditions in different tick species [22–24]. While many studies have attempted to determine the complex nature of saliva or the salivary gland, this is the first study to investigate ncRNAs in *I. ricinus*. We identified many potential lncRNAs from *I. ricinus* and ensured that these potential lncRNAs had the most reliability and reproducibility we could afford by analyzing different transcriptome projects. Initial analysis about lncRNAs expression showed that the expression level of lncRNAs compared to coding RNAs is stable among studies. Additionally, a proof about the reproducibility of differential expression between studies showed that the reproducibility of lncRNAs was similar to the reproducibility of coding RNAs. Next, we analyze the differential expression of the consensus lncRNAs for different feeding treatments in midgut and salivary glands and compared them with the differential expression calculated for a consensus set of coding RNAs. The results obtained showed that even though the percentage of consensus lncRNAs being DE is lower than the one for coding RNAs in both midgut and salivary glands, most of the DE lncRNAs were upregulated in fed ticks, especially in midgut. This suggests that the predicted consensus lncRNAs have a role in the feeding of *I. ricinus* in both midgut and salivary glands. This idea is supported by the KEGG enrichment analysis of miRNAs that target lncRNAs upregulated in fed ticks. This analysis showed that lncRNAs may disrupt the function of host miRNAs that may be harmful for the tick or beneficial for the host immune response. Specifically, the three miRNAs with the highest ratio of targets for lncRNAs upregulated in fed ticks were involved in processes as focal adhesion and PI3K-Akt signaling pathway. Focal adhesion has been related to epithelial motility and mucosal healing of gut due to its involvement in cell motility, proliferation, and survival [25]. Additionally, PI3K-Akt signaling pathway has been reported to control the epithelial proliferation of the gut [26]. Although further research is needed in this field, we found that lncRNAs in midgut are targeted by miRNAs involved in crucial functions for the healing and correct operation of the midgut. Regarding the salivary glands, we identified interesting KEGG pathways as Hippo signaling pathway, which has been reported to maintain the immune homeostasis in immune cells [27]; adherens junction, which is involved in the innate immune response to endotoxins [28]; and biosynthesis of mucin type O-Glycans, which are involved in the inflammatory response [29]. Here, we found that these miRNAs are involved in the immune response of the host, so disrupting its function may be beneficial for the tick interaction with the vertebrate host. All what was discussed here may establish future lines of research in the field of tick–host interactions and its treatment and prevention.

Target and sponge candidate prediction analysis showed that the salivary glands contain a large variety of lncRNAs that may function in controlling host homeostasis and immune responses, most likely by sponging host miRNAs. Ahmad et al. [12] hypothesized that vector lncRNAs are transported to the host in salivary exosomes, where they act as "sponges" of host miRNAs to disrupt natural defense reactions; however, this hypothesis still requires experimental validation. Furthermore, we analyzed human miRNAs or miRNA combinations to understand tick lncRNA functions using a bioinformatics tool. In doing so, we found that miRNA or miRNA combinations likely to be sponged by tick lncRNAs regulate a variety of genes involved in host hemostasis and immune pathways. We observed that the targeted host miRNAs significantly affected the hemostasis-associated pathway, glycosaminoglycan biosynthesis–heparan sulfate/heparin. Heparin is crucial for the catalysis of antithrombin III (a plasma enzyme), enhancing its activity in vertebrates. Antithrombin III inactivates some activated serine proteases of the coagulation cascade, most importantly activated thrombin and factor X [30]. Furthermore, several immune pathways were influenced by the miRNAs, for example, transforming growth factor-beta (TGF-β) signaling, which is a potent cytokine regulator with various effects on hemopoietic cells. The key biological role of TGF-β in the immune system is to maintain tolerance by modulating lymphocyte proliferation, differentiation, and survival [31,32]. WNT is another immune signaling pathway that might be regulated by tick sponge candidates. WNT ligands bind Frizzled and lipoprotein receptor-related protein to induce the canonical WNT signaling pathway, which inactivates the destruction complex, stabilizes, and promotes the nuclear translocation of β-catenin, and subsequently activates T-cell factor/lymphoid enhancing factor (TCF/LEF)-dependent transcription [33,34]. Thus, ticks are likely to "cargo" lncRNAs into the host body via saliva or salivary exosomes, which may then sponge host miRNAs or miRNA combinations important for host hemostasis and the innate and adaptive immune responses, ultimately facilitating tick feeding on the host.

Here, we analyzed lncRNAs and their expression in fed and unfed pathogen-free *I. ricinus* ticks and showed evidence of lncRNAs having an important role in *I. ricinus*–host interaction. This study leads to new research about lncRNAs in ectoparasites and to the comparison of lncRNA behavior between them. Additionally, differences in the expression of lncRNAs between pathogen-infected and pathogen-free ectoparasites has been previously reported in *Ae. aegypti* and *Ae. albopictus* [35,36]. Therefore, a comparison between lncRNA expression in ticks infected by pathogens and pathogen-free ticks and the integration of this analysis with strategies for prediction of sponge candidates may help the research community to have a deeper understanding in the role of lncRNAs in the tick-host interaction and their pathogenicity. Ultimately, this study, for the first time, provides in silico evidence on the putative role of tick lncRNAs as sponge candidates. Future studies will focus on evaluating the biological functions of these sponge candidates, in particular in the context of the ectoparasite–host interaction.

4. Materials and Methods

4.1. Transcriptome Assembly, Expression Analysis and Consensus lncRNA Isolation

This study used RNA-seq data from *Ixodes ricinus* available in NCBI under the BioProject accessions PRJNA716261, PRJNA312361, and PRJNA395009 [37–39]. RNA-seq data from BioProject PRJNA716261 consisted of 88 *Ixodes ricinus* samples exposed to different feeding treatments from two different tissues: 55 samples from the salivary glands, and 33 from the midgut. BioProject PRJNA31236 contained RNA-seq data from 18 salivary gland samples from *Ixodes ricinus*, while BioProject PRJNA395009 contained RNA-seq data from 15 whole body samples of *Ixodes ricinus*. All raw RNA-seq reads were quality filtered and trimmed using Trim Galore [40]. The quality of raw and trimmed reads was determined using FastQC [41]. De novo RNA-seq assemblies from the different BioProjects were constructed separately using Trinity v2.8.6 [42], with a maximum memory usage of 300 gigabytes. Since the RNA-seq data from BioProject PRJNA716261 originated from two different tissues, some additional steps were taken to obtain reliable transcriptomes. First, using Trinity, we created de novo assemblies of the midgut and salivary glands separately. These assemblies were condensed and clustered using CD-HIT v4.8.1 [43]. Contigs with at least 98% identity and at least 80% alignment coverage for the shorter sequences were clustered. Since this study focused on analyzing lncRNAs under different feeding treatments, we calculated the contig expression of all the assemblies for the fed and unfed conditions for both midgut and salivary glands using the RNA-seq data from BioProject PRJNA716261. The contig expression was analyzed using RSEM v1.3.3 [44] and Bowtie2 [45]. Since the dataset coming from midguts and salivary glands is the one with the greatest number of samples (88), which were isolated from single individuals which were siblings, thus reducing genetic variation [37], we decided to give it a major role in the consensus strategy and to make additional and more stringent steps to get from it the most reliable set of lncRNAs and coding RNAs that we could afford. Therefore, to remove possible assembly artifacts from the transcriptome assemblies from the midgut and salivary glands, we applied an expression filter to the contigs. The samples in this BioProject were unfed or fed for different time periods during the slow-feeding phase of the tick. To remain in further analyses, a contig from the transcriptome assembly from the midgut and salivary glands needed to have at least five FPKM in all the samples belonging to the unfed condition or in all the samples that were fed to a given time point.

For every assembled transcriptome, contigs with open reading frame (ORFs) longer than 50 amino acids were isolated using TransDecoder v5.5.0. The ORFs were then aligned to the Swiss-Prot [46] protein database using BlastP v2.10.1+ [47], with an E-value cut-off of 10^{-5}. Contigs with ORFs aligning with the protein database were considered coding RNAs.

To obtain lncRNAs, ncRNAs between 200 bp and 1800 bp were isolated. To avoid possible assembly artifacts and to obtain reliable predicted lncRNAs, consensus lncRNAs were obtained. Using CD-HIT [43], we clustered three sets of lncRNAs: (1) lncRNAs obtained from the midgut and salivary gland (MG-SG lncRNAs); (2) lncRNAs obtained from the salivary gland (SG lncRNAs), and (3) lncRNAs obtained from whole tick bodies (WB lncRNAs). Finally, we retrieved the best representative for each cluster with at least one lncRNA from MG-SG lncRNAs, as this was the most reliable lncRNA set through the application of a stringent expression filter, and at least one lncRNA from the salivary gland or whole body datasets. Thus, we obtained a final set of consensus lncRNAs. For purposes of comparison, we also obtained a consensus set of coding RNAs by following the steps mentioned before. All lncRNAs and cRNAs identified in this study can be downloaded from the Downloads section of IxoriDB [48].

4.2. Differential Expression Analysis

Infecting and feeding on a host are crucial tick activities during which the tick needs to avoid the host immune system [49]. We hypothesized that ticks produce lncRNAs that act as sponges, trapping miRNAs associated with the host immune system or other processes harmful to the tick. To understand this phenomenon, we analyzed the differential expression (DE) of lncRNAs in the unfed and fed states.

To obtain DE under different feeding conditions, we normalized the expected counts provided by RSEM using the TMM method [50]. Then, DE analysis was executed using EdgeR v3.36.0 [51]. Basic statistics and length distribution analysis were conducted for DE lncRNAs and DE coding RNAs using a homemade Python script.

4.3. Target Prediction

The set of mature *H. sapiens* miRNAs was obtained using MirGeneDB v2.1 [17]. Target prediction for the consensus differentially expressed lncRNAs and mature miRNAs from *H. sapiens* was carried out using miRNAconsTarget from sRNAtoolbox [16]. Specifically, the SEED, MIRANDA, and TS (TargetSpy) algorithms were used, and only those differentially expressed lncRNAs confirmed to have targets for miRNAs were retained for further analysis.

4.4. Functional In Silico Analysis of Differentially Expressed lncRNAs

To determine the functional variation between consensus differentially expressed lncRNAs in the unfed and fed conditions, we performed functional in silico analysis based on the miRNAs targeting consensus differentially expressed lncRNAs. For this purpose, we retrieved the set of mature miRNAs from *Homo sapiens* using MirGeneDB v2.1 [17]. Target prediction provided us with the number of targets that a miRNA has for the different sets of DE lncRNAs. We normalized the number of targets in each set using the miRNA with the maximum number of targets in each set. We then calculated the ratio of normalized number of targets for lncRNAs upregulated in fed ticks by dividing it by the number of targets for lncRNAs upregulated in unfed ticks for both midgut and salivary glands. We selected three miRNAs for each tissue with the highest ratio and used them to perform KEGG pathway enrichment analysis using the information provided in DIANA-miRPath v3.0 [18].

4.5. Sponge Candidates and Functional In Silico Analysis

The final objective of this study was to predict lncRNAs from the consensus set of lncRNAs with the highest potential to be sponge candidates. We applied three different approaches to find the best sponge candidates. The first was based on determining which lncRNAs had the highest number of miRNA targets. The second aimed to maximize the number of targets per kb for a miRNA in a lncRNA and was used to normalize the number of targets with the length of the lncRNA. Finally, rather than focusing on lncRNAs with a high number of targets for a single miRNA, we focused on lncRNAs with multiple miRNAs targets. We used the Apriori algorithm to analyze this phenomenon. This algorithm identifies combinations of miRNAs that together target a set of lncRNAs in a statistically significant way. Thus, we obtained sets of miRNAs that consistently targeted lncRNAs together. Lastly, we took as the best sponge candidates the consensus lncRNAs with the greatest number of targets for a miRNA, the greatest number of targets per kb for a given miRNA, and three groups of consensus lncRNAs with no common lncRNAs between them significantly and consistently targeted by a group of miRNAs.

As described above, our functional in silico analysis of lncRNAs is based on the miRNAs that target them. We assigned a function to the miRNA and sponge candidate combinations using miTALOS v2.0 [19]. Using this web tool, we acquired information about the pathways in which the miRNAs participate and the location of the genes they target in those pathways.

5. Conclusions

Here, we analyzed and detected lncRNAs through consensus methods from publicly available RNA-seq data from the salivary glands and midguts (BioProject: PRJNA716261), salivary glands alone (BioProject: PRJNA312361), and whole bodies (BioProject: PR-JNA395009), showing stable overall expression and reproducible response upon feeding challenges. Overall, our bioinformatics analysis suggests that a subset of lncRNAs might act as host microRNA sponges with specific functions in tick–host interactions. We provide here a consensus set of lncRNAs that opens the door for future functional assays to obtain a deeper understanding of the function of these lncRNAs at the tick-vertebrate host interface, which could offer a new mechanistic understanding and may revolutionize the development of novel treatments for tick-borne diseases.

Supplementary Materials: The following supporting information can be downloaded at: https:// www.mdpi.com/article/10.3390/ijms23179761/s1.

Author Contributions: J.M.M. performed data analysis, web development and draft the paper. M.N.A. performed data analysis and drafted the paper. C.B. performed data analysis. M.K. and M.H. designed the project and drafted the paper. All authors have read and agreed to the published version of the manuscript.

Funding: C.B. received the European Union within ESIF in frame of Operational Programme Research, Development and Education (Project No. CZ.02.2.69/0.0/0.0/20_079/0017809). M.H. and J.M.M. received: A-BIO-481-UGR18, financed by the FEDER (Fondo Europeo De Desarrollo Regional–European Regional Development Fund). M.K. received funding from the Grant Agency of the Czech Republic (grant 19-382 07247S) and ERD Funds, project CePaVip OPVVV (No. 384 CZ.02.1.01/0.0/0.0/16_019/0000759). The funders had no role in the design, data collection and analysis, decision to publish, or preparation of the manuscript.

Institutional Review Board Statement: Not applicable. All the presented data were based on available sequence reads and thus no animal experiments were carried out for this publication.

Data Availability Statement: The data presented in this study are available in the article and Supplementary Material. The RNA-seq data used in this study is available in NCBI under the BioProject accessions PRJNA716261, PRJNA312361, and PRJNA395009. The lncRNA data isolated in this study is available in the "Downloads" section of IxoriDB [48].

Conflicts of Interest: The authors declare no conflict of interest.

References

1. Chmelar, J.; Calvo, E.; Pedra, J.H.F.; Francischetti, I.M.B.; Kotsyfakis, M. Tick Salivary Secretion as a Source of Antihemostatics. *J. Proteom.* **2012**, *75*, 3842. [CrossRef] [PubMed]
2. Abbas, M.N.; Chlastáková, A.; Jmel, M.A.; Iliaki-Giannakoudaki, E.; Chmelař, J.; Kotsyfakis, M. Serpins in Tick Physiology and Tick-Host Interaction. *Front. Cell. Infect. Microbiol.* **2022**, *12*, 892770. [CrossRef] [PubMed]
3. Kotál, J.; Langhansová, H.; Lieskovská, J.; Andersen, J.F.; Francischetti, I.M.B.; Chavakis, T.; Kopecký, J.; Pedra, J.H.F.; Kotsyfakis, M.; Chmelař, J. Modulation of Host Immunity by Tick Saliva. *J. Proteom.* **2015**, *128*, 58. [CrossRef] [PubMed]
4. Šimo, L.; Kazimirova, M.; Richardson, J.; Bonnet, S.I. The Essential Role of Tick Salivary Glands and Saliva in Tick Feeding and Pathogen Transmission. *Front. Cell. Infect. Microbiol.* **2017**, *7*, 281. [CrossRef]
5. Chmelař, J.; Kotál, J.; Karim, S.; Kopacek, P.; Francischetti, I.M.B.; Pedra, J.H.F.; Kotsyfakis, M. Sialomes and Mialomes: A Systems-Biology View of Tick Tissues and Tick–Host Interactions. *Trends Parasitol.* **2016**, *32*, 242–254. [CrossRef]
6. De La Fuente, J.; Villar, M.; Estrada-Peña, A.; Olivas, J.A. High Throughput Discovery and Characterization of Tick and Pathogen Vaccine Protective Antigens Using Vaccinomics with Intelligent Big Data Analytic Techniques. *Expert Rev. Vaccines* **2018**, *17*, 569–576. [CrossRef]
7. Wang, H.L.V.; Chekanova, J.A. An Overview of Methodologies in Studying LncRNAs in the High-Throughput Era: When Acronyms ATTACK! *Methods Mol. Biol.* **2019**, *1933*, 1–30. [CrossRef]
8. Dempsey, J.L.; Cui, J.Y. Long Non-Coding RNAs: A Novel Paradigm for Toxicology. *Toxicol. Sci.* **2017**, *155*, 3–21. [CrossRef]
9. Wang, K.C.; Chang, H.Y. Molecular Mechanisms of Long Noncoding RNAs. *Mol. Cell* **2011**, *43*, 904–914. [CrossRef]
10. Mariner, P.D.; Walters, R.D.; Espinoza, C.A.; Drullinger, L.F.; Wagner, S.D.; Kugel, J.F.; Goodrich, J.A. Human Alu RNA Is a Modular Transacting Repressor of MRNA Transcription during Heat Shock. *Mol. Cell* **2008**, *29*, 499–509. [CrossRef]
11. Paraskevopoulou, M.D.; Hatzigeorgiou, A.G. Analyzing MiRNA-LncRNA Interactions. *Methods Mol. Biol.* **2016**, *1402*, 271–286. [CrossRef] [PubMed]
12. Ahmad, P.; Bensaoud, C.; Mekki, I.; Rehman, M.U.; Kotsyfakis, M. Long Non-Coding RNAs and Their Potential Roles in the Vector-Host-Pathogen Triad. *Life* **2021**, *11*, 56. [CrossRef] [PubMed]
13. Kern, C.; Wang, Y.; Chitwood, J.; Korf, I.; Delany, M.; Cheng, H.; Medrano, J.F.; van Eenennaam, A.L.; Ernst, C.; Ross, P.; et al. Genome-Wide Identification of Tissue-Specific Long Non-Coding RNA in Three Farm Animal Species 06 Biological Sciences 0604 Genetics. *BMC Genom.* **2018**, *19*, 684. [CrossRef]
14. Bayer-Santos, E.; Marini, M.M.; da Silveira, J.F. Non-Coding RNAs in Host–Pathogen Interactions: Subversion of Mammalian Cell Functions by Protozoan Parasites. *Front. Microbiol.* **2017**, *8*, 474. [CrossRef]
15. Bensaoud, C.; Hackenberg, M.; Kotsyfakis, M. Noncoding RNAs in Parasite–Vector–Host Interactions. *Trends Parasitol.* **2019**, *35*, 715–724. [CrossRef]
16. Aparicio-Puerta, E.; Lebrón, R.; Rueda, A.; Gómez-Martín, C.; Giannoukakos, S.; Jaspez, D.; Medina, J.M.; Zubkovic, A.; Jurak, I.; Fromm, B.; et al. SRNAbench and SRNAtoolbox 2019: Intuitive Fast Small RNA Profiling and Differential Expression. *Nucleic Acids Res.* **2019**, *47*, W530–W535. [CrossRef]

17. Fromm, B.; Høye, E.; Domanska, D.; Zhong, X.; Aparicio-Puerta, E.; Ovchinnikov, V.; Umu, S.U.; Chabot, P.J.; Kang, W.; Aslanzadeh, M.; et al. MirGeneDB 2.1: Toward a Complete Sampling of All Major Animal Phyla. *Nucleic Acids Res.* **2021**, *50*, D204–D210. [CrossRef]

18. Vlachos, I.S.; Zagganas, K.; Paraskevopoulou, M.D.; Georgakilas, G.; Karagkouni, D.; Vergoulis, T.; Dalamagas, T.; Hatzigeorgiou, A.G. DIANA-MiRPath v3.0: Deciphering MicroRNA Function with Experimental Support. *Nucleic Acids Res.* **2015**, *43*, W460–W466. [CrossRef]

19. Preusse, M.; Theis, F.J.; Mueller, N.S. MiTALOS v2: Analyzing Tissue Specific MicroRNA Function. *PLoS ONE* **2016**, *11*, e0151771. [CrossRef]

20. Sonenshine, E.D. *Biology of Ticks*; Oxford University Press: Oxford, NY, USA, 1991.

21. Sauer, J.R.; McSwain, J.L.; Bowman, A.S.; Essenberg, R.C. Tick Salivary Gland Physiology. *Annu. Rev. Entomol.* **1995**, *40*, 245–267. [CrossRef]

22. Narasimhan, S.; DePonte, K.; Marcantonio, N.; Liang, X.; Royce, T.E.; Nelson, K.F.; Booth, C.J.; Koski, B.; Anderson, J.F.; Kantor, F.; et al. Immunity against Ixodes Scapularis Salivary Proteins Expressed within 24 Hours of Attachment Thwarts Tick Feeding and Impairs Borrelia Transmission. *PLoS ONE* **2007**, *2*, e451. [CrossRef] [PubMed]

23. Francischetti, I.M.B.; Mather, T.N.; Ribeiro, J.M.C. Tick Saliva Is a Potent Inhibitor of Endothelial Cell Proliferation and Angiogenesis. *Thromb. Haemost.* **2005**, *94*, 167–174. [CrossRef]

24. Díaz-Martín, V.; Manzano-Román, R.; Oleaga, A.; Encinas-Grandes, A.; Pérez-Sánchez, R. Cloning and Characterization of a Plasminogen-Binding Enolase from the Saliva of the Argasid Tick Ornithodoros Moubata. *Vet. Parasitol.* **2013**, *191*, 301–314. [CrossRef] [PubMed]

25. Basson, M.D.; Sanders, M.A.; Gomez, R.; Hatfield, J.; VanderHeide, R.; Thamilselvan, V.; Zhang, J.; Walsh, M.F. Focal Adhesion Kinase Protein Levels in Gut Epithelial Motility. *Am. J. Physiol. Gastrointest. Liver Physiol.* **2006**, *291*, G491–G499. [CrossRef] [PubMed]

26. Sheng, H.; Shao, J.; Townsend, C.M.; Evers, B.M. Phosphatidylinositol 3-Kinase Mediates Proliferative Signals in Intestinal Epithelial Cells. *Gut* **2003**, *52*, 1472–1478. [CrossRef] [PubMed]

27. Chen, L.; Chi, H.; Kinashi, T. Editorial: Hippo Signaling in the Immune System. *Front. Immunol.* **2020**, *11*, 587514. [CrossRef] [PubMed]

28. Wang, Y.; Malik, A.B.; Sun, Y.; Hu, S.; Reynolds, A.B.; Minshall, R.D.; Hu, G. Innate Immune Function of the Adherens Junction Protein P120-Catenin in Endothelial Response to Endotoxin. *J. Immunol.* **2011**, *186*, 3180. [CrossRef]

29. Tian, E.; ten Hagen, K.G. Recent Insights into the Biological Roles of Mucin-Type O-Glycosylation. *Glycoconj J.* **2009**, *26*, 325–334. [CrossRef]

30. Gray, E.; Hogwood, J.; Mulloy, B. The Anticoagulant and Antithrombotic Mechanisms of Heparin. *Handb. Exp. Pharm.* **2012**, *207*, 43–61. [CrossRef]

31. Li, M.O.; Wan, Y.Y.; Sanjabi, S.; Robertson, A.K.L.; Flavell, R.A. Transforming Growth Factor-Beta Regulation of Immune Responses. *Annu. Rev. Immunol.* **2006**, *24*, 99–146. [CrossRef]

32. Martínez, V.G.; Hernández-López, C.; Valencia, J.; Hidalgo, L.; Entrena, A.; Zapata, A.G.; Vicente, A.; Sacedón, R.; Varas, A. The Canonical BMP Signaling Pathway Is Involved in Human Monocyte-Derived Dendritic Cell Maturation. *Immunol. Cell Biol.* **2011**, *89*, 610–618. [CrossRef] [PubMed]

33. MacDonald, L.A.; Cohen, A.; Baron, S.; Burchfiel, C.M. Respond to "Search for Preventable Causes of Cardiovascular Disease". *Am. J. Epidemiol.* **2009**, *169*, 1426–1427. [CrossRef] [PubMed]

34. Kim, M.; Choi, S.H.; Jin, Y.B.; Lee, H.J.; Ji, Y.H.; Kim, J.; Lee, Y.S.; Lee, Y.J. The Effect of Oxidized Low-Density Lipoprotein (Ox-LDL) on Radiation-Induced Endothelial-to-Mesenchymal Transition. *Int. J. Radiat. Biol.* **2013**, *89*, 356–363. [CrossRef]

35. Etebari, K.; Asad, S.; Zhang, G.; Asgari, S. Identification of Aedes Aegypti Long Intergenic Non-Coding RNAs and Their Association with Wolbachia and Dengue Virus Infection. *PLoS Negl. Trop. Dis.* **2016**, *10*, e0005069. [CrossRef]

36. Azlan, A.; Obeidat, S.M.; Das, K.T.; Yunus, M.A.; Azzam, G. Genome-Wide Identification of Aedes Albopictus Long Noncoding RNAs and Their Association with Dengue and Zika Virus Infection. *PLoS Negl. Trop. Dis.* **2021**, *15*, e0008351. [CrossRef] [PubMed]

37. Medina, J.M.; Jmel, M.A.; Cuveele, B.; Gómez-Martín, C.; Aparicio-Puerta, E.; Mekki, I.; Kotál, J.; Martins, L.A.; Hackenberg, M.; Bensaoud, C.; et al. Transcriptomic Analysis of the Tick Midgut and Salivary Gland Responses upon Repeated Blood-Feeding on a Vertebrate Host. *Front. Cell. Infect. Microbiol.* **2022**, *12*, 919786. [CrossRef] [PubMed]

38. Perner, J.; Kropáčková, S.; Kopáček, P.; Ribeiro, J.M.C. Sialome Diversity of Ticks Revealed by RNAseq of Single Tick Salivary Glands. *PLoS Negl. Trop. Dis.* **2018**, *12*, e0006410. [CrossRef] [PubMed]

39. Charrier, N.P.; Couton, M.; Voordouw, M.J.; Rais, O.; Durand-Hermouet, A.; Hervet, C.; Plantard, O.; Rispe, C. Whole Body Transcriptomes and New Insights into the Biology of the Tick Ixodes Ricinus. *Parasites Vectors* **2018**, *11*, 364. [CrossRef]

40. Babraham Bioinformatics. Babraham Bioinformatics—Trim Galore. Available online: https://www.bioinformatics.babraham.ac.uk/projects/trim_galore/ (accessed on 4 July 2018).

41. Babraham Bioinformatics. Babraham Bioinformatics—FastQC A Quality Control Tool for High Throughput Sequence Data. Available online: https://www.bioinformatics.babraham.ac.uk/projects/fastqc/ (accessed on 4 July 2018).

42. Grabherr, M.G.; Haas, B.J.; Yassour, M.; Levin, J.Z.; Thompson, D.A.; Amit, I.; Adiconis, X.; Fan, L.; Raychowdhury, R.; Zeng, Q.; et al. Trinity: Reconstructing a Full-Length Transcriptome without a Genome from RNA-Seq Data. *Nat. Biotechnol.* **2011**, *29*, 644. [CrossRef]

43. Fu, L.; Niu, B.; Zhu, Z.; Wu, S.; Li, W. CD-HIT: Accelerated for Clustering the next-Generation Sequencing Data. *Bioinformatics* **2012**, *28*, 3150–3152. [CrossRef]
44. Li, B.; Dewey, C.N. RSEM: Accurate Transcript Quantification from RNA-Seq Data with or without a Reference Genome. *BMC Bioinform.* **2011**, *12*, 323. [CrossRef] [PubMed]
45. Langmead, B.; Salzberg, S.L. Fast Gapped-Read Alignment with Bowtie 2. *Nat. Methods* **2012**, *9*, 357–359. [CrossRef] [PubMed]
46. Boutet, E.; Lieberherr, D.; Tognolli, M.; Schneider, M.; Bairoch, A. UniProtKB/Swiss-Prot. *Methods Mol. Biol.* **2007**, *406*, 89–112. [CrossRef]
47. Altschul, S.F.; Gish, W.; Miller, W.; Myers, E.W.; Lipman, D.J. Basic Local Alignment Search Tool. *J. Mol. Biol.* **1990**, *215*, 403–410. [CrossRef]
48. Medina, J.M.; Jmel, M.A.; Cuveele, B.; Mekki, I.; Kotal, J.; Almeida Martins, L.; Hackenberg, M.; Kotsyfakis, M.; Bensaoud, C. IxoriDB. Available online: https://arn.ugr.es/IxoriDB (accessed on 20 August 2022).
49. Schmid-Hempel, P. Immune Defence, Parasite Evasion Strategies and Their Relevance for 'Macroscopic Phenomena' Such as Virulence. *Philos. Trans. R. Soc. B Biol. Sci.* **2009**, *364*, 85. [CrossRef] [PubMed]
50. Robinson, M.D.; Oshlack, A. A Scaling Normalization Method for Differential Expression Analysis of RNA-Seq Data. *Genome Biol.* **2010**, *11*, R25. [CrossRef]
51. Robinson, M.D.; McCarthy, D.J.; Smyth, G.K. EdgeR: A Bioconductor Package for Differential Expression Analysis of Digital Gene Expression Data. *Bioinformatics* **2010**, *26*, 139. [CrossRef]

International Journal of
Molecular Sciences

Article

Identification of microRNAs in the Lyme Disease Vector *Ixodes scapularis*

Deepak Kumar [1,2]**, Latoyia P. Downs** [2] **, Monica Embers** [3] **, Alex Sutton Flynt** [1,2] **and Shahid Karim** [1,2,*]

[1]   Center for Molecular and Cellular Biosciences, University of Southern Mississippi,
       Hattiesburg, MS 39406, USA; deepak.kumar@usm.edu (D.K.); alex.flynt@usm.edu (A.S.F.)
[2]   School of Biological, Environmental, and Earth Sciences, University of Southern Mississippi,
       Hattiesburg, MS 39406, USA; latoyia.downs@usm.edu
[3]   Division of Immunology, Tulane National Primate Research Center, Covington, LA 70433, USA;
       members@tulane.edu
*   Correspondence: shahid.karim@usm.edu

Abstract: MicroRNAs (miRNAs) are a class of small non-coding RNAs involved in many biological processes, including the immune pathways that control bacterial, parasitic, and viral infections. Pathogens probably modify host miRNAs to facilitate successful infection, so they might be useful targets for vaccination strategies. There are few data on differentially expressed miRNAs in the black-legged tick *Ixodes scapularis* after infection with *Borrelia burgdorferi*, the causative agent of Lyme disease in the United States. Small RNA sequencing and qRT-PCR analysis were used to identify and validate differentially expressed *I. scapularis* salivary miRNAs. Small RNA-seq yielded 133,465,828 (≥18 nucleotides) and 163,852,135 (≥18 nucleotides) small RNA reads from *Borrelia*-infected and uninfected salivary glands for downstream analysis using the miRDeep2 algorithm. As such, 254 miRNAs were identified across all datasets, 25 of which were high confidence and 51 low confidence known miRNAs. Further, 23 miRNAs were differentially expressed in uninfected and infected salivary glands: 11 were upregulated and 12 were downregulated upon pathogen infection. Gene ontology and network analysis of target genes of differentially expressed miRNAs predicted roles in metabolic, cellular, development, cellular component biogenesis, and biological regulation processes. Several Kyoto Encyclopedia of Genes and Genomes (KEGG) pathways, including sphingolipid metabolism; valine, leucine and isoleucine degradation; lipid transport and metabolism; exosome biogenesis and secretion; and phosphate-containing compound metabolic processes, were predicted as targets of differentially expressed miRNAs. A qRT-PCR assay was utilized to validate the differential expression of miRNAs. This study provides new insights into the miRNAs expressed in *I. scapularis* salivary glands and paves the way for their functional manipulation to prevent or treat *B. burgdorferi* infection.

Keywords: *Ixodes scapularis*; salivary glands; microRNA; RNA sequencing; *Borrelia burgdorferi*

Citation: Kumar, D.; Downs, L.P.; Embers, M.; Flynt, A.S.; Karim, S. Identification of microRNAs in the Lyme Disease Vector *Ixodes scapularis*. *Int. J. Mol. Sci.* **2022**, 23, 5565. https://doi.org/10.3390/ijms23105565

Academic Editors: Denis Sereno and Michail Kotsyfakis

Received: 14 April 2022
Accepted: 10 May 2022
Published: 16 May 2022

Publisher's Note: MDPI stays neutral with regard to jurisdictional claims in published maps and institutional affiliations.

1. Introduction

Small non-coding RNAs (sncRNAs) regulate genes at the post-transcriptional level in animals, plants, and arthropods, including ticks [1–5]. MicroRNAs (miRNAs) are a class of sncRNA, between 18 and 25 nucleotides in length, and they are now known to be important in arthropod immunity and host–pathogen interactions through their involvement in several cellular processes, including development, immunity, and pathogen responses in arthropods [2,6–10]. In animals, miRNAs regulate post-transcriptional gene expression, most often by binding to the 3′-untranslated region (3′-UTR) of target genes. While perfect complementarity of 2–8 nucleotides at the 5′ end of the miRNA (seed region) is necessary for miRNA regulation, the remaining sequence can harbor mismatches or bulges [2,11,12]. miRNAs are transcribed as primary miRNA transcripts before processing by Drosha and Pasha proteins into pre-miRNAs. These are then exported to the cytoplasm and processed by Dicer into mature miRNAs, which are loaded onto the microRNA-induced silencing complex (miRISC) before targeting complementary mRNA for degradation [13,14].

Several proteomic and transcriptomic studies have identified differentially expressed transcripts and proteins in uninfected and infected ticks [1,15–18], but few have investigated the role of pathogens in the differential modulation of the post-transcriptional tick and host machinery. Although there are over 800 tick species and despite their importance as vectors of human and animal diseases, ticks are underrepresented in available miRNA resources. For example, the miRbase database contains 49 *Ixodes scapularis* miRNAs and 24 *Rhipicephalus microplus* miRNAs, while MirGeneDB 2.1 contains 64 *Ixodes scapularis* miRNAs. However, small RNA sequencing (RNA-seq) and computational approaches are accelerating the discovery of miRNAs from species with incomplete genome sequencing, assembly, and annotation.

Ixodes scapularis is a primary vector of human pathogens, including the Lyme disease agent *Borrelia burgdorferi*, which infects vertebrates and ticks through evolved complex mechanisms. There is now a good understanding of tick immune pathways and their interactions with *B. burgdorferi* [19–21], but it is still uncertain how *B. burgdorferi* avoids clearance. *B. burgdorferi* must traverse tick salivary glands during transmission [22]. Since saliva/salivary gland proteins can enhance *B. burgdorferi* transmission into the vertebrate host, characterization of molecular interactions at the tick-bite site and the tick salivary glands is expected to facilitate vaccine development [23,24], since promoting immunity against tick salivary proteins could neutralize tick bites and pathogen transmission. Once *B. burgdorferi* is acquired by ticks from infected hosts, it resides in the tick gut and only migrates to the salivary gland during subsequent blood feeding, which generally lasts for 3–7 days [25–28]. While it is known that other tick-borne pathogens, such as *Anaplasma marginale*, must replicate inside salivary glands for efficient transmission [29], the details of *B. burgdorferi* replication are less well understood. Indeed, borrelial spirochetes invade the tick salivary gland via an unknown mechanism [22] and might be carried to the host dermis via tick saliva. Several salivary gland genes are upregulated in *B. burgdorferi*-infected *Ixodes scapularis* nymphs compared with uninfected ones [30], suggesting a significant role for salivary gland gene regulation in *B. burgdorferi* infection and transmission.

To fill the knowledge gap on miRNA expression in *Ixodes scapularis* salivary glands, here, we performed miRNA profiling of partially fed *B. burgdorferi*-infected and uninfected tick salivary glands to identify miRNAs that might play a role in *B. burgdorferi* survival, colonization, transmission, and host immunomodulation. In doing so, we detected 254 miRNAs, of which 25 were high confidence miRNAs and 51 low confidence miRNAs. Forty-one of the identified miRNAs were present as *I. scapularis* miRNAs in miRBase (v22.1). Gene ontology and network analysis of target genes of differentially expressed miRNAs have predicted roles in metabolic and cellular development, cellular component biogenesis, and biological regulation processes. Several KEGG pathways, including sphingolipid metabolism; valine, leucine and isoleucine degradation; lipid transport and metabolism; exosome biogenesis and secretion; and phosphate-containing compound metabolic processes, were predicted as targets of differentially expressed miRNAs.

2. Materials and Methods

2.1. Ethics Statement

All animal experiments were performed in strict accordance with the recommendations in the NIH Guide for the Care and Use of Laboratory Animals. The Institutional Animal Care and Use Committee of the University of Southern Mississippi approved the protocol for blood feeding of field-collected ticks (protocol # 15101501.1).

2.2. Ticks and Tissue Dissections

Ticks were purchased from the Oklahoma State University Tick-Rearing Facility. Adult male and female *I. scapularis* were kept according to standard practices [31] and maintained in the laboratory as described in our previously published work [32,33]. Unfed female adult

I. scapularis were infected with laboratory-grown *B. burgdorferi* strain B31.5A19 using the capillary feeding method at Tulane National Primate Research Center [34]. *Borrelia*-infected and uninfected ticks (n = 45 in each group) were placed on each ear of a rabbit host for tick blood feeding. Blood-fed adult female *I. scapularis* were dissected within 60 min of removal from the rabbit. Tick tissues were dissected and washed in M-199 buffer as described previously [35]. Salivary glands and midguts from individual *I. scapularis* were stored in RNAlater (Life Technologies, Carlsbad, CA, USA) at −80 °C until use.

2.3. RNA Isolation, cDNA Synthesis, and PCR-Based B. burgdorferi Detection in Tick Tissues

The TRIzol method was used for RNA extraction from individually dissected midgut tissues, and cDNA was synthesized as described previously [36,37]. *B. burgdorferi* was detected in tick midguts using the *flaB* gene in a PCR assay [36,38]. After testing for *B. burgdorferi* infection in tick midguts, the corresponding salivary gland tissues from the same uninfected/infected ticks (n = 10 salivary glands from each group) were pooled in separate tubes and RNA isolated using the TRIzol method [37].

2.4. Small RNA Sequencing (RNA-Seq)

Small RNA libraries were made using the Illumina TruSeq Kit following the manufacturer's protocol (Illumina, San Diego, CA, USA). Briefly, short adapter oligonucleotides were ligated to each end of the small RNAs in the sample, cDNA made with reverse transcriptase, and PCR used to add sample-specific barcodes and Illumina sequencing adapters. The final concentration of all sequencing libraries was determined using a Qubit fluorometric assay (Thermo Fisher Scientific, Waltham, MA, USA), and the DNA fragment size of each library was assessed using a DNA 1000 high-sensitivity chip on an Agilent 2100 Bioanalyzer (Agilent Technologies, Santa Clara, CA, USA). After purification by polyacrylamide gel electrophoresis, sample libraries were pooled and sequenced on an Illumina NextSeq 500 (single end 36 base) using the TruSeq SBS kit v3 (Illumina, San Diego, CA, USA) and protocols defined by the manufacturer. Four small RNA libraries of clean and *B. burgdorferi*-infected, partially fed, and pooled salivary glands were sequenced. RNA library preparation and indexing were performed at the University of Mississippi Medical Center (UMMC) sequencing facility. The small RNA sequencing data used in this study was deposited in the National Center for Biotechnology Information (NCBI) under the Sequence Read Archive (SRA) accession number PRJNA837369.

2.5. Data Analysis

A schematic of the experimental plan and data analysis is shown in Supplementary Figure S1. miRDeep2 v.2.0.0.8 [39,40] was used to process RNA-seq data. To predict novel miRNAs, the reads from all samples were combined. The mapper function in miRDeep2 first trims the adapter sequences from the reads and converts the read files from FASTQ to FASTA format. Reads shorter than 18 bases were discarded and the remaining reads mapped to the *I. scapularis* reference genome using default miRDeep2 mapper function parameters. Reads mapping to the genome were used to predict novel miRNAs. The *Drosophila melanogaster* genome was also used as another reference genome, and mapped reads were aligned to available miRNAs of *D. melanogaster* in miRBase (v22) and quantified. Reads were mapped to the reference genomes of *D. melanogaster* and *I. scapularis* and locations of potential miRNA read accumulations identified. The regions immediately surrounding the mapped reads were examined for miRNA biogenesis features including mature miRNAs, star and precursor reads, and stem-loop folding properties. miRDeep2 models the miRNA biogenesis pathway and uses a probabilistic algorithm to score compatibility of the position and frequency of sequencing reads with the secondary structure of the miRNA precursor.

For miRNA expression, a count table was generated using bedtools multicov, which counts alignments from indexed BAM files that overlap intervals in BED files provided from the miRDeep2 analysis.

2.6. ISE6 Cell Culture, Infection with B. burgdorferi, and Validation of Differentially Expressed miRNAs by qRT-PCR

Differentially expressed miRNAs in RNA-seq data were validated by qRT-PCR in the ISE6 cell line derived from *I. scapularis* embryos cultured and maintained as suggested by Munderloh and Kurtti [41]. The *B. burgdorferi* (strain B31) isolate was a kind donation from Dr. Monica Embers (Tulane University, Covington, LA, USA). Cultures were grown and maintained as suggested previously [34]. Once ISE6 cells reached confluency, they were infected with *B. burgdorferi* as described previously [42]. Briefly, ISE6 cells were inoculated with the supernatant of a log phase *B. burgdorferi* culture at a multiplicity of infection of 50 and incubated for 24 h before harvesting. RNA was isolated using the TRIzol method and the expression of miRNAs analyzed by qRT-PCR with the Mir-X miRNA qRT-PCR TB Green kit (Takara Bio Inc, Kusatsu, Shiga, Japan; catalog #638316). qRT-PCR conditions were an initial denaturation at 95 °C for 10 min then 40 cycles of 95 °C for 5 s, 60 °C for 20 s.

2.7. Normalization, Differential Expression (DE), and Statistical Analysis of miRNAs between Uninfected and Infected Salivary Glands

Differential expression (DE) analysis of identified miRNAs was performed using the interactive web interface DeApp [43]. Low-expression genetic features were removed after alignment if the counts per million (CPM) value was ≤ 1 in less than two samples. Sample normalization and a multidimensional scaling (MDS) plot are shown in Supplementary Figure S2. DE analysis was performed with edgeR with a false discovery rate (FDR) adjusted *p*-value of 0.05 and minimum fold-change of 1.5. DeApp displays a dispersion plot showing the overall DE analysis along with statistical significance (*p*-value, FDR adjusted *p*-value) and a volcano plot corresponding to the specified parameters and cutoff values.

2.8. In Silico Mapping of B. burgdorferi-Infected Small RNA Sequences to the Borrelia burgdorferi Genome

In silico mapping of small RNA sequences of *B. burgdorferi*-infected salivary glands to the *Borrelia burgdorferi* genome (GCF_000181575.2_ASM18157v2_genomic.fna) detected 18,165 *Borrelia burgdorferi* sequences, but no miRNAs of *B. burgdorferi* origin were predicted by miRDeep2 analysis.

2.9. Prediction of Target Genes, Proteome Re-Annotation, Gene Ontology (GO), and KEGG Enrichment Analyses

TargetSpy [44], MIRANDA [45], and PITA [46] were used in miRNAconsTarget from sRNAtoolbox to predict genes regulated by tick salivary gland miRNAs up- or downregulated in the in silico analysis [47]. Targets common to all three programs were further considered. In silico target prediction provided a high number of false positives, but cross-species comparisons and combinatorial effects reduced this number [48]. Target gene networks and KEGG pathways significantly enriched for target genes were extracted using the STRING [49] output. PANNZER2 [50] was used to functionally re-annotate the proteome of up- or downregulated genes (targets of detected miRNAs), and WEGO [51] was used to analyze and plot gene ontology (GO) annotations.

3. Results and Discussion

3.1. Profile Characteristics of Small RNA Libraries

There were 216,292,174 raw small RNA reads from *B. burgdorferi*-infected salivary glands and 212,542,697 from uninfected salivary glands. After adapter trimming and removal of short reads (≤ 18 nucleotides (nt)), 133,465,828 small RNA reads were available from *B. burgdorferi*-infected samples and 163,852,135 from uninfected samples for downstream analysis. The read length distribution shows the types of small RNAs present in *B. burgdorferi*-infected and uninfected salivary gland samples. Two main peaks at 22 nt (miRNAs/siRNAs) at 29 nt (piRNAs) were distinguishable in *B. burgdorferi*-infected and uninfected samples (Figure 1A). There were ~5 × 10^7 and ~2.5 × 10^7 22 nt miRNA sequences in uninfected and infected salivary gland samples, respectively. The read length distribution in uninfected samples was comparable to infected samples.

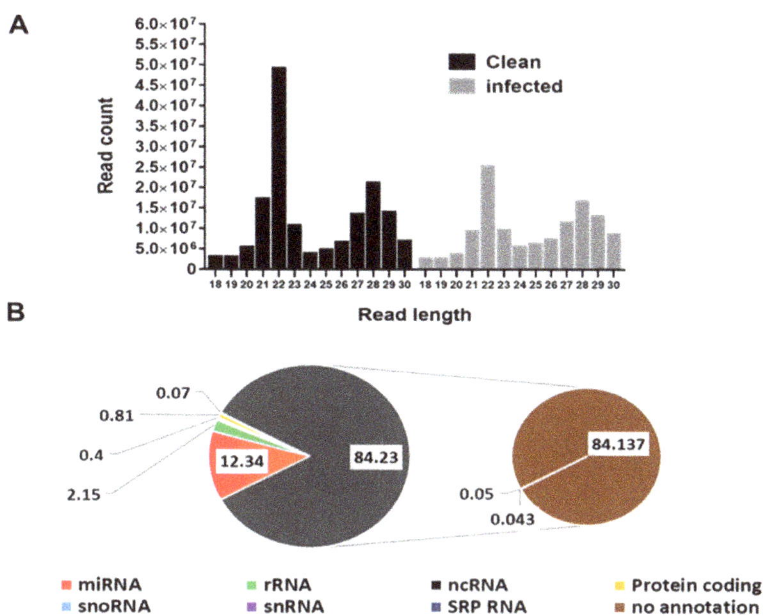

Figure 1. (**A**) Small RNA sequence length distribution in uninfected (clean) and *Borrelia-burgdorferi*-infected *Ixodes scapularis* salivary glands. MicroRNAs are twenty-two (22) nucleotides in length. Clean, uninfected salivary glands; infected, *Borrelia-burgdorferi*-infected salivary glands. (**B**) Summary of the abundance of combined small RNA reads from *Ixodes scapularis* salivary glands matching various small RNA categories. Reads from uninfected and *Borrelia-burgdorferi*-infected samples are combined together for abundance calculation.

3.2. Other Small RNA Categories

A summary of reads matching various small RNA categories from *B. burgdorferi*-infected and uninfected salivary glands is shown in Figure 1B. Other small RNAs include signal recognition particle rRNAs, ncRNAs, protein-coding RNAs, snoRNAs, snRNA SRP RNAs, and not-annotated small RNAs. Of the total small RNA reads from *Ixodes scapularis* salivary glands, 12.3% were miRNAs, 2.2% rRNAs, 0.4% ncRNAs, 0.8% protein coding, 0.07% snoRNAs, 0.05% snRNAs, 0.04% SRP RNAs, and 84.3% reads were not annotated.

3.3. MicroRNA Profiling of Infected and Uninfected Salivary Glands and Identification of Novel Tick miRNAs

Figure 2A shows the basic hairpin-loop structure of an miRNA and other parameters (Dicer cut overhangs, total read count, mature read count, loop read count, total read count, randfold score, and total score) used to determine whether a hairpin-loop-structured RNA is an miRNA. Using miRDeep2, 254 miRNAs were predicted in the salivary gland libraries. By adopting a conservative approach, miRNAs were categorized as high- (n = 25) and low-confidence (n = 51) miRNAs (Figure 2B), based on standard criteria [39,40]. A conservative approach was used to annotate predicted miRNAs obtained from miRDeep2 analysis. The criteria utilized for annotating predicted miRNAs included: (1) a more significant number of deep sequencing reads corresponding to the mature sequence, (2) one or more reads matching to the star sequence, at least a few loop reads, (3) short 3′ duplex overhangs, characteristic of Drosha/Dicer processing, and (4) conservation of 5′ end of the potential mature sequence to a known mature sequence in the miRbase. The score boosts if the 5′

end of the potential mature sequence is identical to a known mature sequence. A high-confidence category will meet all the criteria mentioned above, whereas a low-confidence category will not satisfy all the criteria by missing one or two points. Of 49 *Ixodes scapularis* miRNAs present in miRBase, 41 were detected in our data. Of the 254 *Ixodes scapularis* miRNAs, several had homologs present in *D. melanogaster*. The details of the identified miRNAs and their miRDeep2 scores are presented in Supplementary Table S1.

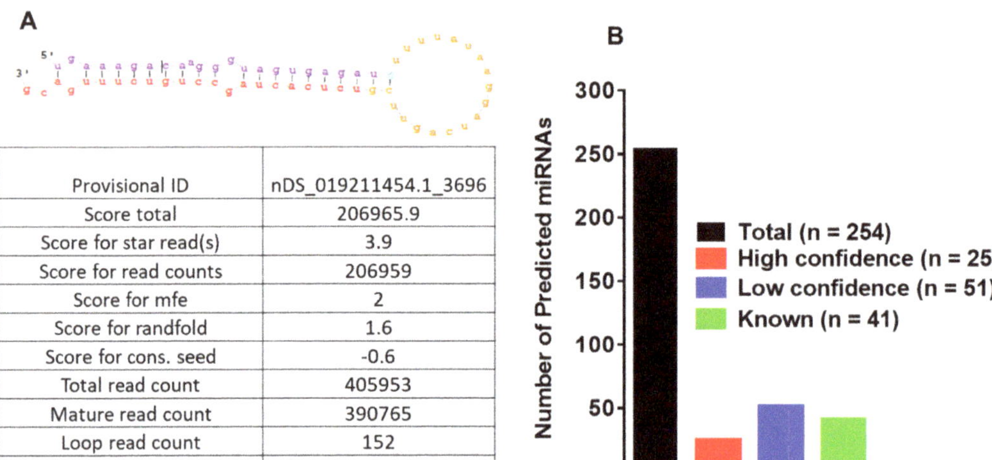

Provisional ID	nDS_019211454.1_3696
Score total	206965.9
Score for star read(s)	3.9
Score for read counts	206959
Score for mfe	2
Score for randfold	1.6
Score for cons. seed	-0.6
Total read count	405953
Mature read count	390765
Loop read count	152
Star read count	15036

Figure 2. (**A**) Basic stem-loop structures of predicted microRNAs. miRDeep2 was used to identify potential miRNA precursors based on nucleotide length, star sequence, stem-loop folding, and homology to the *Ixodes scapularis* reference genome. Shown are the predicted stem-loop structures (yellow), star (pink), and mature sequences of predicted miRNAs (red) in the salivary glands of *Ixodes scapularis* ticks. (**B**) Annotation of predicted miRNAs in partially fed uninfected (clean) and *Borrelia-burgdorferi*-infected *Ixodes scapularis* salivary glands. Further, 254 microRNAs were predicted in infected and uninfected salivary gland tissues, 25 were categorized as high confidence (*n* = 25) and 51 as low confidence (*n* = 51) based on standard criteria and a conservative approach. Out of 49 *Ixodes scapularis* miRNAs available in miRbase (v22.1), 41 were detected in this study.

3.4. In Silico DE Analysis of miRNAs in B. burgdorferi-Infected Salivary Glands

In the DE analysis, 11 miRNAs were upregulated and 12 were downregulated in infected salivary glands compared with uninfected salivary glands (Figure 3 and Table 1). Several of the identified tick miRNAs were conserved in *D. melanogaster* (dme-miR-375-3p, dme-miR-993-5p, dme-miR-12-5p, dme-bantam-3p, dme-miR-100-5p, dme-miR-8-3p, and dme-miR-304-5p). miRNAs downregulated in *B. burgdorferi*-infected salivary glands relative to uninfected salivary glands were isc-miR-153, isc-miR-1 and isc-miR-79, nDS_002871802.1_376, nDS_002633080.1_31953, nDS_002549652.1_43448, nDS_002537755.1_45711, nDS_002537755.1_45739, nDS_002861213.1_1750, nDS_002763926.1_14582, and nDS_002548557.1_43749, while upregulated miRNAs were isc-miR-317, isc-miR5310, isc-miR-2001, isc-miR-5307, isc-miR-71, isc-miR-87, nDS_002784743.1_12165, nDS_002664372.1_28027, nDS_002716687.1_21278, nDS_002620414.1_33611, and nDS_002680650.1_26098.

Table 1. In silico differential expression of *Ixodes scapularis* miRNAs in salivary glands after *Borrelia burgdorferi* infection compared with uninfected salivary glands.

Differentially Expressed miRNA *	Log$_2$FC	logCPM	LR	*p*-Value	FDR
isc-miR-87	4.245604	11.83289	33.20241	8.30×10^{-9}	4.65×10^{-8}
nDS_002784743.1_12165	5.511035	11.48887	29.7299	4.97×10^{-8}	1.55×10^{-7}
isc-miR-71	5.226443	11.95849	48.94389	2.63×10^{-12}	3.69×10^{-11}
isc-miR-5307	2.570363	12.34313	24.71399	6.65×10^{-7}	1.55×10^{-6}
nDS_002664372.1_28027	5.362184	11.43227	26.80077	2.26×10^{-7}	5.84×10^{-7}
nDS_002716687.1_21278	5.511153	11.48887	29.72059	4.99×10^{-8}	1.55×10^{-7}
isc-miR-2001	3.146271	11.78845	19.75379	8.81×10^{-6}	1.64×10^{-5}
nDS_002620414.1_33611	5.361301	11.43227	26.76848	2.29×10^{-7}	5.84×10^{-7}
isc-miR-5310	5.009416	11.31195	20.86089	4.94×10^{-6}	1.06×10^{-5}
nDS_002680650.1_26098	1.567661	15.43032	103.3589	2.80×10^{-24}	7.83×10^{-23}
isc-miR-317	3.146271	11.78845	19.75379	8.81×10^{-6}	1.64×10^{-5}
nDS_002871802.1_376	−0.80359	15.73883	19.22785	1.16×10^{-5}	1.91×10^{-5}
nDS_002633080.1_31953	−4.7431	11.95812	10.11663	0.001469	0.001959
nDS_002549652.1_43448	−4.9974	12.11011	12.13051	0.000496	0.000731
nDS_002537755.1_45711	−6.34111	13.07343	30.24285	3.81×10^{-8}	1.52×10^{-7}
nDS_002537755.1_45739	−6.57147	13.2637	36.56262	1.48×10^{-9}	1.03×10^{-8}
isc-miR-79 dme-miR-9a-5p	−5.35587	12.34273	15.64359	7.65×10^{-5}	0.000119
isc-miR-1 dme-miR-1-3p	−2.44072	12.67077	10.50794	0.001189	0.001664
nDS_002861213.1_1750	−0.80359	15.73883	19.22785	1.16×10^{-5}	1.91×10^{-5}
nDS_002763926.1_14582	−6.57147	13.2637	36.56262	1.48×10^{-9}	1.03×10^{-8}
nDS_002548557.1_43749	−6.34111	13.07343	30.24285	3.81×10^{-8}	1.52×10^{-7}
isc-miR-153	−4.15091	11.64621	6.563386	0.01041	0.013249

* miRNAs with a log2 fold-change expression > |1| and FDR $\leq$ 0.1 were considered significantly differentially expressed. Values highlighted in red indicate significant upregulation and values highlighted in green indicate significant downregulation.

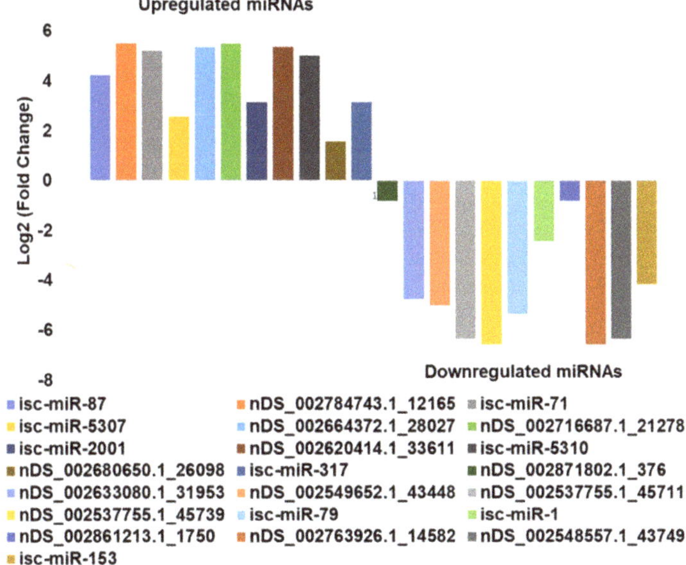

Figure 3. In silico differential expression of predicted miRNAs in *Borrelia-burgdorferi*-infected, partially fed salivary glands relative to partially fed clean salivary glands. EdgeR was used for differential expression analysis. As such, 12 predicted miRNAs were down-regulated, 11 were upregulated, while 39 were unaffected. miRNAs with a log2 fold-change expression > |1| and FDR $\leq$ 0.1 were considered significantly differentially expressed (see Table 1). Further, in Table 2, putative roles and targets of these DE-conserved miRNAs are provided based on available studies in other arthropods.

Table 2. List of differentially expressed microRNAs detected in *Borrelia-burgdorferi*-infected tick salivary glands and their putative roles.

microRNA	Role of Ortholog microRNA	Target Genes	Reference
isc-miR-153	Development, immune response		[52]
isc-miR-1	Stress response, immunity, development, facilitating infection	*HSP60, HSP70, GATA4*	[53–56]
isc-miR-79	Innate immunity, differentiation, apoptosis	*Roundabout protein 2 pathway (robo2), DRAPER, HP2 (Hemolymph protease), P38, pvr, puc*	[57–62]
isc-miR-317	Negative influencer of Toll pathway, host homeostatic response (targeting Gap junctions, TRP channels), mitotic regulation, Fecundity	*Dif-Rc, Thioredoxin peroxidase-2, STAT*	[62–66]
isc-miR-2001	Under extreme environmental conditions		[67]
isc-miR-5307	Propagation of Powassan virus inside vero cells		[68]
isc-miR-71	Stress response, development, facilitates WSSV infection	*cap-1, riok1, Actin1, myD88, IML3*	[62,69–72]
isc-miR-87	Pathogen survival (disruption of Toll and IMD pathway)	*Serine/Threnine Kinase, Toll 1A, Putative TLR 5b, FADD*	[62,73,74]
isc-miR-5310	Pathogen survival by modulating signaling pathways, Feeding behavior		[75–81]

miR-2001 was upregulated in *B. burgdorferi*-infected salivary glands compared with controls. miR-2001 is not present in the *Drosophila* genome but is an evolutionarily conserved miRNA in ticks [81] It has previously been shown to play a role in host immunomodulation, required for pathogen survival. miR-2001 has been detected in *I. ricinus* saliva [63] and *H. longicornis* saliva extracellular vesicles (EVs) [82], along with several other miRNAs, suggesting that EVs containing these miRNAs could be transferred to the host to modulate host cellular functions to facilitate tick and pathogen survival [63,82]. EVs are involved in the intercellular transfer of miRNAs, lipids, and proteins and the disposal of unnecessary cell contents [82,83]. The discovery of EVs in the excretory-secretory products of ectoparasites suggests that EVs are probably taken up by host cells, deliver their cargoes to the host, and favor immunomodulation, pathogen survival, and disease progression [84,85].

miR-1 was downregulated in *B. burgdorferi*-infected tick salivary glands compared with uninfected salivary glands. miR-1 belongs to a family of miRNAs, including miR-7 and miR-34 conserved across fruit flies, shrimps, and humans, where it modulates similar pathways (development, apoptosis) and is upregulated during stress insults [53]. In mosquitoes, miR-1 is upregulated during *Plasmodium* infection [54] and also facilitates West Nile virus infection [55]. miR-1 has been shown to be generally upregulated in response to infection, while we detected its downregulation in *B. burgdorferi*-infected SGs. In *Listeria*-infected macrophages, miR-1 promotes IFN-γ-dependent activation of the innate immune response [56].

Tick saliva and salivary gland extracts reduce IFN-γ and IL-2 production in T cells and inhibit T cell proliferation [86,87], suggesting immune suppression, a possible survival mechanism for tick pathogens. mir-79 was also downregulated in *Borrelia-burgdorferi*-infected salivary glands, and has been shown to participate in immunity and other processes, such as cellular differentiation, neurogenesis, and apoptosis. In *Rhipicephalus haemaphysaloides*, mir-79 was downregulated upon lipopolysaccharide (LPS) induction in female and male ticks, suggesting a role in LPS-mediated stimulation in the innate immune response [57]. The JNK pathway is an immune response pathway against Gram-negative bacteria [58], and mir-79 is known to disrupt JNK signaling by targeting its component genes, *pvr* (CG8222) and *puc* (CG7850) [59]. Our detection of the downregulation of mir-79 in *B. burgdorferi*-infected salivary glands is probably due to stimulation of the JNK pathway as a tick immune response against *B. burgdorferi*. Although *B. burgdorferi* is described as an atypical Gram-negative bacterium due to its double membrane, it lacks classical lipopolysaccharide (LPS). It also has a different cellular organization and membrane composition to other diderms [88]. However, surprisingly, in *A. phagocytophilum* (an intracellular Gram-negative bacterial pathogen)-infected ticks, mir-79 was upregulated to facilitate infection by targeting the Roundabout protein 2 pathway (Robo2) [60], suggesting different roles for mir-79 in ticks when infected with different bacterial pathogens. *Borrelia burgdorferi* (Bb) is extracellular and a Gram-negative bacterial pathogen, but *A. phagocytophilum* is a well-known intracellular, Gram-negative bacterial pathogen. The status of *B. burgdorferi* as extracellular pathogen is true only in ticks, its intracellular localization has been reported in mammalian cells, including fibroblast, endothelial cells, neuron and even epithelial cells [89–93]. Do the extracellular and pseudo-Gram-negative status of *B. burgdorferi* in ticks make it a different bacterial pathogen than Gram negative or positive? More clarity of tick immune response is required in case of *B. burgdorferi* infection.

In Drosophila [35], miR-317 negatively regulates Toll pathway signaling, and its upregulation in *B. burgdorferi*-infected salivary glands may suggest a similar role to facilitate *B. burgdorferi* survival inside tick salivary glands. In silico, miR-317 targets Dif-Rc, an important transcription factor in the Toll pathway in Drosophila [64] and STAT in JAK-STAT signaling in Manduca sexta [62]. An in silico study in *I. ricinus* suggested a combinatorial effect of tick salivary miRNAs on host genes important for maintaining host homeostasis and tick–host interactions, including miR-317 targeting gap junctions and TRP channels, which play significant roles in host homeostatic responses [63]. miR-71 was upregulated in B. burgdorferi-infected salivary glands, and its predicted targets include MyD88, which is activated when ligands bind to the Toll-like receptor (TLR), interleukin 1 receptor (IL-1R), or IFN-γR1 and trigger MyD88-mediated signaling and pro-inflammatory cytokine responses. Another miR-71 target is IML3, an arthropod immunolectin that recognizes LPS on Gram-negative bacteria as a part of arthropod immune defenses. Immunolectins are also predicted targets of miR-87, -276, -9a, and -71 [62]. We hypothesize that miR-71 disrupts tick immune pathways and protects B. burgdorferi in the salivary gland. It has also been shown to prolong the life and regulate stress responses in nematodes, being upregulated in the Dauer larval stage, when food or other life-sustaining resources are scarce [69].

miR-87 was upregulated in *B. burgdorferi*-infected salivary glands. Previous studies in other arthropods, such as *Manduca sexta* and *Aedes albopictus*, suggested a role in disrupting innate immunity, particularly via IMD and Toll signaling pathways [62,73,74]. In *Aedes albopictus*, miR-87's predicted targets are Toll pathway signaling Ser/Thr kinase, Toll-like receptor Toll1A, class A scavenger receptor with Ser-protease domain, galectin [74], and TLR5b [73], while in *Manduca sexta*, its predicted target is FADD, an adaptor protein involved in DISC formation [62]. Our in silico data also showed upregulation of miR-5310 in *B. burgdorferi*-infected salivary glands. miR-5310 is a tick-specific miRNA [81], and a recent study demonstrated its downregulation in *Anaplasma-phagocytophilum*-infected nymphs compared with unfed uninfected nymphs [75]. Previous studies have also indicated modulation of signaling events via miR-5310 upon *A. phagocytophilum* infection [75–80]. In *B. burgdorferi* infection, we speculate that it might modulate signaling events and protect *B. burgdorferi* in the tick salivary glands. miR-5310 might also be involved in tick feeding, as it was found to be downregulated in *Rhipicephalus microplus* tick larvae upon exposure to host odor but not being allowed to feed [81].

3.5. Prediction of Target Genes and Gene Ontology (GO) and Functional Enrichment Analyses of the Target Network

Target proteins were used to build a high-confidence interaction network (interaction scores >0.9). STRING web analysis (Figure 4) showed that the target proteins of 23 DE miRNAs (11 upregulated and 12 downregulated) had similar interactions to those expected for a random set of proteins of similar size sampled from the *I. scapularis* genome (nodes = 687, edges = 79, average node degree = 0.23, average local clustering coefficient

= 0.0966, expected number of edges = 77, PPI enrichment *p*-value = 0.411). This does not necessarily mean that these selected proteins are not biologically meaningful, rather that these tick proteins may not be very well studied and their interactions might not yet be known to STRING.

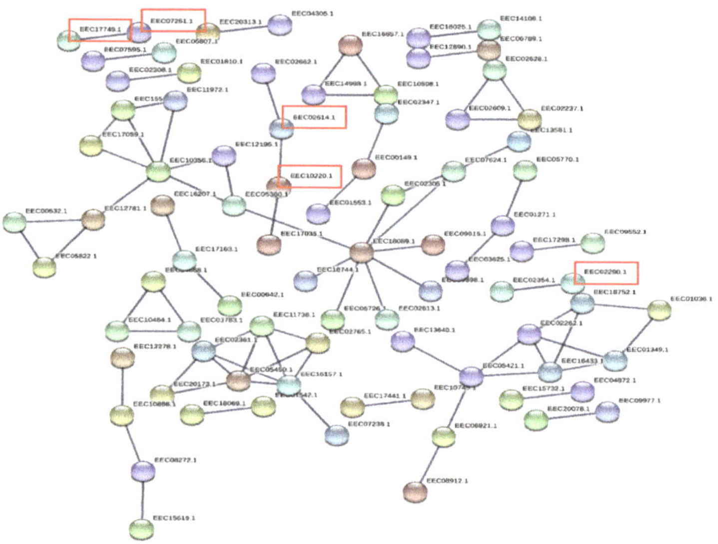

Figure 4. A network built exclusively from *Ixodes scapularis* proteins targeted by in silico differentially expressed miRNAs in *Borrelia-burgdorferi*-infected, partially fed salivary glands relative to partially fed clean salivary glands. Red boxes indicates the proteins involved in significant KEGG pathways such as sphingolipid metabolism (EEC07251.1, KEGG:R02541); valine, leucine, and isoleucine degradation (EEC10220.1, KEGG:R04188); lipid transport and metabolism (EEC02614.1, KEGG:R01178); exosome biogenesis and secretion (EEC07251.1, EEC17749.1, KEGG:R02541); and phosphate-containing compound metabolic process (EEC02290.1, KEGG:R00004).

Many target genes were identified for the DE miRNAs using the sRNAtoolbox miRNAconsTarget program [47]. To minimize false-positive targets, we chose only those targets predicted by all three miRNA target-prediction algorithms (TargetSpy, MIRANDA, and PITA). Forty-one KEGG pathways were enriched for target genes (proteins) of DE miRNAs (Supplementary Table S2) and included sphingolipid metabolism; valine, leucine, and isoleucine degradation; lipid transport and metabolism; exosome biogenesis and secretion; and phosphate-containing compound metabolic processes (Figure 4). Gene ontology (GO) analysis indicated that most target genes of DE miRNAs play significant roles in cellular processes, metabolic processes, biological regulation, developmental processes, and responses to stimuli (Figure 5). Surprisingly, immune response genes were one of the least affected functions.

Lipid metabolism was one of the main KEGG pathways detected by Pannzer and STRING analyses and was predicted to be regulated by tick salivary gland miRNAs miR-1, miR-5310, miR-71, and miR-79. It has previously been shown that the binding of *B. burgdorferi* to host glycosphingolipid can contribute to tissue-specific adhesion of *B. burgdorferi*, and the inflammatory process in Lyme borreliosis might be affected by interactions between *B. burgdorferi* and glycosphingolipid [94]. Therefore, we hypothesize that tick miRNAs (via saliva) promote sphingolipid synthesis inside hosts to promote *Borrelia* adhesion, and indeed, there is evidence that its infection affects lipid metabolism in hosts [95].

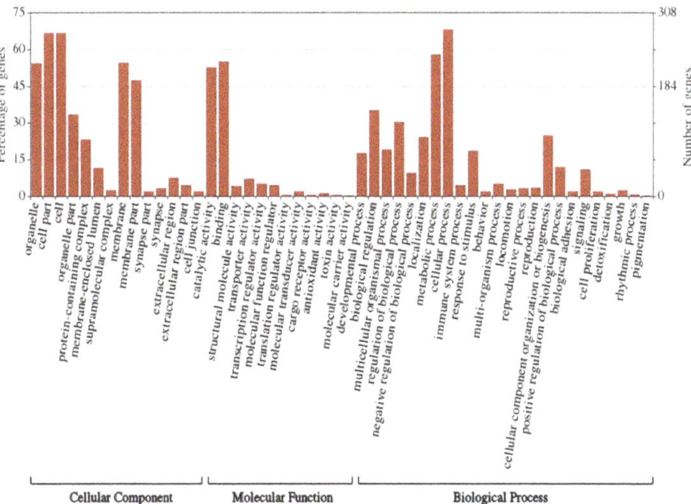

Figure 5. Gene ontology (GO)-derived biological processes related to genes targeted by differentially expressed miRNAs in partially fed *B. burgdorferi*-infected salivary glands relative to partially fed uninfected salivary glands from *Ixodes scapularis* ticks.

3.6. Validation of DE miRNAs by qRT-PCR

The expressions of DE miRNAs were validated in *B. burgdorferi*-infected and uninfected ISE6 cells by qRT-PCR (Figure 6 and Supplementary Table S3), which closely mirrored the RNA-seq data for many targets, although some differences were not statistically significant by qRT-PCR and isc-miR-317 was downregulated rather than upregulated. These differences could be due to the use of different methodologies to quantify miRNA expression [10].

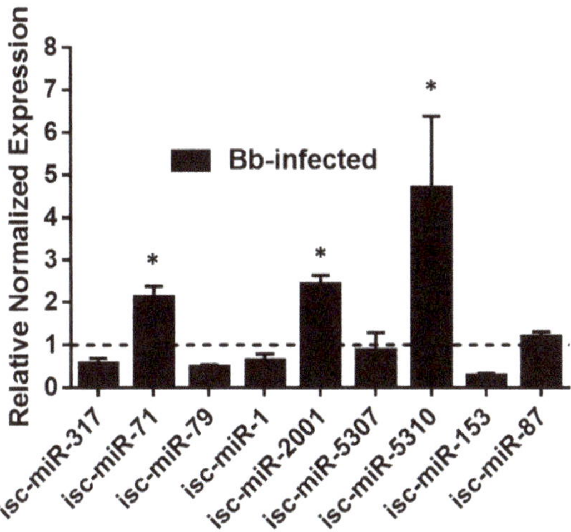

Figure 6. qRT-PCR validation of differentially expressed miRNAs detected in *Borrelia-burgdorferi*-infected, partially fed salivary glands relative to partially fed uninfected salivary glands from *Ixodes scapularis* ticks. qPCR validation was performed in *Borrelia-burgdorferi*-infected and clean ISE6 cells. Expression of miRNAs was normalized to clean ISE6 cells (indicated as 1 on the *y*-axis). Statistical significance for qRT-PCR-based differential expression was determined by the 2-tailed Student's *t*-test, where * is $p < 0.05$.

4. Conclusions

This is the first comprehensive miRNA profiling study of *Ixodes scapularis* salivary glands, with and without *Borrelia burgdorferi* infection. Here, we identified several potential miRNAs targets in tick salivary glands which might play a significant role in *Borrelia* colonization, survival, transmission, and host immunomodulation. Functional validation of these miRNAs is now required. Further characterization of tick salivary gland miRNAs would contribute to a better understanding of the mechanisms underpinning *Borrelia* transmission and propagation inside hosts, not least due to its special status as an extracellular spirochaete and atypical Gram-negative organism that might exploit different survival mechanisms. The impact of *B. burgdorferi* on miRNA expression must also be studied in other tick tissues and hosts to understand cues of its vector competence in ticks and immunomodulation in vertebrates.

Supplementary Materials: The following supporting information can be downloaded at: https://www.mdpi.com/article/10.3390/ijms23105565/s1.

Author Contributions: Conceptualization, D.K. and S.K.; methodology, D.K., L.P.D., M.E., A.S.F. and S.K.; software, D.K. and A.S.F.; validation, D.K. and L.P.D.; formal analysis, D.K. and S.K.; investigation, D.K. and S.K.; resources, S.K.; data curation, D.K. and S.K.; writing—original draft preparation, D.K. and S.K.; writing—review and editing, D.K., L.P.D., M.E., A.S.F. and S.K.; visualization, D.K.; supervision, S.K.; project administration, S.K.; funding acquisition, S.K. All authors have read and agreed to the published version of the manuscript.

Funding: This research was funded by USDA NIFA award #2017-67017-26171; Pakistan-US Science and Technology Cooperation award (US Department of State); the Mississippi INBRE, an institutional Award (IDeA) from the National Institute of General Medical Sciences of the National Institutes of Health under award P20GM103476 and The APC was funded by NIH NIAID award AI135049.

Institutional Review Board Statement: Not applicable.

Data Availability Statement: Publicly available datasets were analyzed in this study. This data can be found here: NCBI (accession: PRJNA837369).

Conflicts of Interest: The authors declare no conflict of interest.

References

1. Bartel, D.P. MicroRNAs: Genomics, Biogenesis, Mechanism, and Function. *Cell* **2004**, *116*, 281–297. [CrossRef]
2. Bartel, D.P. MicroRNAs: Target Recognition and Regulatory Functions. *Cell.* **2009**, *136*, 215–233. [CrossRef] [PubMed]
3. Carrington, J.C.; Ambros, V. Role of microRNAs in plant and animal development. *Science* **2003**, *301*, 336–338. [CrossRef] [PubMed]
4. Griffiths-Jones, S.; Saini, H.K.; van Dongen, S.; Enright, A.J. miRBase: Tools for MicroRNA Genomics. *Nucleic Acids Res.* **2008**, *36*, D154–D158. [CrossRef]
5. Lai, E.C. Two decades of miRNA biology: Lessons and Challenges. *RNA* **2015**, *21*, 675–677. [CrossRef]
6. Bartel, D.P.; Chen, C.Z. Micromanagers of gene expression: The Potentially Widespread Influence of Metazoan MicroRNAs. *Nat. Rev. Genet.* **2004**, *5*, 396–400. [CrossRef]
7. Alvarez-Garcia, I.; Miska, E.A. MicroRNA functions in animal development and human disease. *Development* **2005**, *132*, 4653–4662. [CrossRef]
8. Miesen, P.; Ivens, A.; Buck, A.H.; van Rij, R.P. Small RNA Profiling in Dengue Virus 2-Infected Aedes Mosquito Cells Reveals Viral piRNAs and Novel Host miRNAs. *PLoS Negl Trop Dis.* **2016**, *10*, e0004452. [CrossRef]
9. Momen-Heravi, F.; Bala, S. miRNA regulation of innate immunity. *J. Leukoc. Biol.* **2018**, *103*, 1205–1217. [CrossRef]
10. Saldaña, M.; Etebari, K.; Hart, C.E.; Widen, S.G.; Wood, T.G.; Thangamani, S.; Asgari, S.; Hughes, G.L. Zika virus alters the microRNA expression profile and elicits an RNAi response in Aedes aegypti mosquitoes. *PLoS Negl. Trop. Dis.* **2017**, *11*, e0005760. [CrossRef]
11. Rigoutsos, I. New tricks for animal microRNAS: Targeting of Amino Acid Coding Regions at Conserved and Nonconserved sites. *Cancer Res.* **2009**, *69*, 3245–3248. [CrossRef] [PubMed]
12. Schnall-Levin, M.; Zhao, Y.; Perrimon, N.; Berger, B. Conserved microRNA targeting in Drosophila is as widespread in coding regions as in 3′ UTRs. *Proc. Natl. Acad. Sci. USA* **2010**, *107*, 15751–15756. [CrossRef]
13. Asgari, S. MicroRNAs as regulators of insect host–pathogen interactions and immunity. *Adv. Insect Phys.* **2018**, *55*, 19–45.
14. Flynt, A.S.; Greimann, J.C.; Chung, W.J.; Lima, C.D.; Lai, E.C. MicroRNA biogenesis via splicing and exosome-mediated trimming in Drosophila. *Mol. Cell.* **2010**, *38*, 900–907. [CrossRef] [PubMed]

15. Anderson, J.M.; Sonenshine, D.E.; Valenzuela, J.G. Exploring the mialome of ticks: An Annotated Catalogue of Midgut Transcripts from the Hard Tick, Dermacentor Variabilis (Acari: Ixodidae). *BMC Genom.* **2008**, *9*, 552. [CrossRef]
16. Antunes, S.; Couto, J.; Ferrolho, J.; Sanches, G.S.; Charrez, J.O.M.; Hernández, N.D.L.C.; Mazuz, M.; Villar, M.; Shkap, V.; De La Fuente, J.; et al. Transcriptome and Proteome Response of Rhipicephalus annulatus Tick Vector to Babesia bigemina Infection. *Front. Physiol.* **2019**, *10*, 318. [CrossRef]
17. Popara, M.; Villar, M.; de la Fuente, J. Proteomics characterization of tick-host-pathogen interactions. *Methods Mol. Biol.* **2015**, *1247*, 513–527. [CrossRef]
18. Villar, M.; Ayllón, N.; Alberdi, P.; Moreno, A.; Moreno, M.; Tobes, R.; Mateos-Hernández, L.; Weisheit, S.; Bell-Sakyi, L.; de la Fuente, J. Integrated Metabolomics, Transcriptomics and Proteomics Identifies Metabolic Pathways Affected by Anaplasma phagocytophilum Infection in Tick Cells. *Mol. Cell Proteom.* **2015**, *14*, 3154–3172. [CrossRef]
19. Bechsgaard, J.; Vanthournout, B.; Funch, P.; Vestbo, S.; Gibbs, R.A.; Richards, S.; Sanggaard, K.W.; Enghild, J.J.; Bilde, T. Comparative genomic study of arachnid immune systems indicates loss of beta-1,3-glucanase-related proteins and the immune deficiency pathway. *J. Evol. Biol.* **2016**, *29*, 277–291. [CrossRef]
20. Gulia-Nuss, M.; Nuss, A.; Meyer, J.M.; Sonenshine, D.E.; Roe, R.M.; Waterhouse, R.M.; Sattelle, D.B.; De La Fuente, J.; Ribeiro, J.; Megy, K.; et al. Genomic insights into the Ixodes scapularis tick vector of Lyme disease. *Nat. Commun.* **2016**, *7*, 1–13. [CrossRef]
21. Rosa, R.D.; Capelli-Peixoto, J.; Mesquita, R.D.; Kalil, S.P.; Pohl, P.C.; Braz, G.R.; Fogaça, A.C.; Daffre, S. Exploring the immune signalling pathway-related genes of the cattle tick Rhipicephalus microplus: From Molecular Characterization to Transcriptional Profile upon Microbial Challenge. *Dev. Comp. Immunol.* **2016**, *59*, 1–14. [CrossRef] [PubMed]
22. Pal, U.; Kitsou, C.; Drecktrah, D.; Yaş, Ö.B.; Fikrig, E. Interactions Between Ticks and Lyme Disease Spirochetes. *Curr. Issues Mol. Biol.* **2021**, *42*, 113–144. [CrossRef] [PubMed]
23. Bensaci, M.; Bhattacharya, D.; Clark, R.; Hu, L.T. Oral vaccination with vaccinia virus expressing the tick antigen subolesin inhibits tick feeding and transmission of Borrelia burgdorferi. *Vaccine* **2012**, *30*, 6040–6046. [CrossRef] [PubMed]
24. Narasimhan, S.; Kurokawa, C.; Diktas, H.; Strank, N.O.; Černý, J.; Murfin, K.; Cao, Y.; Lynn, G.; Trentleman, J.; Wu, M.-J.; et al. Ixodes scapularis saliva components that elicit responses associated with acquired tick-resistance. *Ticks Tick Borne Dis.* **2020**, *11*, 101369. [CrossRef]
25. Pal, U.; Fikrig, E. Dynamics of Borrelia burgdorferi transmission by nymphal Ixodes dammini ticks. *J. Infect. Dis.* **2003**, *167*, 1082–1085. [CrossRef]
26. Piesman, J.; Schneider, B.S.; Zeidner, N.S. Use of quantitative PCR to measure density of Borrelia burgdorferi in the midgut and salivary glands of feeding tick vectors. *J. Clin. Microbiol.* **2001**, *39*, 4145–4148. [CrossRef]
27. Pal, U.; Fikrig, E. Adaptation of Borrelia burgdorferi in the vector and vertebrate host. *Microbes Infect.* **2003**, *5*, 659–666. [CrossRef]
28. Fikrig, E.; Narasimhan, S. Borrelia burgdorferi—Traveling incognito? *Microbes Infect* **2006**, *8*, 1390–1399. [CrossRef]
29. Ueti, M.W.; Reagan, J.O., Jr.; Knowles, D.P., Jr.; Scoles, G.A.; Shkap, V.; Palmer, G.H. Identification of midgut and salivary glands as specific and distinct barriers to efficient tick-borne transmission of Anaplasma marginale. *Infect. Immun* **2007**, *75*, 2959–2964. [CrossRef]
30. Ribeiro, J.M.; Alarcon-Chaidez, F.; Francischetti, I.M.B.; Mans, B.; Mather, T.N.; Valenzuela, J.G.; Wikel, S.K. An annotated catalog of salivary gland transcripts from Ixodes scapularis ticks. *Insect Biochem. Mol. Biol.* **2006**, *36*, 111–129. [CrossRef]
31. Patrick, C.D.; Hair, J.A. Laboratory rearing procedures and equipment for multi-host ticks (Acarina: Ixodidae). *J. Med. Entomol.* **1975**, *12*, 389–390. [CrossRef] [PubMed]
32. Karim, S.; Ramakrishnan, V.G.; Tucker, J.S.; Essenberg, R.C.; Sauer, J.R. Amblyomma americanum salivary glands: Double-Stranded RNA-Mediated Gene Silencing of Synaptobrevin Homologue and Inhibition of PGE2 Stimulated Protein Secretion. *Insect Biochem. Mol. Biol.* **2004**, *34*, 407–413. [CrossRef] [PubMed]
33. Kumar, D.; Budachetri, K.; Meyers, V.C.; Karim, S. Assessment of tick antioxidant responses to exogenous oxidative stressors and insight into the role of catalase in the reproductive fitness of the Gulf Coast tick, Amblyomma maculatum. *Insect Mol. Biol.* **2016**, *25*, 283–294. [CrossRef] [PubMed]
34. Embers, M.E.; Grasperge, B.J.; Jacobs, M.B.; Philipp, M.T. Feeding of ticks on animals for transmission and xenodiagnosis in Lyme disease research. *J. Vis. Exp.* **2013**, *78*, e50617. [CrossRef]
35. Karim, S.; Essenberg, R.C.; Dillwith, J.W.; Tucker, J.S.; Bowman, A.S.; Sauer, J.R. Identification of SNARE and cell trafficking regulatory proteins in the salivary glands of the lone star tick, *Amblyomma americanum* (L.). *Insect Biochem. Mol. Biol.* **2002**, *32*, 1711–1721. [CrossRef]
36. Kumar, D.; Embers, M.; Mather, T.N.; Karim, S. Is selenoprotein K required for Borrelia burgdorferi infection within the tick vector Ixodes scapularis? *Parasites Vectors* **2019**, *12*, 289. [CrossRef]
37. Zia, M.F.; Flynt, A.S. Detection and Verification of Mammalian Mirtrons by Northern Blotting. *Methods Mol. Biol.* **2018**, *1823*, 209–219. [CrossRef]
38. Stone, B.L.; Russart, N.M.; Gaultney, R.A.; Floden, A.M.; Vaughan, J.A.; Brissette, C.A. The Western progression of lyme disease: Infectious and Nonclonal Borrelia Burgdorferi Sensu Lato Populations in Grand Forks County, North Dakota. *Appl. Environ. Microbiol.* **2015**, *81*, 48–58. [CrossRef]
39. Friedländer, M.R.; Chen, W.; Adamidi, C.; Maaskola, J.; Einspanier, R.; Knespel, S.; Rajewsky, N. Discovering microRNAs from deep sequencing data using miRDeep. *Nat. Biotechnol.* **2008**, *26*, 407–415. [CrossRef]

40. Friedlander, M.R.; Mackowiak, S.D.; Li, N.; Chen, W.; Rajewsky, N. miRDeep2 accurately identifies known and hundreds of novel microRNA genes in seven animal clades. *Nucleic Acids Res.* **2012**, *40*, 37–52. [CrossRef]
41. Munderloh, U.G.; Kurtti, T.J. Formulation of medium for tick cell culture. *Exp. Appl. Acarol.* **1989**, *7*, 219–229. [CrossRef] [PubMed]
42. Obonyo, M.; Munderloh, U.G.; Fingerle, V.; Wilske, B.; Kurtti, T.J. Borrelia burgdorferi in tick cell culture modulates expression of outer surface proteins A and C in response to temperature. *J. Clin. Microbiol.* **1999**, *37*, 2137–2141. [CrossRef] [PubMed]
43. Li, Y.; Andrade, J. DEApp: An Interactive Web Interface for Differential Expression Analysis of Next Generation Sequence Data. *Source Code Biol. Med.* **2017**, *12*, 2. [CrossRef] [PubMed]
44. Sturm, M.; Hackenberg, M.; Langenberger, D.; Frishman, D. TargetSpy: A Supervised Machine Learning Approach for MicroRNA Target Prediction. *BMC Bioinform.* **2010**, *11*, 292. [CrossRef]
45. John, B.; Enright, A.J.; Aravin, A.; Tuschl, T.; Sander, C.; Marks, D.S. Human MicroRNA targets. *PLoS Biol.* **2004**, *2*, e363. [CrossRef]
46. Kertesz, M.; Iovino, N.; Unnerstall, U.; Gaul, U.; Segal, E. The role of site accessibility in microRNA target recognition. *Nat. Genet.* **2007**, *39*, 1278–1284. [CrossRef]
47. Aparicio-Puerta, E.; Lebrón, R.; Rueda, A.; Gómez-Martín, C.; Giannoukakos, S.; Jáspez, D.; Medina, J.M.; Zubković, A.; Jurak, I.; Fromm, B.; et al. sRNAbench and sRNAtoolbox 2019: Intuitive Fast Small RNA Profiling and Differential Expression. *Nucleic Acids Res.* **2019**, *47*, W530–W535. [CrossRef]
48. Min, H.; Yoon, S. Got target?: Computational Methods for MicroRNA Target Prediction and Their Extension. *Exp. Mol. Med.* **2010**, *42*, 233–244. [CrossRef]
49. Franceschini, A.; Szklarczyk, D.; Frankild, S.; Kuhn, M.; Simonovic, M.; Roth, A.; Lin, J.; Minguez, P.; Bork, P.; von Mering, C.; et al. STRING v9.1: Protein-Protein Interaction Networks, with Increased Coverage and Integration. *Nucleic Acids Res.* **2013**, *41*, D808–D815. [CrossRef]
50. Toronen, P.; Medlar, A.; Holm, L. PANNZER2: A Rapid Functional Annotation Web Server. *Nucleic Acids Res.* **2018**, *46*, W84–W88. [CrossRef]
51. Ye, J.; Zhang, Y.; Cui, H.; Liu, J.; Wu, Y.; Cheng, Y.; Xu, H.; Huang, X.; Li, S.; Zhou, A.; et al. WEGO 2.0: A Web Tool for Analyzing and Plotting GO Annotations, 2018 Update. *Nucleic Acids Res.* **2018**, *46*, W71–W75. [CrossRef] [PubMed]
52. Wu, Z.; He, B.; He, J.; Mao, X. Upregulation of miR-153 promotes cell proliferation via downregulation of the PTEN tumor suppressor gene in human prostate cancer. *Prostate* **2013**, *73*, 596–604. [CrossRef] [PubMed]
53. Huang, T.; Xu, D.; Zhang, X. Characterization of host microRNAs that respond to DNA virus infection in a crustacean. *BMC Genomics* **2012**, *13*, 1–10. [CrossRef] [PubMed]
54. Huang, T.; Zhang, X. Functional analysis of a crustacean microRNA in host-virus interactions. *J. Virol.* **2012**, *86*, 12997–13004. [CrossRef] [PubMed]
55. Hussain, M.; Torres, S.; Schnettler, E.; Funk, A.; Grundhoff, A.; Pijlman, G.P.; Khromykh, A.; Asgari, S. West Nile virus encodes a microRNA-like small RNA in the 3′ untranslated region which up-regulates GATA4 mRNA and facilitates virus replication in mosquito cells. *Nucleic Acids Res.* **2012**, *40*, 2210–2223. [CrossRef]
56. Xu, H.; Jiang, Y.; Xu, X.; Su, X.; Liu, Y.; Ma, Y.; Zhao, Y.; Shen, Z.; Huang, B.; Cau, X. Inducible degradation of lncRNA Sros1 promotes IFN-gamma-mediated activation of innate immune responses by stabilizing Stat1 mRNA. *Nat. Immunol.* **2019**, *20*, 1621–1630. [CrossRef]
57. Wang, F.; Gong, H.; Zhang, H.; Zhou, Y.; Cao, J.; Zhou, J. Lipopolysaccharide-Induced Differential Expression of miRNAs in Male and Female Rhipicephalus haemaphysaloides Ticks. *PLoS ONE.* **2015**, *10*, e0139241. [CrossRef]
58. Bond, D.; Foley, E. A quantitative RNAi screen for JNK modifiers identifies Pvr as a novel regulator of Drosophila immune signaling. *PLoS Pathog.* **2009**, *5*, e1000655. [CrossRef]
59. Fullaondo, A.; Lee, S.Y. Identification of putative miRNA involved in Drosophila melanogaster immune response. *Dev. Comp. Immunol.* **2012**, *36*, 267–273. [CrossRef]
60. Artigas-Jerónimo, S.; Alberdi, P.; Villar Rayo, M.; Cabezas-Cruz, A.; Prados, P.J.E.; Mateos-Hernández, L.; de la Fuente, J. Anaplasma phagocytophilum modifies tick cell microRNA expression and upregulates isc-mir-79 to facilitate infection by targeting the Roundabout protein 2 pathway. *Sci. Rep.* **2019**, *9*, 1–15. [CrossRef]
61. Freitak, D.; Knorr, E.; Vogel, H.; Vilcinskas, A. Gender- and stressor-specific microRNA expression in Tribolium castaneum. *Biol. Lett.* **2012**, *8*, 860–863. [CrossRef] [PubMed]
62. Zhang, X.; Zheng, Y.; Jagadeeswaran, G.; Ren, R.; Sunkar, R.; Jiang, H. Identification of conserved and novel microRNAs in Manduca sexta and their possible roles in the expression regulation of immunity-related genes. *Insect. Biochem. Mol. Biol.* **2014**, *47*, 12–22. [CrossRef]
63. Hackenberg, M.; Langenberger, D.; Schwarz, A.; Erhart, J.; Kotsyfakis, M. In silico target network analysis of de novo-discovered, tick saliva-specific microRNAs reveals important combinatorial effects in their interference with vertebrate host physiology. *RNA* **2017**, *23*, 1259–1269. [CrossRef] [PubMed]
64. Li, R.; Huang, Y.; Zhang, Q.; Zhou, H.; Jin, P.; Ma, F. The miR-317 functions as a negative regulator of Toll immune response and influences Drosophila survival. *Dev. Comp. Immunol.* **2019**, *95*, 19–27. [CrossRef] [PubMed]
65. Pushpavalli, S.N.; Sarkar, A.; Bag, I.; Hunt, C.R.; Ramaiah, M.J.; Pandita, T.K.; Bhadra, U.; Pal-Bhadra, M. Argonaute-1 functions as a mitotic regulator by controlling Cyclin B during DDrosophila early embryogenesis. *FASEB J.* 2014. [CrossRef]

66. Fricke, C.; Green, D.; Smith, D.; Dalmay, T.; Chapman, T. MicroRNAs influence reproductive responses by females to male sex peptide in drosophila melanogaster. *Genetics* **2014**, *198*, 1603–1619. [CrossRef]
67. Zhou, Y.; He, Y.; Wang, C.; Zhang, X. Characterization of miRNAs from hydrothermal vent 20 Rimicaris exoculata. *Mar. Genomics.* **2015**, *24*, 371–378. [CrossRef]
68. Hermance, M.E.; Widen, S.G.; Wood, T.G.; Thangamani, S. Ixodes scapularis salivary gland microRNAs are differentially expressed during Powassan virus transmission. *Sci. Rep.* **2019**, *9*, 1–17. [CrossRef]
69. Boulias, K.; Horvitz, H.R. The C. elegans MicroRNA mir-71 acts in neurons to promote germline-mediated longevity through regulation of DAF-16/FOXO. *Cell Metab.* **2012**, *15*, 439–450. [CrossRef]
70. Mukherjee, K.; Vilcinskas, A. Development and immunity-related microRNAs of the lepidopteran model host Galleria mellonella. *BMC Genom.* **2014**, *15*, 1–12. [CrossRef]
71. Labunskyy, V.M.; Hatfield, D.L.; Gladyshev, V.N. Selenoproteins: Molecular Pathways and Physiological Roles. *Physiol. Rev.* **2014**, *94*, 739–777. [CrossRef] [PubMed]
72. He, Y.; Sun, Y.; Zhang, X. Noncoding miRNAs bridge virus infection and host autophagy in shrimp in vivo. *FASEB J.* **2017**, *31*, 2854–2868. [CrossRef] [PubMed]
73. Avila-Bonilla, R.G.; Yocupicio-Monroy, M.; Marchat, L.A.; De Nova-Ocampo, M.A.; Del Angel, R.M.; Salas-Benito, J.S. Analysis of the miRNA profile in C6/36 cells persistently infected with dengue virus type 2. *Virus Res.* **2017**, *232*, 139–151. [CrossRef] [PubMed]
74. Liu, Y.; Zhou, Y.; Wu, J.; Zheng, P.; Li, Y.; Zheng, X.; Puthiyakunnon, S.; Tu, Z.; Chen, X.-G. The expression profile of Aedes albopictus miRNAs is altered by dengue virus serotype-2 infection. *Cell Biosci.* **2015**, *5*, 1–11. [CrossRef]
75. Ramasamy, E.; Taank, V.; Anderson, J.F.; Sultana, H.; Neelakanta, G. Repression of tick microRNA-133 induces organic anion transporting polypeptide expression critical for Anaplasma phagocytophilum survival in the vector and transmission to the vertebrate host. *PLoS Genet.* **2020**, *16*, e1008856. [CrossRef]
76. Khanal, S.; Sultana, H.; Catravas, J.D.; Carlyon, J.A.; Neelakanta, G. Anaplasma phagocytophilum infection modulates expression of megakaryocyte cell cycle genes through phosphatidylinositol-3-kinase signaling. *PLoS ONE.* **2017**, *12*, e0182898. [CrossRef]
77. Sultana, H.; Neelakanta, G.; Kantor, F.S.; Malawista, S.E.; Fish, D.; Montgomery, R.R.; Fikrig, E. Anaplasma phagocytophilum induces actin phosphorylation to selectively regulate gene transcription in Ixodes scapularis ticks. *J. Exp. Med.* **2010**, *207*, 1727–1743. [CrossRef]
78. Taank, V.; Dutta, S.; Dasgupta, A.; Steeves, T.K.; Fish, D.; Anderson, J.F.; Sultana, H.; Neelakanta, G. Human rickettsial pathogen modulates arthropod organic anion transporting polypeptide and tryptophan pathway for its survival in ticks. *Sci. Rep.* **2017**, *7*, 13256. [CrossRef]
79. Turck, J.W.; Taank, V.; Neelakanta, G.; Sultana, H. Ixodes scapularis Src tyrosine kinase facilitates Anaplasma phagocytophilum survival in its arthropod vector. *Ticks Tick Borne Dis.* **2019**, *10*, 838–847. [CrossRef]
80. Neelakanta, G.; Sultana, H.; Fish, D.; Anderson, J.F.; Fikrig, E. Anaplasma phagocytophilum induces Ixodes scapularis ticks to express an antifreeze glycoprotein gene that enhances their survival in the cold. *J. Clin. Invest.* **2010**, *120*, 3179–3190. [CrossRef]
81. Barrero, R.A.; Keeble-Gagnère, G.; Zhang, B.; Moolhuijzen, P.; Ikeo, K.; Tateno, Y.; Gojobori, T.; Guerrero, F.D.; Lew-Tabor, A.; Bellgrad, M. Evolutionary conserved microRNAs are ubiquitously expressed compared to tick-specific miRNAs in the cattle tick Rhipicephalus (Boophilus) microplus. *BMC Genom.* **2011**, *12*, 1–17. [CrossRef] [PubMed]
82. Nawaz, M.; Malik, M.I.; Zhang, H.; Gebremedhin, M.B.; Cao, J.; Zhou, Y.; Zhou, J. miRNA profile of extracellular vesicles isolated from saliva of Haemaphysalis longicornis tick. *Acta Trop.* **2020**, *212*, 105718. [CrossRef] [PubMed]
83. Colombo, M.; Raposo, G.; Thery, C. Biogenesis, secretion, and intercellular interactions of exosomes and other extracellular vesicles. *Annu. Rev. Cell. Dev. Biol.* **2014**, *30*, 255–289. [CrossRef] [PubMed]
84. Coakley, G.; Maizels, R.M.; Buck, A.H. Exosomes and Other Extracellular Vesicles: The New Communicators in Parasite Infections. *Trends Parasitol.* **2015**, *31*, 477–489. [CrossRef] [PubMed]
85. Lambertz, U.; Oviedo Ovando, M.E.; Vasconcelos, E.J.; Unrau, P.J.; Myler, P.J.; Reiner, N.E. Small RNAs derived from tRNAs and rRNAs are highly enriched in exosomes from both old and new world Leishmania providing evidence for conserved exosomal RNA Packaging. *BMC Genomics.* **2015**, *16*, 151. [CrossRef]
86. Ramachandra, R.N.; Wikel, S.K. Effects of Dermacentor andersoni (Acari: Ixodidae) salivary gland extracts on Bos indicus and B. taurus lymphocytes and macrophages: In vitro cytokine elaboration and lymphocyte blastogenesis. *J. Med. Entomol.* **1995**, *32*, 338–345. [CrossRef]
87. Urioste, S.; Hall, L.R.; Telford, S.R., 3rd; Titus, R.G. Saliva of the Lyme disease vector, Ixodes dammini, blocks cell activation by a nonprostaglandin E2-dependent mechanism. *J. Exp. Med.* **1994**, *180*, 1077–1085. [CrossRef]
88. Kung, F.; Anguita, J.; Pal, U. Borrelia burgdorferi and tick proteins supporting pathogen persistence in the vector. *Future Microbiol.* **2013**, *8*, 41–56. [CrossRef]
89. Ma, Y.; Sturrock, A.; Weis, J.J. Intracellular localization of Borrelia burgdorferi within human endothelial cells. *Infect. Immun.* **1991**, *59*, 671–678. [CrossRef]
90. Georgilis, K.; Peacocke, M.; Klempner, M.S. Fibroblasts protect the Lyme disease spirochete, Borrelia burgdorferi, from ceftriaxone in vitro. *J. Infect. Dis.* **1992**, *166*, 440–444. [CrossRef]
91. Klempner, M.S.; Noring, R.; Rogers, R.A. Invasion of human skin fibroblasts by the Lyme disease spirochete, Borrelia burgdorferi. *J. Infect. Dis.* **1993**, *167*, 1074–1081. [CrossRef] [PubMed]

92. Wawrzeniak, K.; Gaur, G.; Sapi, E.; Senejani, A.G. Effect of *Borrelia burgdorferi* Outer Membrane Vesicles on Host Oxidative Stress Response. *Antibiotics* **2020**, *9*, 275. [CrossRef] [PubMed]
93. Gaur, G.; Sawant, J.Y.; Chavan, A.S.; Khatri, V.A.; Liu, Y.-H.; Zhang, M.; Sapi, E. Effect of Invasion of *Borrelia burgdorferi* in Normal and Neoplastic Mammary Epithelial Cells. *Antibiotics* **2021**, *10*, 1295. [CrossRef] [PubMed]
94. Kaneda, K.; Masuzawa, T.; Yasugami, K.; Suzuki, T.; Suzuki, Y.; Yanagihara, Y. Glycosphingolipid-binding protein of Borrelia burgdorferi sensu lato. *Infect. Immun.* **1997**, *65*, 3180–3185. [CrossRef]
95. O'Neal, A.J.; Butler, L.R.; Rolandelli, A.; Gilk, S.D.; Pedra, J.H. Lipid hijacking: A unifying theme in vector-borne diseases. *Elife* **2020**, *9*, e61675. [CrossRef]

International Journal of
Molecular Sciences

MDPI

Article

A Deeper Insight into the Tick Salivary Protein Families under the Light of Alphafold2 and Dali: Introducing the TickSialoFam 2.0 Database

Ben J. Mans [1,2,3], John F. Andersen [4] and José M. C. Ribeiro [4,*]

[1] Epidemiology, Parasites and Vectors, Agricultural Research Council—Onderstepoort Veterinary Research, Onderstepoort, Pretoria 0110, South Africa
[2] Department of Life and Consumer Sciences, University of South Africa, Pretoria 0002, South Africa
[3] Department of Veterinary and Tropical Diseases, University of Pretoria, Pretoria 0002, South Africa
[4] Laboratory of Malaria and Vector Research, National Institute of Allergy and Infectious Diseases, Rockville, MD 20852, USA
* Correspondence: jribeiro@niaid.nih.gov

Abstract: Hard ticks feed for several days or weeks on their hosts and their saliva contains thousands of polypeptides belonging to dozens of families, as identified by salivary transcriptomes. Comparison of the coding sequences to protein databases helps to identify putative secreted proteins and their potential functions, directing and focusing future studies, usually done with recombinant proteins that are tested in different bioassays. However, many families of putative secreted peptides have a unique character, not providing significant matches to known sequences. The availability of the Alphafold2 program, which provides in silico predictions of the 3D polypeptide structure, coupled with the Dali program which uses the atomic coordinates of a structural model to search the Protein Data Bank (PDB) allows another layer of investigation to annotate and ascribe a functional role to proteins having so far being characterized as "unique". In this study, we analyzed the classification of tick salivary proteins under the light of the Alphafold2/Dali programs, detecting novel protein families and gaining new insights relating the structure and function of tick salivary proteins.

Keywords: medical entomology; tick; salivary glands; structure; classification

Citation: Mans, B.J.; Andersen, J.F.; Ribeiro, J.M.C. A Deeper Insight into the Tick Salivary Protein Families under the Light of Alphafold2 and Dali: Introducing the TickSialoFam 2.0 Database. *Int. J. Mol. Sci.* **2022**, 23, 15613. https://doi.org/10.3390/ijms232415613

Academic Editor: Michail Kotsyfakis

Received: 8 November 2022
Accepted: 6 December 2022
Published: 9 December 2022

Publisher's Note: MDPI stays neutral with regard to jurisdictional claims in published maps and institutional affiliations.

1. Introduction

The saliva of bloodsucking animals contains a vast array of compounds that inhibit their hosts' hemostasis (which consists of platelet or thrombocyte aggregation, blood clotting and vasoconstriction) and can also have immunomodulatory properties and counteract their hosts' tissue repair mechanisms [1]. Hard ticks feed for several days or weeks on their hosts and their saliva contains thousands of polypeptides belonging to dozens of families, as identified by salivary transcriptomes [2]. Bioinformatic methods help to catalog these families, after identification of signal peptides indicative of secretion [3], identification of transmembrane domains [4] help to exclude transcripts that might code for proteins with a signal peptide domain, but the product may be a membrane protein having extracellular domains. Comparison of the coding sequences to protein databases through the blastp tool, or to motif databases using the rpsblast tool against the Conserved Domains Database (CDD), Pfam, Smart and Kog databases helps to identify putative secreted proteins and their potential functions, directing and focusing future studies, usually done with recombinant proteins that are tested in different bioassays. However, many families of putative secreted peptides have a unique character, not providing significant matches to known sequences. Recently, a database named TickSialoFam (TSFam) classified tick salivary proteins based on psiblast-generated position-specific scoring matrices (PSSM) [5], listing 136 groups of tick salivary proteins, from which only 24 had at least one member biochemically characterized.

Of these 136 groups, 63 are tick specific, or "unique" as indicated by their primary structure having no significant matches to known proteins.

The availability in 2021 of the Alphafold2 program [6], which provides in silico predictions of the 3D polypeptide structure, coupled with the Dali program [7], which uses the atomic coordinates of a structural model to search the Protein Data Bank (PDB) [8] allowing another layer of investigation to annotate and ascribe a functional role to proteins having so far being characterized as "unique". In this study, we analyzed the TSFam database classification under the light of the Alphafold2/Dali programs, having made changes in the names and putative function of 15 protein families, and updating the TSFam database to a 2.0 version.

2. Results

The Alphafold2/Dali analysis of the tick salivary proteins allowed for changes in our previous classification of these proteins as proposed in the TickSialoFam database [5]. New protein groups were created, such as metalloproteoid, while some protein groups of unknown function or structure were incorporated into previously known groups, and the predicted structures of formerly known groups are here described, together with evolutionary considerations.

2.1. The Metalloproteoid Group

2.1.1. The 28 kDa Protein Family—Changed to New Group Metalloproteoid, Metastriate

The AlphaFold2-Dali (AF-Dali) pipeline identified 58 sequences of the TSFam protein family named "28 kDa", which appears exclusively in the Metastriata, as producing significant matches (Z values ranging from 7.4 to 14.8) to zinc metalloproteases (spreadsheet S1). Alignment of 28 kDa sequences with their Dali matching metalloproteases, mostly from snake venom, indicate only 2.5% identity and 16.7% similarity (Figure S1). Notice that the zinc-binding histidine motif HEXXHXXGXXH (shown by a rectangle on Figure S1) required for enzymatic activity by the metalloproteases [9] is absent in the tick proteins. Even so, the predicted tertiary structure of the tick proteins matches well with all the beta sheets and most of the alpha helices of the metalloproteases (Figure 1 and movie on Figure S2). We thus rename the 28 kDa family as the Metalloproteoid family, subfamily 28 kDa. The substrate specificity of snake metalloproteases covers a wide range, including coagulation factors, platelet membrane receptors or von Willebrand factor [10]. It is possible that tick metalloproteoids function by binding to specific host proteins without hydrolyzing them, but possibly inhibiting their function, thus acting as kratagonists.

2.1.2. Amb-25-357 Changed to Metalloproteoid, Amblyomma Koch, 1844 (Acari: Ixodidae)

The Alphafold2 predictions of the "unknown" protein family restricted to the *Amblyomma* named Amb-25-357 were matched, with Z values above 14, to metalloproteases (Spreadsheet S1 and Figure S3). Similar to the 28 kDa family above, the alignment of members of this family with metalloproteases shows the absence of the histidine rich domain indicative of zinc ion binding. This family is thus reclassified to the Metalloproteoid group, subgroup *Amblyomma*.

2.1.3. Lipocalin P32 Antigen and Lipocalin, 26 kDa_a—Changed to Metalloproteoid, Subfamilies P32 and 26 kDa_a

The TSFam 1.0 group Lipocalins, sub-groups P32 antigen and 26 kDa_a revealed by the AF-Dali pipeline to be members of the Metalloproteoid group as their AF structure predictions matches metalloproteases with Z values ranging from 10 to 15 for the proteins having MW larger than 17 kDa. All sequences from these groups originated from the *Ixodes* Latreille, 1795 genus.

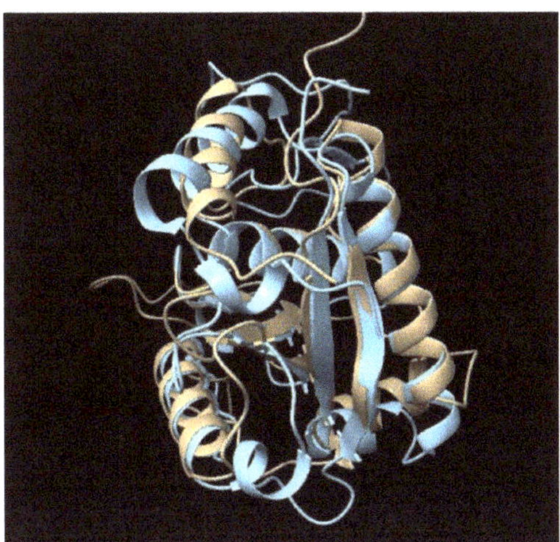

Figure 1. Metalloproteoid JAA55829.1 predicted Alphafold structure (Brown color) superimposed to the metalloprotease leucurolysin-a (PDB 4q1l-A) (Blue color).

2.2. Reevaluation of the Metalloprotease Family

Having discovered the Metalloproteoid family (see above), we decided to reanalyze the Metalloprotease family identified by the TSFam1.0 database, in particular to certify whether each sequence had the Zn binding domain. We thus searched the motif H-E-x(2)-H-x(2)-G-x(2)-H using the tool ps_scan.pl [11] and discovered that the TSFam1.0 PSMM models 30-49, 30-561, 30-93, 35-121, 35-167, 35-414, 35-415, 35-768, 35-77 and 50-341 were identifying Metalloproteoid sequences. Figure S4 displays three tick protein sequences previously classified as metalloproteases aligned to three snake metalloproteases. Notice the lack of the Zn binding domain on the tick proteins, now classified as members of the Metalloproteoid family.

2.3. New Lipocalin Acquisitions

Hyp 94 Protein Family—Changed to Lipocalin, Hyp94

The Hyp 94 protein family, which appears exclusive to the genus *Amblyomma*, showed Dali similarities of AF-predicted structures to tick lipocalin proteins named Jacalin and female-specific histamine binding protein [12,13] with Z values ranging from 12 to 15. Thus, it is being reclassified within the Lipocalin group, subgroup Hyp 94. A 3D structure comparison is shown in Figure S5.

2.4. New Niemann-Pick Acquisitions

2.4.1. Group 17.7 kDa—Changed to Niemann-Pick, Argasidae

This family of proteins, apparently exclusive to soft ticks, presents AF-Dali similarities with Z values above 12 to Niemann-Pick proteins [14] having a ML domain [15] and accordingly its group is added to the previously existing group "Niemann Pick" forming a new subgroup "argasidae". These proteins recognize lipids, including pathogen motifs. If secreted in tick saliva they may act as kratagonists. It is interesting that the primary sequence of these proteins does not provide for matches to the pfam02221, E1_DerP2_DerF2, ML domain when searched with the rpsblast tool. However, there is a striking match of the 3D structure between the 17.7 kDa family members and the crystal structure of canonical ML domain containing proteins (movie on Figure S6).

2.4.2. Lipocalin Metastriate—Changed to Niemann-Pick, Metastriate

Similar to the 17.7 kDa group, the Lipocalin Metastriate subgroup was transferred to the Niemann-Pick group.

2.4.3. Cytotoxin Sequences Changed to Niemann-Pick, Der f 7 Antigen and Sulfotransferase Groups

The TSFam1.0 database identified 137 sequences as belonging to the Cytotoxin group. Following AF-DALI results, 94 of these sequences matched the Der f 7 antigen and 32 matched bacterial toxin-like proteins. Two proteins matched protein tyrosine sulfotransferases and 9 did not provide significant structural matches. The two sequences matching sulfotransferases matched by blastp sequences annotated as sulfotransferases had CDD matches to sulfotransferases, and accordingly they were wrongly included in the TSFam Cytotoxin group. The description of the 2 families that were newly identified as characterized by the AF-Dali pipeline follow below.

2.5. Cytotoxin

The AF-DALI pipeline identified 32 sequences from ticks of the genus *Ixodes* that matched, with Z values ranging from 10 to 16, bacterial proteins with pdb accessions 2d42-A (NON-TOXIC CRYSTAL PROTEIN), 4rhz-A (CRY23AA1), 3zjx-A (EPSILON-TOXIN), 2ztb-B (CRYSTAL PROTEIN),7ml9-A (INSECTICIDAL PROTEIN), 6lh8-A (AEROLYSIN-LIKE PROTEIN) and 1w3a-A (HEMOLYTIC LECTIN LSLA). These proteins are related to bacterial pore-forming proteins. The presence of these proteins in ticks suggest they were acquired by horizontal transfer, as was the case of the DAE antimicrobial proteins. Figure S7A shows the AF predicted structure for *Ixodes ricinus* (Linnaeus, 1758) JAB71021.1. and in 7B shows the comparison of JAB71021.1 with the dimeric bacterial protein PDB:2d42. Similarly, the DAP-36 protein identified as a *Dermacentor andersoni* Stiles, 1908 immunosuppressive protein [16] matched bacterial toxins. Accordingly, its classification changed from Group DAP-36 to Group Cytotoxin, subgroup DAP-36.

2.6. Der f 7 Allergen/JHBP

Protein sequences from both prostriate and metastriate ticks were found matching the structure of Der f 7 allergen with Z scores above 10, and also matching, usually with smaller Z scores, structures annotated as Juvenile Hormone Binding Proteins (JHBP) (Figure S8). The former group Mys-30-94 was identified as a member of the Der f 7 allergen, as were the former group OneOfEach, which were reclassified accordingly. The Der f7 antigen is an allergenic protein derived from the mite *Dermatophagoides farina* Hughes, 1961 (Acari: Pyroglyphidae). While the function of Der f 7 allergen is unknown, JHBP are well known carriers of the lipidic juvenile hormone found in arthropods. It is possible this family acts as lipid kratagonists.

2.7. Additional Changes

Following AF-DALI analysis, the following TSFam motifs were changed as indicated below:

40-800, Evasin → changed to Kunitz
40-584, Evasin → changed to acid tail
35-53, Kunitz → changed to Lipocalin
Rapp-25-325 → changed to cystatin
25-159 Cytochrome_P450 → changed to Cytochrome_B2

2.8. The Search for Salivary Disintegrins

Disintegrin was the name given to soluble snake venom toxins consisting of peptides varying in length from 49-84 amino acids which were able to prevent platelet aggregation by inhibiting the cross-linking of activated platelets by fibrinogen, which binds to the platelet integrin $\alpha_{IIb}\beta_3$ (fibrinogen receptor) [17]. These inhibitors contain an amino acid triad (usually KGD or RGD) flanked, a few amino acids up or down the triad, by cysteines that form a bridge and stabilizes the hairpin structure. The environment around the disintegrin loop has an important role in the curvature and flexibility of this loop, determining the integrin specificity: some RGD disintegrins do interact with $\alpha_3\beta_1$, $\alpha_6\beta_1$, and $\alpha_7\beta_1$ thus interfering with laminin-based cell adhesion while the RGD in jarastatin targets $\alpha_M\beta_2$ in neutrophils. KTS and RTS disintegrins, on the other hand, are selective inhibitors of the $\alpha_1\beta_1$ integrin, a receptor for collagen IV. Later it was shown that snake disintegrins were derived from limited proteolysis of proteases of the ADAM (A Disintegrin And Metalloprotease) and ADAMTS (A Disintegrin And Metalloprotease with Thrombospondin motifs), which are widely found in both invertebrates and vertebrates [9,10,18–20].

Saliva of ticks contain disintegrins inhibiting platelet aggregation [21], as well inhibitors of neutrophil aggregation, which were identified by their similarity to ADAMTS metalloproteases [22]. Saliva of sand flies also contain a typical disintegrin [23]. Interestingly, saliva of a horse fly contains a protein named Tablysin-15 which is a member of the widespread antigen 5 family [24] that, however, acquired an RGD domain resulting in a high affinity binding for platelet $\alpha_{IIb}\beta_3$ and endothelial cell $\alpha_V\beta_3$, but not for $\alpha_5\beta_1$ or $\alpha_2\beta_1$ integrins [25] thus showing that the disintegrin function can be found in non-canonical protein families. To predict a disintegrin function on a protein it is important that the disintegrin triad is shown as a hairpin protruding from the remaining protein backbone. We thus used ps_scan to scan our tick salivary protein database (Spreadsheet S1) against a disintegrin motifs database (Supplementary File S1), resulting in 804 of the 15,797 sequences, or 5%, displaying at least one disintegrin motif. These results are mapped to column AM of Spreadsheet S1. After searching for the hallmark loop on 100 of these 804 candidate sequences, we found 18 that were strong candidates for further testing of disintegrin activity, eight of which are shown in Figure S9, belonging to different tick salivary families. A complete description of all the predicted disintegrin sequences is out of the scope of this work.

2.9. Secreted cyP450 Enzymes

Cytochrome P450 enzymes are ubiquitous heme-containing proteins that catalyze oxidative reactions in steroids, fatty acids, prostaglandins, leukotrienes, biogenic amines, pheromones and plant metabolites [26]. In eukaryotes these enzymes are mostly found bound to membranes. However, we found many tick proteins from salivary transcriptomes that present a clear signal peptide indicative of secretion, no transmembrane domains (outside the signal peptide) or GPI anchors. Because most reactions catalyzed by p450 enzymes involves the oxidative transfer of electrons from the porphyrinic iron to the substrate, a mechanism to regenerate the reduced iron is needed, and this task is associated with the p450 reductase or cytochrome b5 enzymes, which rely on cellular reserves of NADPH. This poses a problem for extracellular p450 enzymes, as they would not be able to access their iron reducing partners [2]. However, p450 enzymes can participate in peroxidase reactions. For example, prostaglandin H_2 is converted to thromboxane and prostacyclin by two P450s [27,28]. These unusual P450 reactions do not need electrons or O_2 [29].

The alignment of tick salivary secreted proteins of the p450 family with matching sequences from the PDB database produces the phylogenetic tree depicted in Figure S10, which shows several clusters of sequences, including the large cluster I. The ClustalX alignment of cluster I (Supplementary Figure S11A,B) shows the conserved motif FxxGx(H/R)xCxG associated with the prosthetic heme group (marked by the pink amino acids in Figure S11B). The absolutely conserved cysteine is the proximal ligand to the heme iron. This sulfur is the origin of the characteristic name-giving 450 nm Soret absorbance observed [30]. An additional conserved motif, [AG]-G-x-[ED]-T (indicated by green amino acids in Figure S11B), contains the highly conserved threonine preceded by an acidic residue which is positioned in the active site and believed to be involved in catalysis [30].

To the extent that these peptides coding for P450 enzymes are secreted in saliva, they may function to convert platelet derived PGH_2 to prostacyclin or PGE_2, which are inhibitors of platelet aggregation and vasodilators, or they can act solely as kratagonists of lipidic agonists of hemostasis. These possibilities remain untested.

2.10. The Salp15/Ixostatin Group

Ixostatin was the name given to a group of cysteine-rich protein sequences found in the sialotranscriptome of *Ixodes pacificus* Cooley and Kohls, 1943 [31] displaying remarkable similarities to the cysteine-rich domain of ADAMST-4 (aggrecanase). Due to their similarities to these domains, they could be involved in disruption of platelet aggregation or neutrophil function, cell–matrix interactions, or inhibition of angiogenesis [31]. A few proteins of this group were characterized functionally. Two recombinant proteins from *Ixodes scapularis* Say, 1821 of the Ixostatin group, namely ISL929 (CAX36743.1) and ISL1373 (CAX36742.1), were shown to reduce expression of β_2 integrins and impair adherence of polymorphonuclear leukocytes [22]. The ixostatin protein from *I. scapularis* named Salp15 (AAK97817.1) was shown to inhibit CD4+ T cell activation. Repression of calcium fluxes triggered by TCR ligation resulted in lower production of interleukin 2. Salp15 also inhibited the development of CD4+ T cell-mediated immune responses in vivo [32].

Psiblast search of the NR database starting with members of this group, for example, CAX36743.1, finds only tick sequences in the initial blastp, but in subsequent iterations it identifies metalloproteases from insects, without converging to a unique group of sequences. Using more limited psiblast parameters (-j 2 and -h 1e-15) when starting the search with typical members of the group, we created 41 models that identified 632 sequences as belonging to the Salp15/Ixostatin group, out of a total of 15,796 sequences. The vast majority are found in the *Ixodes* genus, but a few sequences are also found in metastriate ticks. No member of this group was found in the Argasidae. Alignment and phylogenetic analysis of selected 197 protein sequences of the Salp15/Ixostatin group shows a shallow distribution of branches, most with low bootstrap support (Figure S12). This is indicative of a scenario of fast evolution with possible events of recombination. The genes coding for these proteins have been proposed to be evolving under positive selection pressure [33]. Alignment of 73 sequences most similar to Salp15 (AAK97817.1), all from the *Ixodes* genus (Figure S13), shows seven conserved cysteines within 19% identities in 109 sites of the mature proteins. Alphafold2 predictions for 80 peptide sequences of the Salp15 family containing seven cysteines shows, on the monomeric prediction mode, three disulfide bonds with the structure | C:1 C:3 | C:2 C:5 | C:4 C:6 | (Figure S14). The most carboxyterminal cysteine remains in the reduced state. The NMR structure of a monomeric Salp15 has been recently published [34] and compared to the Alphafold2 prediction. The authors concluded that "the global fold of Salp15 likely consists of a disordered N-terminal region and a globular domain with an α-helix, a ß-sheet, and regions with non-regular structure. Only C135, the C-terminal residue, is reduced, according to its 13C chemical shifts". Perhaps the free carboxyterminal cysteine could be joined with another in a dimer structure. Submission of these sequences to Alphafold2 in dimer and tetramer mode failed to recover multimers. The disordered N terminal region may impair folding predictions.

Alignment of 43 sequences most similar to Ixostatins ISL929 (CAX36743.1) and ISL1373 (CAX36742.1) reveals 8 conserved cysteines (Figure S15). Alphafold2 structure predictions in the monomeric mode indicate that six of the cysteines are involved in disulfide bonds as A:2 A:4 | A:3 A:6 | A:5 A:7 (Figure S16A), with the first and last cysteines remaining in the reduced state. However, dimer predictions (Figure S16B) assign these residues to be linked as A:1 | B:8 and A:8 | B:1 in 38 of these 43 sequences, suggesting that a dimer conformation is more probable in this subgroup of the Salp15/Ixostatin group.

The most cys rich subgroup of the Salp15/Ixostatin group, named 20-Cys has 20 or 21 cysteines (Alignment on Figure S17). The Alphafold2 predictions (Figure S18 and spreadsheet S1) indicates the 20 cysteines to be connected as | A:1 A:3 | A:2 A:5 | A:4 A:15 | A:6 A:8 | A:7 A:10 | A:9 A:20 | A:11 A:13 | A:12 A:17 | A:14 A:16 | A:18 A:19. Within the sequences containing 21 cysteines, the most carboxy terminal cysteine remains reduced. Alphafold2 predictions in the multimer mode do not predict these unpaired cysteines to be involved in disulfide bridges.

2.11. The 10 k Da-WC Group

The 10 kDa-WC family is a large secretory family thus far restricted to the genus *Ixodes* [35]. It is named based on the WC motif found in the C-terminal end of the protein. PSI-BLAST analysis of the non-redundant database retrieves 107 sequences after an exhaustive search, while the current supplemental spreadsheet S1 contains 353 proteins annotated for this family. At sequence level, the family is characterized by 7 conserved cysteines (although this may vary for individual family members) (Figure S19). Not all members possess the WC motif as can be seen for some sequences. The sequences in the alignment show 27–88% sequence similarity (median 40%). However, a stable compact fold is predicted for all sequences composed of N- and C-terminal interacting alpha-helices packed against a central three-stranded anti-parallel beta-sheet (Figure S21). All modelled structures show 3 intact disulfide bonds with a disulfide bond pattern of C1–C6, C2–C4, C3–C5 (Figure S20). Interestingly, the cysteine involved in the WC motif sits at the C-terminal end and does not have a disulfide bond partner. Its solvent exposure suggests that this cysteine is reactive and would either form dimers with itself or other members of this family or other proteins. A homodimer model predicted by AlphaFold2 result in disulfide bond formation between the respective C7 cysteines with stacking of the tryptophans, suggesting that this may indeed be the natural quaternary conformation of family members that possess the WC motif. The dimer is formed by a two- fold rotation of the alpha helices to form a cleft that may be involved in binding host target proteins or even ligands (Figure S21). The possibility also exists that family members may form heterodimers thereby increasing the potential functional repertoire of this family. Structural alignment of the different monomer models indicates RMSD values that range from 0.59–1.97 Å ± 0.34 Å for alignment of the secondary structure elements. The Dali search algorithm found for the best hit a Z-score of 4.3 and RMSD of 2.6 Å over 90% of the sequence alignment. This match is for a domain of unknown function (DUF2470) found in glutamyl-tRNA reductase binding protein and heme utilization gene Z (HugZ), both belonging to the Heme-binding protein family found in bacteria [36,37]. DUF2470 is involved in binding to glutamyl-tRNA reductase and partially shields the heme-binding pocket for the heme-binding proteins. Potential functions for this family in ticks may include anti-microbial activity by targeting heme-degradation by bacterial heme oxygenases [38]. The binding cleft formed by the dimers may also be functional in a range of other binding activities.

2.12. The 5.3 kDa Family

The 5.3 kDa family is restricted to hard ticks with extensive expansions in the genus *Ixodes* [38]. An exhaustive PSI-BLAST search of the non-redundant database retrieves 35 sequences that all belong to *I. scapularis* or *I. ricinus*. The current supplemental spreadsheet S1 report 137 sequences. The family is characterized by 6 conserved cysteine residues with some members having an additional pair or single N-terminal cysteine (Figure S22). Sequence similarity for the presented family members ranges from 11–89% (median 33%). AlphaFold2 predicts a triple-stranded anti-parallel beta-sheet with a cystine knot formed by two of the disulfide bonds (Figure S23). Structural alignment of the different monomer models indicates RMSD values that range from 0.41–2.25 Å ± 0.42 Å for alignment of the secondary structure elements. The disulfide bond structure is the same as other cystine knot family members namely, C1–C4, C2–C5, C3–C6 [39], with some members exhibiting an additional disulfide bond (Figure S22). Recently, holocyclotoxin was shown to exhibit the cystine knot fold [40] and previously the relationship between the 5.3 kDa and holocyclotoxin was noted [2,35]. Expanding on this, the assignment of the 5.3 kDa family to the cysteine knot fold and their conserved disulfide bond pattern was also confirmed [41]. Potential functions for the 5.3 kDa family may be anti-microbial [42] or targeting of ion channels [43]. Of interest is the high structural similarity found with a potato carboxypeptidase inhibitor that also present a knottin fold [44] that gives a RMSD of 1.15 Å. There is therefore a possibility that the 5.3 kDa family may also perform this function.

2.13. Seven-Disulfide Bond Family (7 DB)

The seven-disulfide bond family is to date unique to soft ticks and was first described in *Argas monolakensis* Schwan, Corwin & Brown, 1992 (Acari: Argasidae), *Ornithodoros coriaceus* Koch 1844 and *Ornithodoros parkeri* Cooley, 1936 [45–47]. PSI-BLAST analysis of the non-redundant database retrieves 2000 proteins after 8 iterations that span all Kingdoms of Life except for Arthropods, indicating that the 7 DB members may belong to a large and extensive protein family. It also suggests that this family may have been acquired via horizontal gene transfer, probably from a vertebrate host. The current database contains 7 members.

It was previously suggested that the 7 DB family is composed of two BTSP domains [47]. Alphafold2 modeling indicates, however, that the 7 DB-family shows an integrated fold where the two BTSP-like domains are not independent entities but are formed by the integration of the various beta-sheets to form a globular domain that cannot be differentiated into sub-domains (Figure S24). The structure starts with a two-stranded anti-parallel beta-sheet as the BTSP domain, but while the first strand of the second three-stranded anti-parallel beta-sheet follows from the first beta-sheet, it does not turn back on itself to complete the beta-sheet but moves into a new three-stranded beta-sheet. The second beta strand of this sheet leaves the sheet to form an alpha helix that connects a new two-stranded antiparallel beta-sheet, which moves to a beta-strand that completes the previous three-stranded beta-sheet. The alpha chain then moves back to the "first" domain to form two anti-parallel beta-sheets that stack against the first beta-strand to form the three-stranded anti-parallel beta-sheet of the "first domain". Pairwise structural comparison of the members shows RMSD values from 0.66 Å −1.7 Å + −0.27 Å. As expected, the disulfide bond structure also presents an inter-mixing of disulfide bonds between the domains giving the pattern: C1–C5, C2–C4, C3–C13, C6–C14, C7–C10, C8–C12, C9–C11 (Figure S25).

Remarkably, the BTSP Barium sulphate adsorption protein1 (BSAP1) can be structurally fitted to the "first BTSP domain" with an RMSD value of 0.91 Å and to the "second domain" with an RMSD value of 1.57 Å (Figure S24). However, fitting to the "second domain" is in reverse order, with the three-stranded beta-sheet from the second domain preceding the two-stranded beta-sheet. This intermixing of domains also results in a disulfide bond pattern that is different from that of the BTSP fold. Conserved disulfide bonds in "domain 1"include C1–C5, C2–C4 and in "domain 2" C7–C10, C8–C12, C9–C11. The last two conserved disulfide bonds are inter-domain and include C3–C13 and C6–C14.

Structural comparison using the DALI server indicates similarity to Evasin-1 (Z-scores: 3.1–4.5 and RMSD 1.5 Å −2.2 Å) and Evasin-4 (Z-scores: 2.5–3.8 and RMSD 3.3 Å −4.6 Å). Evasin-1 and Evasin-4 have been shown to be evolutionary related [48]. Pairwise comparison of Evasin-1 to the 7 DB family indicates a fit of 1.51 Å to the "first domain" and 0.82 Å to the "second domain" (Figure S26). The fit to the first domain is only for some of the beta-sheets while the overall fit to the second domain is better. It is clear that an evolutionary relationship exists between all of these families, related to their beta-sheet structure (Figure S27). The first domain of the 7 DB fold seems to be closer related to the BTSP family and the second domain to the evasins. In the case of the 7 DB family, strand switching between the domains may have resulted in a more overall stable structure. Strand switching is not common in protein structures but has been observed at least once in blood-feeding bugs, where triabin, a lipocalin, have a switched beta-stranded barrel that does not conform to the canonical continuous eight stranded anti-parallel beta barrel characteristic of the lipocalin family A-H, but presents the order ACBDEFGH [49]. To date, no function has been associated with the 7 DB fold.

2.14. Basic Tail Secretory Family (BTSP)

The basic tail secretory proteins were first described from *I. pacificus* and *I. scapularis* as a family of small disulfide-rich proteins [31,35]. Since then, BTSP members have been found in all tick salivary gland transcriptomes sequenced to date and are one of the most abundant secretory protein families [47]. Recently, the structure of BSAP1, a member of the BTSP, was determined using NMR and shown to consist of an N-terminal two-stranded and a C-terminal three-stranded anti-parallel beta-sheet arranged end-to-end and stabilized by three conserved disulfide bonds [50,51]. An exhaustive PSI-BLAST analysis using the sequence of BSAP1, retrieved 142 proteins from the non-redundant database. The current supplemental spreadsheet S1 report 484 sequences. The canonical single BTSP domain is characterized by 6 cysteines followed by a lysine-rich tail. The cysteine spacing may be variable and the basic tail may be exchanged for an acidic tail or no tail. The disulfide bond structure is C1–C3, C2–C5, C4–C6 (Figure S28). Alphafold2 prediction of BSAP1 resulted in a structure with a more defined secondary structure that suggests that the NMR structure represents a more labile structure in solution. Even so, the core structure gave an RMSD of 1.47 Å between the NMR and Alphafold2 predicted structure (Figure S29). For comparative purposes we used the Alphafold2 predicted structure. Structural comparison shows that the core domain is conserved as predicted by Alphafold2 with RMSD values ranging from 0.68–1.95 ± 0.24 Å. The C-terminal tail region is unstructured for most proteins although alpha-helices are sometimes predicted. In some cases, proteins have double-BTSP domains. In regard to this, a special case is the related seven disulfide bond family (7 DB) discussed abovee that does not present the same fold. It was suggested that the BTSP family core domain is similar to EGF domains [51]. DALI searches indicate, however, that the closest hits of the core structure resemble the tick evasin-4 fold [52], and an RMSD of 1.01 Å is obtained between BSAP1 and Evasin-4 for the core two-stranded beta-sheet. The general disulfide bond pattern within this core structure is also conserved. An evolutionary relationship, therefore, exists between the BTSP and evasin-4 family. To date the BTSP family has been implicated in fXa inhibition [53], modulation of fibrinolysis [54], and the complement pathway mediated by lectin [51].

2.15. Evasins

The evasins are a well-characterized group of proteins that consist of at least two evolutionary independent families named Evasin A and Evasin B with the latest count from the non-redundant database finding 271 Class A evasins and 128 Class B evasins [48]. The Evasin A group contains Evasin-1 and Evasin-4 for which structures have been determined [52,55]. The structure for the Class A evasins presents a discontinuous N-terminal domain comprised of the N-terminus linked by a disulfide bond to two antiparallel beta-sheets (β3-β4) and a C-terminal domain comprised of 2 anti-parallel beta-sheets with 5 anti-parallel beta-strands and an alpha helix in the order β1-β2-β5-α1-β6-β7 (Figures S30 and S31). The core structure consisting of the various beta-sheets and the alpha helix may be absent from some structures. Its disulfide bond pattern is C1–C3, C2–C6, C4–C7, C5–C8 (Figure S30). The Evasin B group contain Evasin-3 for which a structure has been determined [56,57]. It possesses a three disulfide bond knottin fold composed of an anti-parallel beta-sheet with the topology of β1-β3-β2. The disulfide bond pattern is C1–C4, C2–C5, C3–C6 (Figure S32). Fitting of the Alphafold2 modeled structures indicate an RMSD range of 0.8 Å–2.04 Å ± 0.36 Å that shows a largely correctly predicted fold. The evasin A family in general targets CC chemokines, while the evasin B family targets CXC chemokines [48].

2.16. Evolutionary Considerations for the BTSP-Evasin A-7 DB Superfamily

The structural similarities of the BTSP, Evasin A and 7 DB families suggest a common ancestral structural fold from which these families evolved that included at least two beta-sheets stabilized by three to four disulfide bonds. Exhaustive PSI-BLAST analysis of the Evasin A and the BTSP families only retrieve family specific members from ticks and their relationships to each other was only detected once Alphafold2 models were searched against the PDB database using the DALI server. This would partly be due to the small size and high sequence divergence observed in these families but could also suggest that these are proteins that originated exclusively within the tick family. If this is the case, then the ability of the 7 DB family to retrieve distant homologs from all kingdoms of life may suggest that this was the ancestral fold for this superfamily. Gene duplication of N- or C-terminal beta-sheet domains would have then resulted in beta-sheets that had to form continuous structures resulting in either the core BTSP or Evasin A fold (Figure S27). It is of interest that the BTSP family shows the best fit to the N-terminal domain, while the Evasin A family shows the best fit to the C-terminal domain, suggesting that the N-terminal domain gave rise to the BTSP family and the C-terminal domain to the Evasin A family (Scenario 1). Alternatively, the Evasin A fold may have originated from the BTSP fold (Scenario 2). The viability of these scenarios would have depended on a horizontal gene transfer even in the ancestral tick lineage with subsequent duplication of the BTSP or Evasin A folds from the 7 DB fold before divergence of the tick families, with subsequent loss of the 7 DB fold in ixodids.

2.17. The 13 kDa-Basic Group

The 13 kDa-basic family is a small secretory family found in prostriate and metastriate ticks. An exhaustive PSI-BLAST search of the non-redundant database retrieved 47 sequences, 16 tick proteins with the rest derived from mites and spiders, while the current supplemental spreadsheet S1 contains 15 proteins annotated for this family. This family is highly conserved with sequence similarity ranging from 63–100% (median 73%). The family is characterized by 6 conserved cysteines with a predicted disulfide bond pattern of C1–C6, C2–C3, C4–C5 (Figure S31). Alphafold2 predicts a compact globular structure formed by six alpha-helices, with all structures presenting intact disulfide bonds (Figure S32). Structural alignment indicates RMSD values that range from 0.2–0.78 Å $\pm$ 0.14 Å for alignment of the secondary structure elements. The Dali search algorithm found as structural homologs members of the odorant-binding, pheromone-binding or venom 2 allergen families with average Z-scores of 4.0 $\pm$ 0.4 and RMSD of 3.5 $\pm$ 0.5 Å over 90% of the sequence alignment [58,59]. Odorant-binding proteins have not yet been found in ticks as prominent ligand scavengers involved in tick-host interactions since the lipocalins generally perform this function [2]. The fact that single family members have been found in prostriate, metastriate and other arachnids may point to a role as a odorant or pheromone-binding protein. A major difference between the various odorant-binding families and the tick proteins is that the odorant-binding proteins from *Anopheles gambiae* Giles, 1902 (Diptera: Culicidae) have a disulfide bond pattern of C1–C3, C2–C5, C4–C6, while the venom allergen 2 protein from the fire ant has a disulfide bond pattern of C1–C2, C3–C4, C5–C6. There is therefore a possibility that these families converged on the same general fold, explaining the absence of odorant-binding proteins in ticks [60,61]. Conversely, an ancient relationship between arachnid and insect odorant-binding proteins was postulated [62]. In this regard, the odorant-binding fold of this family has been previously identified in a study that focused on the tick foreleg tarsi transcriptome and was postulated to play a role in chemosensation [63]. The disulfide bond patterns predicted by the Swiss Model and RaptorX servers in this latter study were identical to the patterns found in the current study and also indicated homology to the odorant-binding proteins.

2.18. Lipocalin Family

The lipocalin family is one of the largest and most diverse secretory families found in tick salivary glands [64,65]. PSI-BLAST analysis using the histamine-binding protein 2 (O77421) sequence from the hard tick *Rhipicephalus appendiculatus* (Neumann, 1901) [12] against the non-redundant database retrieves 971 proteins that are restricted to ticks. Similarly, moubatin (A46618) from the soft tick *Ornithodoros moubata* (Murray, 1877) [66,67] retrieves ~1081 proteins restricted to ticks. In both cases, PSI-BLAST analysis was run for

15 iterative cycles with no sign of convergence and of those sequences retrieved only 807 (64%) was shared between the two searches. As such, the PSI-BLAST analysis cannot retrieve all members since it cannot be run to exhaustion, indicating that the position-specific scoring matrices derived from alignment do not cover all possible tick lipocalin sequence variations. This indicates that the family is likely too divergent to be represented by any one position-specific scoring matrix. To address this, the current TSFam database contains 125 PSSM models representing various lipocalin clades that should cover the majority of diverse lipocalins in tick sialomes. The current TSFam database contains 2229 members assigned to the lipocalin group, while other in-house databases that also have representatives from transcriptomes present in the small read archives contain more than 11,000 proteins annotated as lipocalins. Figure S33 displays the alignment of representative members of this family with their conserved disulfide bonds.

Canonical lipocalins possess an N-terminal 3_{10}-helix, a central eight-stranded antiparallel β-barrel and a C-terminal α-helix [68]. The β-barrel may or may not possess a cavity that can be augmented by the inter β-strand loops to form a binding cavity. In some cases, the inter strand-loops allow formation of a second binding cavity near the opening of the barrel [69]. The N-terminal 3_{10}-helix closes of one side of the barrel, while the C-terminal helix packs against the side of the barrel, giving the structure a distinct resemblance to a coffee cup [68]. Lipocalins may have 2–4 conserved disulfide bonds that stabilize the β-barrel and pin the C-terminal α-helix to the side of the barrel. The binding cavity allow tick lipocalins to scavenge a variety of bio-active molecules including biogenic amines, leukotrienes, and cholesterol [12,13,64,70–73]. Lipocalins also target complement C5 and properdin via protein–protein interactions and can modulate dendritic cell responses [74–77]. These functional properties allow ticks to modulate host defense responses such as inflammation, platelet aggregation, complement activation and immune responses.

For the current analysis, 175 lipocalin sequences were modelled with Alphafold2. Dali searches indicate that for the majority of proteins, the best targets are known tick lipocalins, with average Z-scores of 15 ± 3 (range of 2.8–23). The majority of models shows the core lipocalin fold, with deviations found for N- or C-terminal extensions. Pairwise comparison with the structure of the female histamine-binding protein HBP2 (1QFTA) indicated an average RMSD of 1.43 Å $\pm$ 0.15 Å. Some lipocalins presented as double lipocalin-domains. The high sequence diversity observed in the lipocalins preclude the construction of a single multiple sequence alignment. However, for the purposes of illustration a representative alignment of those lipocalins with known conserved functions are presented with some examples of structures and their bound ligands (Figure S34).

2.19. Kunitz-BPTI Family

The Kunitz-BPTI (basic pancreatic trypsin inhibitor) family is a ubiquitous family of serine protease inhibitors [78]. Ticks possess a large, expanded family of Kunitz-BPTI proteins that exist as single domain proteins, or multiple domain proteins ranging from two domains to more than seven Kunitz-BPTI domains linked in tandem [2]. Their basic single domain structure comprises an N-terminal 3_{10}-helix, followed by an unordered strand that turn back to form a substrate-binding presenting loop, followed by a two-stranded antiparallel β-sheet terminating in a C-terminal α-helix packed against the β-sheet. Generally, three conserved disulfide bonds stabilize the structure, the first between the 3_{10}-helix and the C-terminal portion of the α-helix (C1–C6), the second at the start of the substrate-binding presenting loop linking to the turn that follows after the β-sheet (C2–C4) and the third linking the second β-sheet and the N-terminal portion of the α-helix (C3–C5) pinning the secondary structures together (Figure S35). In ticks, some BPTI proteins have been found that lack the second disulfide bond [79,80]. This presumably allows greater freedom of movement for the substrate-binding presenting loop and may have functional significance in those proteins that lacks this disulfide bond.

Members of this family mainly target serine proteases by binding to the enzyme active site via the substrate-binding presenting loop with additional interaction conferred by the C-terminal α-helix [78]. While some tick inhibitors would certainly use this mechanism to target serine proteases, significant deviations from this standard mode of inhibition is observed, notably that inhibitors from both hard and soft ticks insert their N-terminal residues into the active site of particularly fXa and thrombin, while interacting with substrate-binding exosites via their C-terminal α-helices [81–83]. In soft ticks, thrombin inhibitors are related to ornithodorin and seems to be evolutionary conserved in the Argasidae [84], while orthologs to boophilin are found in most metastriate ticks characterized to date [85,86]. Other Kunitz-BPTI inhibitors that target the blood clotting cascade include ixolaris and penthalaris that both target fX and fXa [79,87].

Inhibitors that target various receptors have also been characterized. This include integrins and ion channels [21,80,88]. The RGD integrin recognition motif is presented on the substrate-binding presenting loop of the savignygrins that target the platelet fibrinogen receptor $\alpha_{IIb}\beta_3$ [21]. The RGD-motif may also be replaced by the RED motif that still allow targeting of the $\alpha_{IIb}\beta_3$ receptor [84,88]. Kunitz-BPTI inhibitors with the RGD motif have been identified in all soft tick salivary gland transcriptomes sequenced to date [45–47,89].

For the current database 87 proteins have been modeled by Alphafold2. These included single domain Kunitz-BPTI proteins as well as three, four, five and seven domain proteins. These modeled structures correspond with manual assignment of the BPTI domains based on conserved disulfide bond structure. The double-domain proteins presented two types of structures, notably extended to produce a dumbbell shaped protein as observed for tick thrombin inhibitors [90], or two domains packed against each other to form a globular protein as observed in bikunin and ixolaris [91,92]. The modeling of multi-domain proteins ($n \geq 3$) as observed for the models produced by Alphafold2 is a new and novel feature creating interesting possibilities to study the structure-function relationships of these molecules.

2.20. The 8.9 kDa Family

The 8.9 kDa family occurs exclusively in hard ticks and takes on a highly variable structure with most forms containing one or two domains. Also present are duplicated structures having two complete two-domain modules contained in a single polypeptide. Conserved cysteines are distributed throughout the structure with the simplest, most compact types having four cysteines (two disulfide bonds), with 6-, 8-cysteine proteins found in the two domain forms and 12- and 16 cysteines in the duplicated forms. Considerable variation in the disulfide bonding pattern of conserved cysteines is apparent. The common feature of the 8.9 kDa family is an N-terminal domain containing a four (or sometimes three)-stranded β-sheet having a Greek key topology [93]. The segment between strands 1 and 2 is extended and usually forms a 2- or 3-stranded antiparallel β-sheet which lies over the larger sheet, forming a sandwich. Most forms have a C-terminal domain of varying structure. The fold is like the type C domain of von Willebrand factor with most of the disulfide bonds being conserved [94]. Prostriate and metastriate genera contain two-domain, eight-cysteine types that resemble the C5-binding complement inhibitor CirpT1, but overall, the C-terminal domain is extremely variable and displays a number of different disulfide bond arrangements strongly suggesting that these proteins interact with many different targets and play a variety of biological roles.

2.20.1. Taxonomic Distribution

Ixodes

The transcriptome of *I. ricinus* contains approximately 70 unique transcripts varying in identity from 14-80%. A large group of single domain peptides is present that are truncated at the C-terminal end of the N-terminal β-sheet structure (Figures S36A and S37A). These contain two disulfide bonds (four cysteines) with the first (DS1) linking the third strand from the larger β-sheet (β-sheet 1) with the second strand of the smaller sheet (β-sheet 2) and the second bond linking the loop between the second and third β-strands of β-sheet 1 with some part of the C-terminal region of the protein (DS-2). In this group of proteins β-strand A of β-sheet 1 is absent in the models and the N-terminus extends away from the protein and forms a short α-helical segment (Figure S37A). This group exhibits high overall amino acid sequence conservation ranging between 60 and 85% in amino acid identity.

Also present in *I. ricinus* is a larger group of peptides containing both the N- and C-terminal domains and having four disulfide bonds (eight cysteines). These are very diverse in amino acid sequence (17–70% identity) but are all similar in overall structure to the CirpT-type [93] complement inhibitors (Figures S36B and S37B). The N-terminal domain is much like the single-domain sequences discussed above. Many of these sequences are modeled by AlphaFold2 as lacking the first strand of β-sheet 1. A segment of unstructured sequence is present at the N-terminus in these cases suggesting that the full four-stranded β-sheet structure may be present (Figure S37B). The C-terminal domain of this group contains three disulfide bonds, most notably a cluster of two (DS2 and DS3) that is characteristic of all members of the 8.9 kDa family except those *Ixodes* variants that do not contain a C-terminal domain. This cluster contains two sequential cysteine residues, with the first linking to the loop between β-strands B and C of the N-terminal domain (DS2) and the second linking to the C- end of β-strand D of the N-terminal domain (DS3). A fourth disulfide bond (DS4) links the extreme C-terminus with a second C-terminal domain cysteine residue linking the strands forming a small two-stranded β-sheet (β-sheet 3) that is normally present in this domain (Figure S37B).

Amblyomma

Ticks in the genus *Amblyomma* produce a diverse set of peptides from the 8.9 kDa family containing variations in the structure of the C-terminal domain as well as forms in which the entire two domain structure is duplicated or partially duplicated to give proteins that could have multivalent binding properties. The *Amblyomma* forms contain 6, 8, 12 or 16 cysteine residues (3, 4, 6 or 8 disulfide bonds).

Six-cysteine forms from *Amblyomma* species are variable in sequence (20–80% amino acid sequence identity within the group) and contain the N-terminal domain seen in *Ixodes* proteins as well as a C-terminal domain that is truncated directly C-terminal to the "CC" double cysteine motif involved in DS2 and DS3 of the eight-cysteine forms described above (Figures S38A and S39A). Because of the shortened C-terminal domain in these forms, they do not contain DS4. Eight-cysteine variants are also present in *Amblyomma* which resemble those described in *Ixodes* and are quite diverse in sequence, showing about 20–80% amino acid identity within the group (Figures S38B and S39B). This group contains the C5 complement inhibitors and is similar in general structure to the eight-cysteine forms from *I. ricinus* (Figure S4B).

In addition to the six- and eight-cysteine forms of the 8.9 kDa family, *A. maculatum* (Kock, 1844) contains extended variants that have extra domains attached to the eight-cysteine structure. The simplest type has a three-domain structure resembling the four-cysteine, single domain forms described in *I. ricinus* attached to the C-terminal end of the eight-cysteine, two-domain module (Figures S40 and S41A). The chain continues from the N-terminal module into the extra domain forming a hairpin loop corresponding to β-sheet 2 of the *I. ricinus* single domain protein, then into a three-stranded, antiparallel β-sheet corresponding to β-sheet 1 of *I. ricinus* four-cysteine proteins and the eight-cysteine proteins from all species (Figure S41A). This type contains twelve cysteine residues forming six disulfide bonds which are conserved in position relative to those from the previously described proteins.

Also present in *A. maculatum* are variants with four domains made up of two complete eight-cysteine units fused end-to-end into a single polypeptide (Figures S40 and S41B). Like the three-domain forms described above, the second structural module of these proteins contains a modified N-terminal domain β-sheet structure, but also has a fully elaborated

C-terminal domain containing a β-sheet 3 type structure (Figure S41B). These proteins contain sixteen conserved cysteine residues forming eight disulfide bonds.

Rhipicephalus

Like *Amblyomma*, *Rhipicephalus* (Kock, 1844) species contain 8.9 kDa family members having two domains with 6 or 8 cysteine residues as well as four-domain proteins containing 16 cysteine residues. An unusual group of 8-cysteine proteins with a shortened C-terminal domain is present. It contains an N-terminal domain with β-sheet 1 modeled as having three strands, and β-sheet 2 as having two or three strands (Figures S42A and S43A). The C-terminal domain is truncated relative to the 8-cysteine forms of *Ixodes* or *Amblyomma* but contains the "CC" double cysteine motif that makes up part of the two-disulfide bond cluster (Figure S43A). All members of this group are modeled by AlphaFold2 as having single free cysteines near the N-terminus of the protein and near the C-terminus, which are not in proximity to one another (Figure S43A). This suggests that they may form multimers or contain an unmodeled model structure in which these two unpaired cysteines are in proximity to form a disulfide bond.

A group of "conventional" 8-cysteine forms are also found in *Rhipicephalus* in which the domain structure and disulfide bonding pattern of most closely resembles comparable proteins from *Amblyomma* or *Ixodes*. This not to say they are highly similar at the sequence level or would be expected to be functionally homogeneous. Within this single species these proteins exhibit a range in amino acid identity of 17–52% (Figures S42B and S43B).

Finally, *R. appendiculatus* contains a set of extended variants like those of *Amblyomma* which contain four protein domains and sixteen cysteine residues. As in the *Amblyomma* forms, these are derived from the end-to-end fusion of two eight-cysteine forms into a single polypeptide. In this case, the N-terminus of the first eight-cysteine unit forms part of the C-terminal domain of the unit by forming one strand in in β-sheet 3, resulting in it having three strands rather than the usual two. In a second model, the C-terminus of the protein integrates into β-sheet 1 of the C-terminal eight-cysteine unit, making it a 4-stranded antiparallel sheet. The disulfide bonding pattern of this group is that expected from the fusion of two eight-cysteine units.

2.20.2. C5-Binding Anticomplement Proteins

The only established function of the 8.9 kDa family is inhibition of complement by binding to component C5 and preventing its activation [93]. Orthologous 8-cysteine 8.9 kDa family members from *R. pulchellus* (Gerstäcker, 1873), *Dermacentor andersoni*, *R. sanguineus* and *A. americanum* have been found to function similarly. Using the crystal structure of CirpT1, the variant from *R. pulchellus*, complexed with the macroglobulin domain 7 of human complement C5 and the cryo electron microscopy structure of CirpT1 complexed with C5 along with the inhibitors OmCI and RaCI1, we identified a block of eight residues in the interaction interface (Figure S44). Of the selected set of sequences from the TickSialofam database, only those eight-cysteine forms closely resembling the previously described inhibitors contained even a small number of amino acid identities within the selected sequence block. Few partial matches or weakly similar sequences were found in the set of *Ixodes* transcripts suggesting that C5 inhibitors of this "type one" clade of the 8.9 kDa family and are restricted to metastriate species. As anticipated, due to the high variability and low degree of sequence identity within the groups described here, only orthologs of CirpT1 appear to have a C5 binding function. The structural diversity revealed by Alphafold2 modeling therefore suggests that the 8.9 kDa salivary protein family of the eight-cysteine type can be expected to perform multiple functions.

2.21. The 8-kDa Family

The 8-kDa family occurs in metastriate hard ticks and contain a cysteine knot structure characterized by a three- or four-stranded antiparallel β-sheet folded into an open sided barrel stabilized by three disulfide bonds (Figures S45 and S46) [95]. The disulfides are clustered at the end of the β-sheet containing both the N- and C-termini. Two alternative disulfide bonding patterns are seen, one with a pattern (based on relative cysteine positions) of: 1–6, 2–4, 3–5, and the second having a pattern of 1–4, 2–5, 3–6, which is also seen in the evasins. This family contains the RaCI complement inhibitors from *R. appendiculatus* which target complement factor C5 (Figures S45 and S46) [96]. These bind at a surface of C5 containing elements of the MG1, MG2 and C5d domains and prevent its cleavage by the C5 convertase. RaCI proteins from *R. appendiculatus*, *R. microplus* (Canestrini, 1888) and *D. andersoni* have been analyzed and found to act in a similar manner but have somewhat variable sequences in regions interacting with C5. This is explained by the large number of backbone interactions involved in C5 binding. RaCI peptides contain a lengthened loop between β-strands 1 and 2 that inserts into a pocket lying between the MG1 and MG2 domains of C5. This loop is not extended in other members of the family suggesting that these binding interactions cannot occur and that these proteins do not bind C5.

2.22. The 15-kDa Basic Family

The 15-kDa basic family found in *Amblyomma* sp. is a cysteine knot variant similar in structure to the 8-kDa family [95]. It contains an antiparallel β-sheet domain with two or three strands stabilized by three disulfide bonds in the pattern (based on relative cysteine positions) 1–4, 2–5, 3–6. Some members contain two additional cysteine residues forming a potential fourth disulfide bond as part of a disordered N-terminal coil (Figures S47 and S48). C-terminal to the cysteine knot domain is a length of disordered coil, followed by one or two helical segments containing greater than five turns which are then followed by a second intrinsically disordered region.

2.23. Complement-Binding Family

Members of the complement binding family are large proteins containing lectin or von Willebrand A (vWA) domains linked to strings of all β-sheet sushi-like domains, mostly stabilized by one or two disulfide bonds, that are reminiscent of complement control protein (CCP) domains. The lectin and vWA domains occur at the N-terminal end of the protein with the CCP domains extending out from them. One member of the group (JAR89651) from *I. ricinus* contains a fucose-binding lectin domain followed by a C-type lectin domain leading to ten repeated sushi domains. A second protein (JAR90946), also from *I. ricinus*, contains an N-terminal vWA domain followed by eight repeated modular domains Figures S49–S51). The proteins contain large numbers of cysteine residues, and all were predicted by AlphaFold2 to participate in disulfide bonds. These structures suggest that in blood feeding, the N-terminal parts may bind to exposed carbohydrate or collagen patches and the repeated domains may function as modulators of complement function in the mode of factor H. Other members that can be categorized as belonging to this group include JAB71472 from *I. ricinus* that contain only the repeated domains without the apparent lectin or vWA "anchors". Interestingly, these proteins (see alignments of JAR89651 and JAR90946, Figures S49–S51) show high degrees of similarity to the N-terminal parts and complete conservation of cysteine residues of these large proteins from a wide variety of arthropods such as mites, crustaceans and horseshoe crabs that do not feed on vertebrate hosts suggesting that they have functions in endogenous systems not involving vertebrate blood such as immune surveillance. They are related to the sushi, von Willebrand factor type A, EGF, and pentraxin domain-containing (SVEP1) and CUB and sushi modular domain proteins (CSMD1) of vertebrates. CSMD1 is known to inhibit complement by a mechanism in which it serves as a cofactor to factor I in the degradation of C4b and C3b from the classical and alternative pathways of complement, respectively [97].

2.24. Dae-2 Family

The Dae-2 (domesticate amidase effector 2) proteins are a group of cysteine peptidases whose genes have been acquired by tick species by lateral transfer from bacterial genomes [98,99]. These serve an antimicrobial function by proteolytically cleaving cell wall peptidoglycan. The tick salivary Dae-2 proteins have been shown to have broader substrate specificity than microbial forms and are thought to act by controlling growth of skin microbes at the site of feeding. The catalytic cysteine and histidine residues (Cys 23 and His 73 in the bacterial Tae-2 numbering system) are conserved in all tick forms (Figure S52). Differences in surface structure, particularly along the substrate binding groove, are considered to be determinants of selectivity for peptidoglycans from different bacterial forms. Unlike the bacterial Tae-2 which contain no disulfide bonds, tick Dae-2 proteins contain a conserved disulfide linking Cys 58 and Cys 89 (in JAC30591 numbering, Figure S53). There is also a free cysteine at position 4 (JAC30591 numbering) and in most tick sequences this is paired with a cysteine at position 31 to form a second disulfide (Compare JAC30591 and AEO34830 in Supplementary Figures S52 and S53).

3. Methods

3.1. TSFam Database

From the TSFam database [5], 15,796 sequences obtained from tick salivary transcriptomes were selected based on the presence of a signal peptide indicative of secretion, and no transmembrane domains outside the signal peptide. The original database was clustered by blastp and paired-joining the sequences to attain 25, 30... 90, 95% identity in at least 70% of the longest sequence. Thus, for each degree of identity there are n clusters, sorted by their decreasing abundance. Accordingly, each particular cluster can be determined by two numbers, the first being the identity threshold and the second the cluster number for that identity (Supplemental spreadsheet S1). The sequences from each cluster were used to construct a psiblast generated PSSM, and these were combined, after proper annotation, within the database where tick sequences can be searched using the rpsblast tool (Supplementary TickSialoFam 2.0 database).

3.2. Alphafold2 Program

The Alphafold2 program [6] was run locally on the NIH Biowulf cluster in a Linux environment using 8 cpu's, one v100x GPU and 60 GB RAM, using the monomer or multimer mode. Prediction of disulfide bonds were made by calculating the distances between all sulfur atoms from cysteine residues, available from the pdb files, and assigning a disulfide bond for those pairs that had a distance smaller than 3 Å.

3.3. Dali

The Dali program [7], available online: http://ekhidna2.biocenter.helsinki.fi/dali/README.v5.html (accessed on 1 July 2022) was used to compare the Alphafold2 predictions to the structures available in the PDB database. The program was run locally in the NIH Biowulf cluster. The program generates statistical analyses of the comparisons. According to the manual, a Z score above 20 indicates that two structures are homologous, between 8 and 20 that two structures are probably homologous, between 2 and 8 is a gray area, and a Z-score below 2 is not significant.

3.4. Disintegrin Searches

We have previously scanned salivary proteins from blood sucking arthropods for disintegrin motifs [19] after building prosite blocks (Supplementary File S1—Prosite disintegrin motifs) which were used to search the tick salivary protein database (Supplemental spreadsheet S1) using the program ps_can.pl [100] Available online: https://ftp.expasy.org/databases/prosite/.

4. Conclusions

The addition of structural fold prediction algorithms in the classification of secretory salivary gland protein families adds a powerful dimension that allows confirmation and validation of various protein families, groups or folds. It also allowed assignment of distantly related families or groups to well-known families or to predict novel folds not yet determined by conventional structural biological methodologies. The models provided by Alphafold2 also allow identification of potential homodimers and insights into the quaternary folds of proteins with multiple domains. In addition, the models provide insight into the potential functions and mechanisms of various families and provide a basis for assessment of structural integrity via disulphide bond predictions. Alphafold2-based classification as utilized for the TSFam2.0 database has already improved the original database, while adding significant information that can be used in hypothesis driven research on protein family function and evolution. The TSFam2.0 database is therefore a significant improvement on the original TSFam database that will with subsequent refinement and addition of more tick protein families and structures result in a comprehensive classification of tick protein families.

Supplementary Materials: The following supporting information can be downloaded at: https://www.mdpi.com/article/10.3390/ijms232415613/s1, (1) Supplemental Figures: PowerPoint file with manuscript figures and movies. (2) Supplemental File S1-Disintegrin motifs in prosite format used to scan tick salivary proteins using the program ps_scan.pl Available online: https://github.com/ebi-pf-team/interproscan/blob/master/core/jms-implementation/support-mini-x86-32/bin/prosite/ps_scan.pl. (3) Supplemental spreadsheet S1–Hyperlinked spreadsheet containing putative tick salivary proteins linked to comparisons to several databases and AlphaFold predicted structures. Clusterization of the proteins allowed for extraction of reversed-position specific motifs collected into the TSFam 2.0 database. The spreadsheet has links to pdb files, which need programs that are able to open them. We suggest the use of ChimeraX Available online: (https://www.cgl.ucsf.edu/chimerax/download.html) or Swiss-PDBViewer Available online: https://spdbv.unil.ch/. (4) Supplementary TickSialoFam 2.0 database–Include RPS models and a formatted database which should be used to query protein sequences by means of the rpsblast program from the NCBI Blast suite of programs Available online: https://ftp.ncbi.nlm.nih.gov/blast/executables/blast+/LATEST/.

Author Contributions: B.J.M. Study design: Data analysis; Manuscript draft; Manuscript revision. J.F.A.: Study design: Data analysis; Manuscript draft; Manuscript revision. J.M.C.R.: Study design: Software design; Data analysis; Manuscript draft; Manuscript revision. All authors have read and agreed to the published version of the manuscript.

Funding: J.M.C.R. and J.F.A. were supported by the Intramural Research Program of the National Institute of Allergy and Infectious Diseases (Vector-Borne Diseases: Biology of Vector Host Relationship, Z01 AI000810-21). B.J.M. was supported by the National Research Foundation of South Africa (Grant Numbers: 137966).

Data Availability Statement: Publicly available datasets were analyzed in this study. This data can be found here: https://proj-bip-prod-publicread.s3.amazonaws.com/transcriptome/TickSialoFam/TSF2.0/SupSpreadsheet+1.xlsx.

Acknowledgments: This work utilized the computational resources of the NIH HPC Biowulf cluster (http://hpc.nih.gov).

Conflicts of Interest: The authors declare no conflict of interest.

References

1. Ribeiro, J.M. Francischetti IM: Role of arthropod saliva in blood feeding: Sialome and post-sialome perspectives. *Annu. Rev. Entomol.* **2003**, *48*, 73–88. [CrossRef] [PubMed]
2. Francischetti, I.M.; Sa-Nunes, A.; Mans, B.J.; Santos, I.M.; Ribeiro, J.M. The role of saliva in tick feeding. *Front. Bio-Sci. A J. Virtual Libr.* **2009**, *14*, 2051. [CrossRef] [PubMed]
3. Nielsen, H. Predicting secretory proteins with SignalP. In *Protein Function Prediction*; Kihara, D., Ed.; Humana Press: New York, NY, USA, 2017; pp. 59–73. [CrossRef]
4. Sonnhammer, E.L.; Von Heijne, G.; Krogh, A. A hidden Markov model for predicting transmembrane helices in protein sequences. In Proceedings of the International Conference on Intelligent Systems for Molecular Biology, Montréal, QC, Canada, 28 June–1 July 1998; Volume 6, pp. 175–182.
5. Ribeiro, J.M.C.; Mans, B.J. TickSialoFam (TSFam): A Database That Helps to Classify Tick Salivary Proteins, a Review on Tick Salivary Protein Function and Evolution, With Considerations on the Tick Sialome Switching Phenomenon. *Front. Cell. Infect. Microbiol.* **2020**, *10*, 374. [CrossRef] [PubMed]

6. Jumper, J.; Evans, R.; Pritzel, A.; Green, T.; Figurnov, M.; Ronneberger, O.; Tunyasuvunakool, K.; Bates, R.; Žídek, A.; Potapenko, A.; et al. Highly accurate protein structure prediction with AlphaFold. *Nature* **2021**, *596*, 583–589. [CrossRef] [PubMed]

7. Holm, L.; Kääriäinen, S.; Rosenström, P.; Schenkel, A. Searching protein structure databases with DaliLite v. 3. *Bioinformatics* **2008**, *24*, 2780–2781. [CrossRef]

8. Rose, Y.; Duarte, J.M.; Lowe, R.; Segura, J.; Bi, C.; Bhikadiya, C.; Chen, L.; Rose, A.S.; Bittrich, S.; Burley, S.K.; et al. RCSB Protein Data Bank: Architectural Advances Towards Integrated Searching and Efficient Access to Macromolecular Structure Data from the PDB Archive. *J. Mol. Biol.* **2021**, *433*, 166704. [CrossRef]

9. Markland, F.S., Jr.; Swenson, S. Snake venom metalloproteinases. *Toxicon* **2013**, *62*, 3–18. [CrossRef]

10. Matsui, T.; Fujimura, Y.; Titani, K. Snake venom proteases affecting hemostasis and thrombosis. *Biochim. Et Biophys. Acta-Protein Struct. Mol. Enzymol.* **2000**, *1477*, 146–156. [CrossRef]

11. De Castro, E.; Sigrist, C.J.; Gattiker, A.; Bulliard, V.; Langendijk-Genevaux, P.S.; Gasteiger, E.; Bairoch, A.; Hulo, N. ScanProsite: Detection of PROSITE signature matches and ProRule-associated functional and structural residues in proteins. *Nucleic Acids Res.* **2006**, *34*, W362–W365. [CrossRef]

12. Paesen, G.; Adams, P.; Harlos, K.; Nuttall, P.; Stuart, D. Tick Histamine-Binding Proteins: Isolation, Cloning, and Three-Dimensional Structure. *Mol. Cell* **1999**, *3*, 661–671. [CrossRef]

13. Roversi, P.; Johnson, S.; Preston, S.G.; Nunn, M.A.; Paesen, G.C.; Austyn, J.M.; Nuttall, P.A.; Lea, S.M. Structural basis of cholesterol binding by a novel clade of dendritic cell modulators from ticks. *Sci. Rep.* **2017**, *7*, 16057. [CrossRef] [PubMed]

14. Friedland, N.; Liou, H.-L.; Lobel, P.; Stock, A.M. Structure of a cholesterol-binding protein deficient in Niemann–Pick type C2 disease. *Proc. Natl. Acad. Sci. USA* **2003**, *100*, 2512–2517. [CrossRef]

15. Inohara, N.; Nuñez, G. ML—A conserved domain involved in innate immunity and lipid metabolism. *Trends Biochem. Sci.* **2002**, *27*, 219–221. [CrossRef] [PubMed]

16. Alarcon-Chaidez, F.J.; Müller-Doblies, U.U.; Wikel, S. Characterization of a recombinant immunomodulatory protein from the salivary glands of Dermacentor andersoni. *Parasite Immunol.* **2003**, *25*, 69–77. [CrossRef] [PubMed]

17. McLane, M.A.; Marcinkiewicz, C.; Vijay-Kumar, S.; Wierzbicka-Patynowski, I.; Niewiarowski, S. Viper venom disintegrins and related molecules. *Proc. Soc. Exp. Biol. Med.* **1998**, *219*, 109–119. [CrossRef]

18. Calvete, J.J. The continuing saga of snake venom disintegrins. *Toxicon* **2013**, *62*, 40–49. [CrossRef]

19. Assumpcao, T.C.F.; Ribeiro, J.M.C.; Francischetti, I.M.B. Disintegrins from Hematophagous Sources. *Toxins* **2012**, *4*, 296–322. [CrossRef]

20. Francischetti, I.M. Platelet aggregation inhibitors from hematophagous animals. *Toxicon* **2010**, *56*, 1130–1144. [CrossRef]

21. Mans, B.J.; Louw, A.I.; Neitz, A.W. Savignygrin, a platelet aggregation inhibitor from the soft tick Ornithodoros savignyi, presents the RGD integrin recognition motif on the Kunitz-BPTI fold. *J. Biol. Chem.* **2002**, *277*, 21371–21378. [CrossRef]

22. Guo, X.; Booth, C.J.; Paley, M.A.; Wang, X.; DePonte, K.; Fikrig, E.; Narasimhan, S.; Montgomery, R.R. Inhibition of Neutrophil Function by Two Tick Salivary Proteins. *Infect. Immun.* **2009**, *77*, 2320–2329. [CrossRef]

23. Kato, H.; Gomez, E.A.; Fujita, M.; Ishimaru, Y.; Uezato, H.; Mimori, T.; Iwata, H.; Hashiguchi, Y. Ayadualin, a novel RGD peptide with dual antihemostatic activities from the sand fly Lutzomyia ayacuchensis, a vector of Andean-type cutaneous leishmaniasis. *Biochimie* **2015**, *112*, 49–56. [CrossRef] [PubMed]

24. Gibbs, G.M.; Roelants, K.; O'bryan, M.K. The CAP superfamily: Cysteine-rich secretory proteins, antigen 5, and pathogene-sis-related 1 proteins—Roles in reproduction, cancer, and immune defense. *Endocr. Rev.* **2008**, *29*, 865–897. [CrossRef] [PubMed]

25. Liu, H.; Yang, X.; Andersen, J.F.; Wang, Y.; Tokumasu, F.; Ribeiro, J.M.C.; Ma, D.; Xu, X.; An, S.; Francischetti, I.M.B.; et al. A novel family of RGD-containing disintegrins (Tablysin-15) from the salivary gland of the horsefly Tabanus yao targets αIIbβ3 or αVβ3 and inhibits platelet aggregation and angiogenesis. *Thromb. Haemost.* **2011**, *105*, 1032–1045. [CrossRef] [PubMed]

26. Nebert, D.W.; Gonzalez, F.J. P450 genes: Structure, evolution, and regulation. *Annu. Rev. Biochem.* **1987**, *56*, 945–993. [CrossRef]

27. Haurand, M.; Ullrich, V. Isolation and characterization of thromboxane synthase from human platelets as a cytochrome P-450 enzyme. *J. Biol. Chem.* **1985**, *260*, 15059–15067. [CrossRef]

28. Hara, S.; Miyata, A.; Yokoyama, C.; Inoue, H.; Brugger, R.; Lottspeich, F.; Ullrich, V.; Tanabe, T. Isolation and molecular cloning of prostacyclin synthase from bovine endothelial cells. *J. Biol. Chem.* **1994**, *269*, 19897–19903. [CrossRef]

29. Hecker, M.; Ullrich, V. On the mechanism of prostacyclin and thromboxane A2 biosynthesis. *J. Biol. Chem.* **1989**, *264*, 141–150. [CrossRef]

30. Denisov, I.G.; Makris, T.M.; Sligar, S.G.; Schlichting, I. Structure and Chemistry of Cytochrome P450. *Chem. Rev.* **2005**, *105*, 2253–2278. [CrossRef]

31. Francischetti, I.M.; Pham, V.M.; Mans, B.J.; Andersen, J.F.; Mather, T.N.; Lane, R.S.; Ribeiro, J.M. The transcriptome of the salivary glands of the female western black-legged tick Ixodes pacificus (Acari: Ixodidae). *Insect Biochem. Mol. Biol.* **2005**, *35*, 1142–1161. [CrossRef]

32. Anguita, J.; Ramamoorthi, N.; Hovius, J.W.; Das, S.; Thomas, V.; Persinski, R.; Conze, D.; Askenase, P.W.; Rincón, M.; Kantor, F.S.; et al. Salp15, an Ixodes scapularis Salivary Protein, Inhibits CD4+ T Cell Activation. *Immunity* **2002**, *16*, 849–859. [CrossRef]

33. Schwalie, P.C.; Schultz, J. Positive Selection in Tick Saliva Proteins of the Salp15 Family. *J. Mol. Evol.* **2009**, *68*, 186–191. [CrossRef]

34. Chaves-Arquero, B.; Persson, C.; Merino, N.; Tomás-Cortazar, J.; Rojas, A.L.; Anguita, J.; Blanco, F.J. Structural Analysis of the Black-Legged Tick Saliva Protein Salp15. *Int. J. Mol. Sci.* **2022**, *23*, 3134. [CrossRef] [PubMed]

35. Ribeiro, J.M.; Alarcon-Chaidez, F.; Francischetti, I.M.B.; Mans, B.J.; Mather, T.N.; Valenzuela, J.G.; Wikel, S.K. An annotated catalog of salivary gland transcripts from Ixodes scapularis ticks. *Insect Biochem. Mol. Biol.* **2006**, *36*, 111–129. [CrossRef] [PubMed]

36. Hu, Y.; Jiang, F.; Guo, Y.; Shen, X.; Zhang, Y.; Zhang, R.; Guo, G.; Mao, X.; Zou, Q.; Wang, D.-C. Crystal Structure of HugZ, a Novel Heme Oxygenase from Helicobacter pylori. *J. Biol. Chem.* **2011**, *286*, 1537–1544. [CrossRef]

37. Zhao, A.; Fang, Y.; Chen, X.; Zhao, S.; Dong, W.; Lin, Y.; Gong, W.; Liu, L. Crystal structure of Arabidopsis glutamyl-tRNA reductase in complex with its stimulator protein. *Proc. Natl. Acad. Sci. USA* **2014**, *111*, 6630–6635. [CrossRef] [PubMed]

38. Lyles, K.V.; Eichenbaum, Z. From Host Heme To Iron: The Expanding Spectrum of Heme Degrading Enzymes Used by Pathogenic Bacteria. *Front. Cell. Infect. Microbiol.* **2018**, *8*, 198. [CrossRef] [PubMed]

39. Pallaghy, P.K.; Norton, R.S.; Nielsen, K.J.; Craik, D. A common structural motif incorporating a cystine knot and a triple-stranded β-sheet in toxic and inhibitory polypeptides. *Protein Sci.* **1994**, *3*, 1833–1839. [CrossRef]

40. Vink, S.; Daly, N.L.; Steen, N.; Craik, D.J.; Alewood, P.F. Holocyclotoxin-1, a cystine knot toxin from Ixodes holocyclus. *Toxicon* **2014**, *90*, 308–317. [CrossRef]

41. Pienaar, R.; Neitz, A.W.H.; Mans, B.J. Tick Paralysis: Solving an Enigma. *Veter. Sci.* **2018**, *5*, 53. [CrossRef]

42. Pichu, S.; Ribeiro, J.M.; Mather, T.N. Purification and characterization of a novel salivary antimicrobial peptide from the tick, Ixodes scapularis. *Biochem. Biophys. Res. Commun.* **2009**, *390*, 511–515. [CrossRef]

43. Norton, R.S.; Pallaghy, P.K. The cystine knot structure of ion channel toxins and related polypeptides. *Toxicon* **1998**, *36*, 1573–1583. [CrossRef] [PubMed]

44. Rees, D.; Lipscomb, W. Refined crystal structure of the potato inhibitor complex of carboxypeptidase A at 2.5 Å resolution. *J. Mol. Biol.* **1982**, *160*, 475–498. [CrossRef] [PubMed]

45. Francischetti, I.M.; Mans, B.J.; Meng, Z.; Gudderra, N.; Veenstra, T.D.; Pham, V.M.; Ribeiro, J.M. An insight into the sialome of the soft tick, Ornithodorus parkeri. *Insect Biochem. Mol. Biol.* **2008**, *38*, 1–21. [CrossRef]

46. Francischetti, I.M.; Meng, Z.; Mans, B.J.; Gudderra, N.; Hall, M.; Veenstra, T.D.; Pham, V.M.; Kotsyfakis, M.; Ribeiro, J.M. An insight into the salivary transcriptome and proteome of the soft tick and vector of epizootic bovine abortion, Ornithodoros coriaceus. *J. Proteom.* **2008**, *71*, 493–512. [CrossRef]

47. Mans, B.J.; Andersen, J.F.; Francischetti, I.M.; Valenzuela, J.G.; Schwan, T.G.; Pham, V.M.; Garfield, M.K.; Hammer, C.H.; Ribeiro, J.M. Comparative sialomics between hard and soft ticks: Implications for the evolution of blood-feeding behavior. *Insect Biochem. Mol. Biol.* **2008**, *38*, 42–58. [CrossRef] [PubMed]

48. Bhattacharya, S.; Nuttall, P.A. Phylogenetic Analysis Indicates That Evasin-Like Proteins of Ixodid Ticks Fall Into Three Distinct Classes. *Front. Cell. Infect. Microbiol.* **2021**, *11*, 991. [CrossRef]

49. Fuentes-Prior, P.; Noeske-Jungblut, C.; Donner, P.; Schleuning, W.-D.; Huber, R.; Bode, W. Structure of the thrombin complex with triabin, a lipocalin-like exosite-binding inhibitor derived from a triatomine bug. *Proc. Natl. Acad. Sci. USA* **1997**, *94*, 11845–11850. [CrossRef]

50. Denisov, S.S.; Ippel, J.H.; Mans, B.J.; Dijkgraaf, I.; Hackeng, T.M. SecScan: A general approach for mapping disulfide bonds in synthetic and recombinant peptides and proteins. *Chem. Commun.* **2018**, *55*, 1374–1377. [CrossRef]

51. Denisov, S.S.; Ippel, J.H.; Castoldi, E.; Mans, B.J.; Hackeng, T.M.; Dijkgraaf, I. Molecular basis of anticoagulant and anticomplement activity of the tick salivary protein Salp14 and its homologs. *J. Biol. Chem.* **2021**, *297*, 100865. [CrossRef]

52. Denisov, S.S.; Ramírez-Escudero, M.; Heinzmann, A.C.A.; Ippel, J.H.; Dawson, P.E.; Koenen, R.R.; Hackeng, T.M.; Janssen, B.J.C.; Dijkgraaf, I. Structural characterization of anti-CCL5 activity of the tick salivary protein evasin-4. *J. Biol. Chem.* **2020**, *295*, 14367–14378. [CrossRef]

53. Narasimhan, S.; Koski, R.; Beaulieu, B.; Anderson, J.; Ramamoorthi, N.; Kantor, F.; Cappello, M.; Fikrig, E. A novel family of anti-coagulants from the saliva of Ixodes scapularis. *Insect Mol. Biol.* **2002**, *11*, 641–650. [CrossRef] [PubMed]

54. Assumpcao, T.C.; Mizurini, D.M.; Ma, D.; Monteiro, R.Q.; Ahlstedt, S.; Reyes, M.; Kotsyfakis, M.; Mather, T.N.; Andersen, J.F.; Lukszo, J.; et al. Ixonnexin from tick saliva promotes fibrinolysis by interacting with plasminogen and tissue-type plasminogen acti-vator, and prevents arterial thrombosis. *Sci. Rep.* **2018**, *8*, 4806. [CrossRef] [PubMed]

55. Dias, J.M.; Losberger, C.; Déruaz, M.; Power, C.A.; Proudfoot, A.E.I.; Shaw, J.P. Structural Basis of Chemokine Sequestration by a Tick Chemokine Binding Protein: The Crystal Structure of the Complex between Evasin-1 and CCL3. *PLoS ONE* **2009**, *4*, e8514. [CrossRef] [PubMed]

56. Denisov, S.S.; Ippel, J.H.; Heinzmann, A.C.A.; Koenen, R.R.; Ortega-Gomez, A.; Soehnlein, O.; Hackeng, T.M.; Dijkgraaf, I. Tick saliva protein Evasin-3 modulates chemotaxis by disrupting CXCL8 interactions with glycosaminoglycans and CXCR2. *J. Biol. Chem.* **2019**, *294*, 12370–12379. [CrossRef] [PubMed]

57. Lee, A.W.; Deruaz, M.; Lynch, C.; Davies, G.; Singh, K.; Alenazi, Y.; Eaton, J.R.O.; Kawamura, A.; Shaw, J.; Proudfoot, A.E.I.; et al. A knottin scaffold directs the CXC-chemokine–binding specificity of tick evasins. *J. Biol. Chem.* **2019**, *294*, 11199–11212. [CrossRef]

58. Borer, A.S.; Wassmann, P.; Schmidt, M.; Hoffman, D.R.; Zhou, J.-J.; Wright, C.; Schirmer, T.; Marković-Housley, Z. Crystal structure of Sol i 2: A major allergen from fire ant venom. *J. Mol. Biol.* **2012**, *415*, 635–648. [CrossRef]

59. Ziemba, B.P.; Murphy, E.J.; Edlin, H.T.; Jones, D.N.M. A novel mechanism of ligand binding and release in the odorant binding protein 20 from the malaria mosquito Anopheles gambiae. *Protein Sci.* **2012**, *22*, 11–21. [CrossRef]

60. Carr, A.L.; DMitchell, R., III; Dhammi, A.; Bissinger, B.W.; Sonenshine, D.E.; Roe, R.M. Tick Haller's organ, a new paradigm for arthropod olfaction: How ticks differ from insects. *Int. J. Mol. Sci.* **2017**, *18*, 1563. [CrossRef]
61. Josek, T.; Walden, K.K.; Allan, B.F.; Alleyne, M.; Robertson, H.M. A foreleg transcriptome for Ixodes scapularis ticks: Candidates for chemoreceptors and binding proteins that might be expressed in the sensory Haller's organ. *Ticks Tick-Borne Dis.* **2018**, *9*, 1317–1327. [CrossRef]
62. Vizueta, J.; Frías-López, C.; Macías-Hernández, N.; Arnedo, M.; Sánchez-Gracia, A.; Rozas, J. Evolution of chemosensory gene families in arthropods: Insight from the first inclusive comparative transcriptome analysis across spider appendages. *Genome Biol. Evol.* **2016**, *9*, 178–196. [CrossRef]
63. Renthal, R.; Manghnani, L.; Bernal, S.; Qu, Y.; Griffith, W.P.; Lohmeyer, K.; Guerrero, F.D.; Borges, L.M.; Pérez De León, A. The chemosensory appendage proteome of *Amblyomma americanum* (Acari: Ixodidae) reveals putative odorant-binding and other chemoreception-related proteins. *Insect Sci.* **2017**, *24*, 730–742. [CrossRef] [PubMed]
64. Mans, B.J.; Ribeiro, J.M. Function, mechanism and evolution of the moubatin-clade of soft tick lipocalins. *Insect Biochem. Mol. Biol.* **2008**, *38*, 841–852. [CrossRef] [PubMed]
65. Mans, B.J.; de Castro, M.H.; Pienaar, R.; de Klerk, D.; Gaven, P.; Genu, S.; Latif, A.A. Ancestral reconstruction of tick lineages. *Ticks Tick-Borne Dis.* **2016**, *7*, 509–535. [CrossRef] [PubMed]
66. Keller, P.; Waxman, L.; Arnold, B.; Schultz, L.; Condra, C.; Connolly, T. Cloning of the cDNA and expression of moubatin, an inhibitor of platelet aggregation. *J. Biol. Chem.* **1993**, *268*, 5450–5456. [CrossRef] [PubMed]
67. Waxman, L.; Connolly, T. Isolation of an inhibitor selective for collagen-stimulated platelet aggregation from the soft tick Ornithodoros moubata. *J. Biol. Chem.* **1993**, *268*, 5445–5449. [CrossRef]
68. Flower, D.R.; North, A.C.; Sansom, C.E. The lipocalin protein family: Structural and sequence overview. *Biochim. Et Biophys. Acta Protein Struct. Mol. Enzymol.* **2000**, *1482*, 9–24. [CrossRef]
69. Paesen, G.C.; Adams, P.L.; Nuttall, P.A.; Stuart, D.L. Tick histamine-binding proteins: Lipocalins with a second binding cavity. *Biochim. Et Biophys. Acta Protein Struct. Mol. Enzymol.* **2000**, *1482*, 92–101. [CrossRef]
70. Sangamnatdej, S.; Paesen, G.C.; Slovak, M.; Nuttall, P.A. A high affinity serotonin-and histamine-binding lipocalin from tick saliva. *Insect Mol. Biol.* **2002**, *11*, 79–86. [CrossRef]
71. Mans, B.J.; Ribeiro, J.M.C.; Andersen, J.F. Structure, Function, and Evolution of Biogenic Amine-binding Proteins in Soft Ticks. *J. Biol. Chem.* **2008**, *283*, 18721–18733. [CrossRef]
72. Mans, B.J.; Ribeiro, J.M. A novel clade of cysteinyl leukotriene scavengers in soft ticks. *Insect Biochem. Mol. Biol.* **2008**, *38*, 862–870. [CrossRef]
73. Beaufays, J.; Adam, B.; Menten-Dedoyart, C.; Fievez, L.; Grosjean, A.; Decrem, Y.; Prévôt, P.-P.; Santini, S.; Brasseur, R.; Brossard, M.; et al. Ir-LBP, an Ixodes ricinus Tick Salivary LTB4-Binding Lipocalin, Interferes with Host Neutrophil Function. *PLoS ONE* **2008**, *3*, e3987. [CrossRef] [PubMed]
74. Nunn, M.A.; Sharma, A.; Paesen, G.C.; Adamson, S.; Lissina, O.; Willis, A.C.; Nuttall, P.A. Complement Inhibitor of C5 Activation from the Soft Tick *Ornithodoros moubata*. *J. Immunol.* **2005**, *174*, 2084–2091. [CrossRef] [PubMed]
75. Roversi, P.; Lissina, O.; Johnson, S.; Ahmat, N.; Paesen, G.C.; Ploss, K.; Boland, W.; Nunn, M.A.; Lea, S.M. The Structure of OMCI, a Novel Lipocalin Inhibitor of the Complement System. *J. Mol. Biol.* **2007**, *369*, 784–793. [CrossRef] [PubMed]
76. Preston, S.G.; Majtán, J.; Kouremenou, C.; Rysnik, O.; Burger, L.F.; Cruz, A.C.; Guzman, M.C.; Nunn, M.A.; Paesen, G.C.; Nuttall, P.A.; et al. Novel Immunomodulators from Hard Ticks Selectively Reprogramme Human Dendritic Cell Responses. *PLOS Pathog.* **2013**, *9*, e1003450. [CrossRef]
77. Braunger, K.; Ahn, J.; Jore, M.M.; Johnson, S.; Tang, T.T.L.; Pedersen, D.V.; Andersen, G.R.; Lea, S.M. Structure and function of a family of tick-derived complement inhibitors targeting properdin. *Nat. Commun.* **2022**, *13*, 317. [CrossRef]
78. Laskowski, M., Jr.; Kato, I. Protein inhibitors of proteinases. *Annu. Rev. Biochem.* **1980**, *49*, 593–626. [CrossRef]
79. Francischetti, I.M.B.; Valenzuela, J.G.; Andersen, J.F.; Mather, T.N.; Ribeiro, J.M.C. Ixolaris, a novel recombinant tissue factor pathway inhibitor (TFPI) from the salivary gland of the tick, *Ixodes scapularis*: Identification of factor X and factor Xa as scaffolds for the inhibition of factor VIIa/tissue factor complex. *Blood* **2002**, *99*, 3602–3612. [CrossRef]
80. Paesen, G.C.; Siebold, C.; Dallas, M.L.; Peers, C.; Harlos, K.; Nuttall, P.A.; Nunn, M.A.; Stuart, D.I.; Esnouf, R.M. An ion-channel mod-ulator from the saliva of the brown ear tick has a highly modified Kunitz/BPTI structure. *J. Mol. Biol.* **2009**, *389*, 734–747. [CrossRef]
81. Van De Locht, A.; Stubbs, M.T.; Bode, W.; Friedrich, T.; Bollschweiler, C.; Höffken, W.; Huber, R. The ornithodorin-thrombin crystal structure, a key to the TAP enigma? *EMBO J.* **1996**, *15*, 6011–6017. [CrossRef]
82. Wei, A.; Alexander, R.S.; Duke, J.; Ross, H.; Rosenfeld, S.A.; Chang, C.-H. Unexpected binding mode of tick anticoagulant peptide complexed to bovine factor Xa. *J. Mol. Biol.* **1998**, *283*, 147–154. [CrossRef]
83. Macedo-Ribeiro, S.; Almeida, C.; Calisto, B.M.; Friedrich, T.; Mentele, R.; Stürzebecher, J.; Fuentes-Prior, P.; Pereira, P.J.B. Isolation, Cloning and Structural Characterisation of Boophilin, a Multifunctional Kunitz-Type Proteinase Inhibitor from the Cattle Tick. *PLoS ONE* **2008**, *3*, e1624. [CrossRef] [PubMed]
84. Mans, B.J.; Andersen, J.F.; Schwan, T.G.; Ribeiro, J.M. Characterization of anti-hemostatic factors in the argasid, Argas monolakensis: Implications for the evolution of blood-feeding in the soft tick family. *Insect Biochem. Mol. Biol.* **2008**, *38*, 22–41. [CrossRef] [PubMed]

85. Lai, R.; Takeuchi, H.; Jonczy, J.; Rees, H.H.; Turner, P.C. A thrombin inhibitor from the ixodid tick, Amblyomma hebraeum. *Gene* **2004**, *342*, 243–249. [CrossRef] [PubMed]
86. Liao, M.; Zhou, J.; Gong, H.; Boldbaatar, D.; Shirafuji, R.; Battur, B.; Nishikawa, Y.; Fujisaki, K. Hemalin, a thrombin inhibitor iso-lated from a midgut cDNA library from the hard tick Haemaphysalis longicornis. *J. Insect Physiol.* **2009**, *55*, 165–174. [CrossRef]
87. Mather, T.N.; Ribeiro, J.M.C.; Francischetti, I.M.B. Penthalaris, a novel recombinant five-Kunitz tissue factor pathway inhibitor (TFPI) from the salivary gland of the tick vector of Lyme disease, Ixodes scapularis. *Thromb. Haemost.* **2004**, *91*, 886–898. [CrossRef]
88. Karczewski, J.; Endris, R.; Connolly, T. Disagregin is a fibrinogen receptor antagonist lacking the Arg-Gly-Asp sequence from the tick, Ornithodoros moubata. *J. Biol. Chem.* **1994**, *269*, 6702–6708. [CrossRef]
89. Reck, J.; Webster, A.; Dall'Agnol, B.; Pienaar, R.; De Castro, M.H.; Featherston, J.; Mans, B.J. Transcriptomic analysis of salivary glands of Ornithodoros brasiliensis Aragão, 1923, the agent of a neotropical tick-toxicosis syndrome in humans. *Front. Physiol.* **2021**, *1224*. [CrossRef]
90. Mans, B.; Louw, A.; Neitz, A. Amino acid sequence and structure modeling of savignin, a thrombin inhibitor from the tick, Ornithodoros savignyi. *Insect Biochem. Mol. Biol.* **2002**, *32*, 821–828. [CrossRef]
91. Xu, Y.; Carr, P.D.; Guss, M.; Ollis, D.L. The crystal structure of bikunin from the inter-α-inhibitor complex: A serine protease inhibitor with two kunitz domains. *J. Mol. Biol.* **1998**, *276*, 955–966. [CrossRef]
92. De Paula, V.S.; Sgourakis, N.G.; Francischetti, I.M.; Almeida, F.C.; Monteiro, R.Q.; Valente, A.P. NMR structure determination of Ixolaris and factor X (a) interaction reveals a noncanonical mechanism of Kunitz inhibition. *Blood J. Am. Soc. Hematol.* **2019**, *134*, 699–708. [CrossRef]
93. Reichhardt, M.P.; Johnson, S.; Tang, T.; Morgan, T.; Tebeka, N.; Popitsch, N.; Deme, J.C.; Jore, M.M.; Lea, S.M. An inhibitor of com-plement C5 provides structural insights into activation. *Proc. Natl. Acad. Sci. USA* **2020**, *117*, 362–370. [CrossRef] [PubMed]
94. Zhou, Y.-F.; Eng, E.T.; Zhu, J.; Lu, C.; Walz, T.; Springer, T.A. Sequence and structure relationships within von Willebrand factor. *Blood* **2012**, *120*, 449–458. [CrossRef] [PubMed]
95. Isaacs, N.W. Cystine Knots. *Curr. Opin. Struc. Biol.* **1995**, *5*, 391–395. [CrossRef] [PubMed]
96. Jore, M.M.; Johnson, S.; Sheppard, D.; Barber, N.M.; Li, Y.I.; Nunn, M.A.; Elmlund, H.; Lea, S.M. Structural basis for therapeutic in-hibition of complement C5. *Nat. Struct. Mol. Biol.* **2016**, *23*, 378–386. [CrossRef] [PubMed]
97. Escudero-Esparza, A.; Kalchishkova, N.; Kurbasic, E.; Jiang, W.G.; Blom, A.M. The novel complement inhibitor human CUB and Sushi multiple domains 1 (CSMD1) protein promotes factor I-mediated degradation of C4b and C3b and inhibits the membrane attack complex assembly. *FASEB J.* **2013**, *27*, 5083–5093. [CrossRef]
98. Chou, S.; Daugherty, M.D.; Peterson, S.B.; Biboy, J.; Yang, Y.; Jutras, B.L.; Fritz-Laylin, L.K.; Ferrin, M.A.; Harding, B.N.; Jacobs-Wagner, C.; et al. Transferred interbacterial antagonism genes augment eukaryotic innate immune function. *Nature* **2014**, *518*, 98–101. [CrossRef]
99. Hayes, B.M.; Radkov, A.D.; Yarza, F.; Flores, S.; Kim, J.; Zhao, Z.; Lexa, K.W.; Marnin, L.; Biboy, J.; Bowcut, V. Ticks resist skin com-mensals with immune factor of bacterial origin. *Cell* **2020**, *183*, 1562–1571.e1512. [CrossRef]
100. Gattiker, A.; Gasteiger, E.; Bairoch, A. ScanProsite: A reference implementation of a PROSITE scanning tool. *Appl. Bioinform.* **2002**, *1*, 107–108.

International Journal of
Molecular Sciences

Article

Ixodes ricinus Salivary Serpin Iripin-8 Inhibits the Intrinsic Pathway of Coagulation and Complement

Jan Kotál [1,2], Stéphanie G. I. Polderdijk [3], Helena Langhansová [1], Monika Ederová [1], Larissa A. Martins [2], Zuzana Beránková [1], Adéla Chlastáková [1], Ondřej Hajdušek [4], Michail Kotsyfakis [1,2], James A. Huntington [3] and Jindřich Chmelař [1,*]

1. Department of Medical Biology, Faculty of Science, University of South Bohemia in České Budějovice, Branišovská 1760c, 37005 České Budějovice, Czech Republic; jan.kotal@nih.gov (J.K.); hlanghansova@prf.jcu.cz (H.L.); ederom01@prf.jcu.cz (M.E.); beranz02@prf.jcu.cz (Z.B.); chlasa00@prf.jcu.cz (A.C.); kotsyfakis@paru.cas.cz (M.K.)
2. Laboratory of Genomics and Proteomics of Disease Vectors, Institute of Parasitology, Biology Center CAS, Branišovská 1160/31, 37005 České Budějovice, Czech Republic; larissa.martins@nih.gov
3. Cambridge Institute for Medical Research, Department of Haematology, University of Cambridge, The Keith Peters Building, Hills Road, Cambridge CB2 0XY, UK; stephanie.polderdijk@cantab.net (S.G.I.P.); jah52@cam.ac.uk (J.A.H.)
4. Laboratory of Vector Immunology, Institute of Parasitology, Biology Center CAS, Branišovská 1160/31, 37005 České Budějovice, Czech Republic; hajdus@paru.cas.cz
* Correspondence: chmelar@prf.jcu.cz

Citation: Kotál, J.; Polderdijk, S.G.I.; Langhansová, H.; Ederová, M.; Martins, L.A.; Beránková, Z.; Chlastáková, A.; Hajdušek, O.; Kotsyfakis, M.; Huntington, J.A.; et al. *Ixodes ricinus* Salivary Serpin Iripin-8 Inhibits the Intrinsic Pathway of Coagulation and Complement. *Int. J. Mol. Sci.* **2021**, *22*, 9480. https://doi.org/10.3390/ijms22179480

Academic Editor: József Tőzsér

Received: 23 June 2021
Accepted: 26 August 2021
Published: 31 August 2021

Abstract: Tick saliva is a rich source of antihemostatic, anti-inflammatory, and immunomodulatory molecules that actively help the tick to finish its blood meal. Moreover, these molecules facilitate the transmission of tick-borne pathogens. Here we present the functional and structural characterization of Iripin-8, a salivary serpin from the tick *Ixodes ricinus*, a European vector of tick-borne encephalitis and Lyme disease. Iripin-8 displayed blood-meal-induced mRNA expression that peaked in nymphs and the salivary glands of adult females. Iripin-8 inhibited multiple proteases involved in blood coagulation and blocked the intrinsic and common pathways of the coagulation cascade in vitro. Moreover, Iripin-8 inhibited erythrocyte lysis by complement, and Iripin-8 knockdown by RNA interference in tick nymphs delayed the feeding time. Finally, we resolved the crystal structure of Iripin-8 at 1.89 Å resolution to reveal an unusually long and rigid reactive center loop that is conserved in several tick species. The P1 Arg residue is held in place distant from the serpin body by a conserved poly-Pro element on the P′ side. Several PEG molecules bind to Iripin-8, including one in a deep cavity, perhaps indicating the presence of a small-molecule binding site. This is the first crystal structure of a tick serpin in the native state, and Iripin-8 is a tick serpin with a conserved reactive center loop that possesses antihemostatic activity that may mediate interference with host innate immunity.

Keywords: blood coagulation; crystal structure; *Ixodes ricinus*; parasite; saliva; serpin; tick

1. Introduction

Ticks are blood-feeding ectoparasites and vectors of human pathogens, including agents of Lyme disease and tick-borne encephalitis. *Ixodes ricinus* is a species of European tick in the Ixodidae (hard tick) family found also in northern Africa and the Middle East [1]. *I. ricinus* ticks feed only once in each of their three developmental stages (larva, nymph, imago), and their feeding course can last over a week in adult females [2]. In order to stay attached to the host for such extended periods of time, ticks counteract host defense mechanisms that would otherwise lead to tick rejection or death.

Insertion of tick mouthparts into host skin causes mechanical injury that immediately triggers the hemostatic mechanisms of blood coagulation, vasoconstriction, and platelet aggregation to prevent blood loss [3]. Consequently, innate immunity is activated as noted by inflammation with edema formation, inflammatory cell infiltration, and itching at tick feeding sites. Long-term feeding and/or repeated exposures of the host to ticks also activate adaptive immunity [4]. As an adaptation to host defenses, ticks modulate and suppress host immune responses and hemostasis by secreting a complex cocktail of pharmacoactive substances via their saliva into the host. For further information on this topic, we refer readers to several excellent reviews describing the impact of saliva and salivary components on the host [4–8].

Blood coagulation is a cascade driven by serine proteases that leads to the production of a fibrin clot. It can be initiated via the extrinsic or intrinsic pathway [9]. The extrinsic pathway starts with blood vessel injury and complex formation between activated factor VII (fVIIa) and tissue factor (TF). The TF/fVIIa complex then activates factor X (fX) either directly or via activation of factor IX (fIX), which in turn activates fX. The intrinsic pathway is triggered by the activation of factor XII (fXII) via kallikrein. Activated fXII (fXIIa) activates factor XI (fXI), which next activates fIX and results in the activation of fX, followed by a common pathway that terminates the coagulation process through the activation of thrombin (fII) and the cleavage of fibrinogen to fibrin, the primary component of the clot [9,10].

Similar to blood coagulation, the complement cascade is based on serine proteases. Complement represents a fast and robust defense mechanism against bacterial pathogens, which are lysed or opsonized by complement to facilitate their killing by other immune mechanisms [11,12]. Complement can be activated via three pathways: the classical pathway, responding to antigen–antibody complexes; the lectin pathway, which needs a lectin to bind to specific carbohydrates on the pathogen surface; and the alternative pathway, which is triggered by direct binding of C3b protein to a microbial surface [12]. All three pathways result in the cleavage of C3 by C3 convertases to C3a and C3b fragments. C3b then triggers a positive feedback loop to amplify the complement response and opsonize pathogens for phagocytosis. Together with other complement components, C3b forms C5 convertase, which cleaves C5 to C5a and C5b fragments. C5b initiates membrane attack complex (MAC) formation, leading to lysis of a target cell. Small C3a and C5a subunits promote inflammation by recruiting immune cells to the site of injury [11].

Both processes, coagulation and complement, are detrimental to feeding ticks, so their saliva contains many anticoagulant and anticomplement molecules, often belonging to the group of serine protease inhibitors (serpins) [13–16]. Serpins form the largest and most ubiquitous family of protease inhibitors in nature and can be found in viruses, prokaryotes, and eukaryotes [17,18]. Serpins are irreversible inhibitors with a unique inhibitory mechanism and highly conserved tertiary structure [19,20] classified in the I4 family of the MEROPS database [21]. Similar to other serine protease inhibitors, the serpin structure contains a reactive center loop (RCL) that serves as bait for the protease. The RCL amino acid sequence determines serpin's inhibitory specificity [22].

Arthropod serpins have mostly homeostatic and immunological functions. They regulate hemolymph coagulation or activation of the phenoloxidase system in insects [23]. Additionally, serpins from blood-feeding arthropods can modulate host immunity and host hemostasis [23]. Indeed, over 20 tick salivary serpins have been functionally characterized with described effects on coagulation or immunity [13]. However, according to numerous transcriptomic studies, the total number of tick serpins is significantly higher [13,24–27]. In *I. ricinus*, at least 36 serpins have been identified based on transcriptomic data, but only 3 of them have been characterized at the biochemical, immunomodulatory, anticoagulatory, or antitick vaccine levels [13,28–32].

Interestingly, one serpin has a fully conserved RCL across various tick species [24]. Homologs of this serpin have been described in *Amblyomma americanum* as AAS19 [33], *Rhipicephalus haemaphysaloides* as RHS8 [34], *Rhipicephalus microplus* as RmS-15 [35], and *I. ricinus* as IRS-8 [30], and it can also be found among transcripts of other tick species in which the serpins have not yet been functionally characterized.

Here we present the functional characterization of Iripin-8, the serpin from *I. ricinus* previously referred to as IRS-8 [30,34], whose RCL is conserved among several tick species. We demonstrate its inhibitory activity against serine proteases involved in coagulation and direct the inhibition of the intrinsic coagulation pathway in vitro. Moreover, we report for the first time the inhibition of complement by a tick serpin. Finally, we provide the structure of Iripin-8 in its native, uncleaved form, revealing an unusual RCL conserved among several tick serpins.

2. Results

2.1. Iripin-8 Is Predominantly a Salivary Protein with Increased Expression during Tick Feeding

Analysis of Iripin-8 mRNA expression levels revealed its highest abundance in tick nymphs with a peak during the first day of feeding (Figure 1A). In salivary glands, increased Iripin-8 transcription positively correlated with the length of tick feeding on its host. A similar increasing trend was also observed in tick midguts; however, the total number of Iripin-8 transcripts was lower than in the salivary glands. Iripin-8 transcript levels were lowest in the ovaries of all the tested tissues/stages.

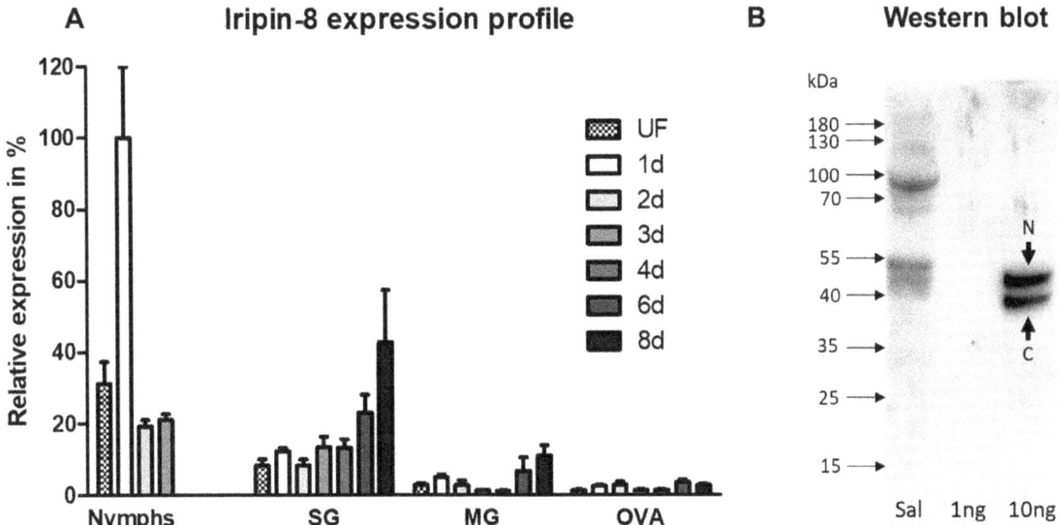

Figure 1. Iripin-8 expression in ticks and its presence in tick saliva. (**A**) Pools of *I. ricinus* salivary glands, midguts, and ovaries from female ticks and whole bodies from nymphs were dissected under RNase-free conditions. cDNA was subsequently prepared as a template for qRT-PCR. Iripin-8 expression was normalized to elongation factor 1α and compared between all values with the highest expression set to 100% (*y*-axis). The data show an average of three biological replicates for adult ticks and six replicates for nymphs (±SEM). SG = salivary glands; MG = midguts; OVA = ovaries; UF = unfed ticks; 1 d, 2 d, 3 d, 4 d, 6 d, 8 d = ticks after 1, 2, 3, 4, 6, or 8 days of feeding. For nymphs, the last column represents fully fed nymphs. All feeding points for each development stage/tissue are compared with the unfed ticks of the respective group. (**B**) Iripin-8 can be detected in tick saliva by Western blotting. Saliva from ticks after 6 days of feeding and recombinant Iripin-8 protein were visualized by Western blotting using serum from naïve and Iripin-8-immunized rabbits. Sal = tick saliva; 1 ng, 10 ng = Iripin-8 recombinant protein at 1 ng and 10 ng load. N: native Iripin-8, C: cleaved Iripin-8.

Next, we performed Western blot analysis and confirmed the presence of Iripin-8 protein in tick saliva (Figure 1B). We detected two bands of the recombinant protein, representing the full-length native serpin (N) and a molecule cleaved in its RCL near the C-terminus, likely due to bacterial protease contamination (C). The proteolytic cleavage of RCL has previously been documented for serpins from various organisms, including ticks [36–38]. The ~5 kDa difference in molecular weight observed between native and recombinant Iripin-8 was probably due to glycosylation, since two N-glycosylation sites are predicted to exist in this serpin. The signal at ~90 kDa in saliva was also detected when using serum from a naïve rabbit (data not shown) and is probably caused by nonspecific antibody binding.

Based on these results, we proceeded to test how Iripin-8 affects host defense mechanisms as a component of tick saliva. Despite the highest expression being observed in the salivary glands, activity in other tissues cannot be ruled out.

2.2. Sequence Analysis and Production of Recombinant Iripin-8

The full transcript encoding Iripin-8 was obtained using cDNA from tick salivary glands. Following sequencing, we found a few amino acid mutations (K10 → E10, L36 → F36, P290 → T290, and F318 → S318) compared with the sequence of Iripin-8 (IRS-8) published as a supplement in our previous work [30] (GenBank No. DQ915845.1; ABI94058.1), probably as a result of intertick variability. The RCL was identical to other homologous tick serpins [34], with arginine at the P1 position (Supplementary Figure S1A); however, the remainder of the sequence had undergone evolution, separating species-specific sequences in strongly supported groups (Supplementary Figure S1B). Iripin-8 has a predicted MW of 43 kDa and a pI of 5.85, with two predicted N-linked glycosylation sites.

Iripin-8 was expressed in 2 L of medium with a yield of 45 mg of protein at >90% purity, as analyzed by pixel density analysis in ImageJ software, where a majority was formed from the native serpin and a fraction from a serpin cleaved at its RCL (Supplementary Figure S2). This mixed sample of native and cleaved serpin was used for all subsequent analyses because the molecules were inseparable by common chromatographic techniques. Proper folding of Iripin-8 was verified by CD spectroscopy (Supplementary Data) [39,40] and subsequently by activity assays against serine proteases, as presented below. Recombinant Iripin-8 protein solution was tested for the presence of LPS, which was detected at 0.038 endotoxin unit/mL, below the threshold for a pyrogenic effect [41,42].

2.3. Iripin-8 Inhibits Serine Proteases Involved in Coagulation

Based on sequence analysis of Iripin-8 and the presence of arginine in the RCL P1 position, we focused on analyzing its inhibitory specificity towards serine proteases related to blood coagulation. Considering the covalent nature of the serpin mechanism of inhibition, we analyzed by SDS-PAGE whether Iripin-8 forms covalent complexes with selected proteases. Figure 2 shows covalent inhibitory complex formation between Iripin-8 and 10 out of 11 tested proteases: thrombin, fVIIa, fIXa, fXa, fXIa, fXIIa, plasmin, APC, kallikrein, and trypsin. We did not detect complexes between Iripin-8 and chymotrypsin. All inhibited proteases could also partially cleave Iripin-8 as indicated by a C-terminal fragment and a stronger signal of cleaved serpin molecule. Chymotrypsin cleaved Iripin-8 in its RCL completely. Inhibition rates of Iripin-8 against these proteases were subsequently determined and are shown in Table 1. Among the tested proteases, plasmin was inhibited significantly faster than other proteases, with a second-order rate constant (k_2) of >200,000 M^{-1} s^{-1}. Trypsin, kallikrein, fXIa, and thrombin were inhibited with a k_2 in the tens of thousands range and the other proteases with lower k_2 values.

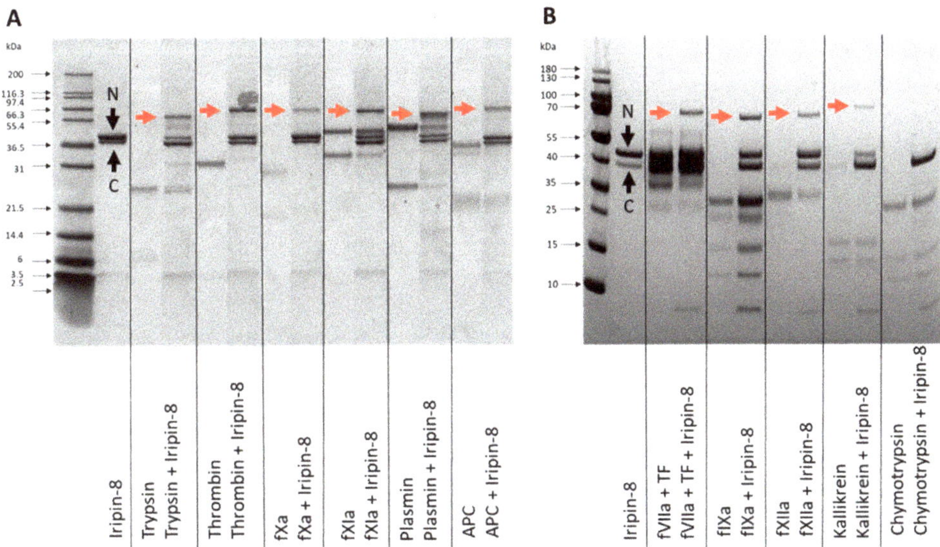

Figure 2. Formation of covalent complexes between Iripin-8 and serine proteases. Iripin-8 and selected serine proteases were incubated for 1 h and subsequently analyzed for complex formation by reducing SDS-PAGE. Protein separation differs between (**A**,**B**) due to the use of gels with different polyacrylamide contents. Gels show the profile of Iripin-8 serpin alone, various serine proteases alone, and proteases incubated with Iripin-8. Complex formation between fVIIa and Iripin-8 was tested in the presence of tissue factor (TF) at an equimolar concentration. Covalent complexes between Iripin-8 and protease are marked with a red arrow. N: native Iripin-8, C: cleaved Iripin-8.

Table 1. Inhibition rate of Iripin-8 against selected serine proteases.

Protease	k_2 (M^{-1} s^{-1})	±SE
Plasmin	225,064	14,183
Trypsin	29,447	3508
Kallikrein	16,682	1119
fXIa	16,328	948
Thrombin	13,794	1040
fXIIa	3324	409
fXa	2088	115
APC	523	35
fVIIa + TF	456	35
fIXa	N/A	N/A

2.4. Iripin-8 Inhibits the Intrinsic and Common Pathways of Blood Coagulation

Given the in vitro inhibition of coagulation proteases by Iripin-8, we tested its activity in three coagulation assays. The prothrombin time (PT) assay simulates the extrinsic pathway of coagulation, the activated partial thromboplastin time (aPTT) represents the intrinsic (contact) pathway, and thrombin time (TT) represents the final common stage of coagulation. Iripin-8 had no significant effect on PT, which increased from 15.3 to 16.7 s in the presence of 6 µM serpin (not shown). Iripin-8 extended aPTT in a dose-dependent manner, with a statistically significant increase already apparent at 375 nM. With 6 µM Iripin-8, the aPTT was delayed over five times from 31.8 ± 0.4 s to 167.9 ± 3.2 s (Figure 3A). Iripin-8 also inhibited TT in a dose-dependent manner and blocked fibrin clot formation completely at concentrations of 800 nM and higher (Figure 3B). The other serpins presented for comparison in Figure 3C did not have any effect on blood coagulation except the inhibition of PT by Iripin-3, which we published elsewhere [32].

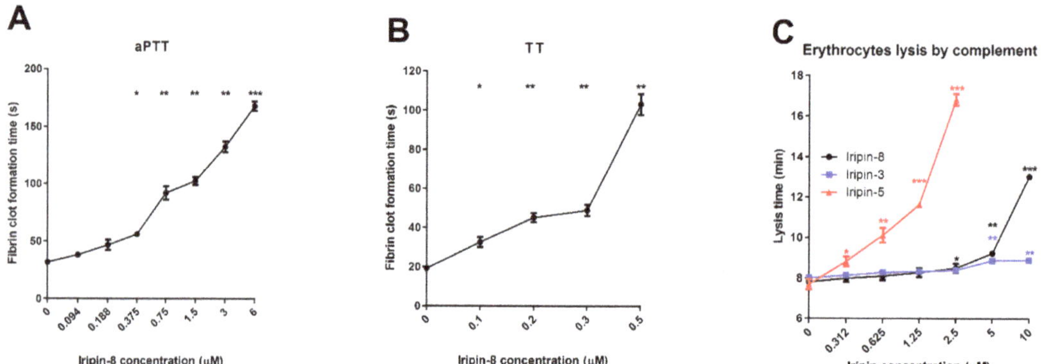

Figure 3. Inhibition of complement and coagulation pathways by Iripin-8. (**A**) Iripin-8 inhibits the intrinsic coagulation pathway. Human plasma was preincubated with increasing concentrations of Iripin-8 (94 nM–6 μM). Coagulation was triggered by the addition of Dapttin® reagent and CaCl₂, and clot formation time was measured. A sample without Iripin-8 was used as a control for statistical purposes. (**B**) Iripin-8 delays fibrin clot formation in a thrombin time assay in a dose-dependent manner. Coagulation of human plasma was initiated by thrombin reagent preincubated with various concentrations of Iripin-8, and thrombin time was measured. Samples without Iripin-8 were used as a control for statistical purposes. (**C**) Iripin-8 inhibits erythrocyte lysis by human complement. Human plasma was preincubated with increasing concentrations of Iripin-3, 5, and 8 (312 nM–10 μM). After the addition of rabbit erythrocytes, their lysis time by complement was measured. Values represent prolongation of time needed for erythrocyte lysis compared with the control group. * $p \leq 0.05$; ** $p \leq 0.01$; *** $p \leq 0.001$.

2.5. Anticomplement Activity of Iripin-8

The complement pathway readily lyses erythrocytes from various mammals, and those from rabbits were found to be the best complement activators [43]. We used human serum and rabbit erythrocytes to test the effect of tick protease inhibitors on the activity of human complement in vitro. Since the complement cascade is driven by serine proteases, we tested the potential effect of Iripin-8 as a complement regulator. There was a statistically significant reduction in complement activity against erythrocytes when human plasma was incubated with Iripin-8 at concentrations of 2.5 μM and higher (Figure 3C). We also compared Iripin-8 with other two tick salivary serpins: Iripin-3 [32] showed very weak anticomplement activity and was used as a control; compared with Iripin-5 [44], Iripin-8 had lower activity.

2.6. Iripin-8 Knockdown Influences Tick Feeding but Not Borrelia Transmission

Since Iripin-8 is predominantly expressed in tick nymphs (Figure 1A), we decided to investigate its importance in tick feeding by RNA interference (RNAi) in the nymphal stage. Knockdown efficiency was 87% for transcript downregulation. Ticks with downregulated Iripin-8 expression showed a significantly lower feeding success rate and higher mortality, with only 51.0% (25/49) finishing feeding compared with 94.1% (48/51) in the control group. Moreover, in ticks that finished feeding, the feeding time was longer compared with control nymphs (Figure 4A). Despite this promising phenotype, we did not observe any effect of Iripin-8 RNAi on the weight of fully engorged nymphs (Figure 4B) or on *B. afzelii* transmission from infected nymphs to mice in any of the tested mouse tissues (Figure 4C).

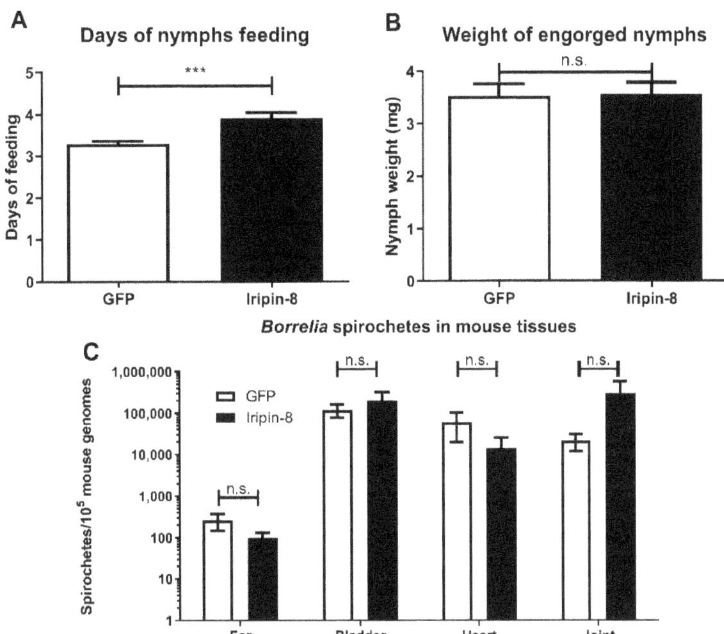

Figure 4. Effect of RNAi on tick fitness and *Borrelia* transmission. (**A**) RNAi of Iripin-8 prolonged the length of *I. ricinus* nymph feeding compared with the control group (GFP). (**B**) Weight of fully engorged nymphs with Iripin-8 knockdown was not different from the control group (GFP). (**C**) Presence of *B. afzelii* spirochetes in mouse tissues after infestation with infected *I. ricinus* nymphs. There were no significant differences between Iripin-8 knockdown and GFP control groups in any of the tested tissues. *** $p \leq 0.001$; n.s., not significant.

2.7. Role of Iripin-8 in Modulating Host Immunity

Next, we evaluated a possible role for Iripin-8 in the modulation of the host immune response to tick feeding via two assays (OVA antigen-specific CD4$^+$ T cell proliferation model using splenocytes isolated from OT-II mice and neutrophil migration towards the chemoattractant (fMLP), in which we previously observed effects with other *I. ricinus* salivary protease inhibitors (the serpin Iripin-3 [32] and the cystatin Iristatin [45]). However, there was no inhibition by Iripin-8 in either assay (Supplementary Figure S4).

2.8. Structural Features of Iripin-8

The crystallographic asymmetric unit contained a single molecule of native Iripin-8 (details in Supplementary Table S2). Electron density was of sufficient quality to model all residues from Ser5 to the C-terminus, including the entire RCL. Iripin-8 has the typical native serpin fold with a Cα RMSD (root-mean-square deviation) of 1.93 Å compared with the archetypal serpin alpha-1-antitrypsin (A1AT, 342 of 352 residues, Figure 5A). The most remarkable feature of Iripin-8 is its long RCL (11 residues longer on the P' side), which extends away from the body of the serpin, moving the P1 Arg364 17.8 Å further than the P1 residue of A1AT. This extended conformation is not the result of a crystal contact; rather it forms the basis of a crystal contact with the RCL of a symmetry-related molecule (Supplementary Figure S5). The P' extension contains a stretch of proline residues that form a type II polyproline helix, conferring rigidity and extending the P1 residue away from the body of Iripin-8 (Figure 5B). We can infer from this that the extended RCL is a feature of native Iripin-8 in solution and that it has functional consequences in determining protease specificity and/or inhibitory promiscuity.

We also observed several molecules of PEG (polyethylene glycol) originating from the crystallization buffer bound to Iripin-8. One of the binding sites was a deep 109 Å^3 cavity in the core structure between helixes A, B, and C (Supplementary Figure S6). This observation suggests that Iripin-8 can bind small molecules, which may have functional implications. The coordinates and structure factors are deposited in the Protein Data Bank under accession code XXX (note: will be submitted before publication).

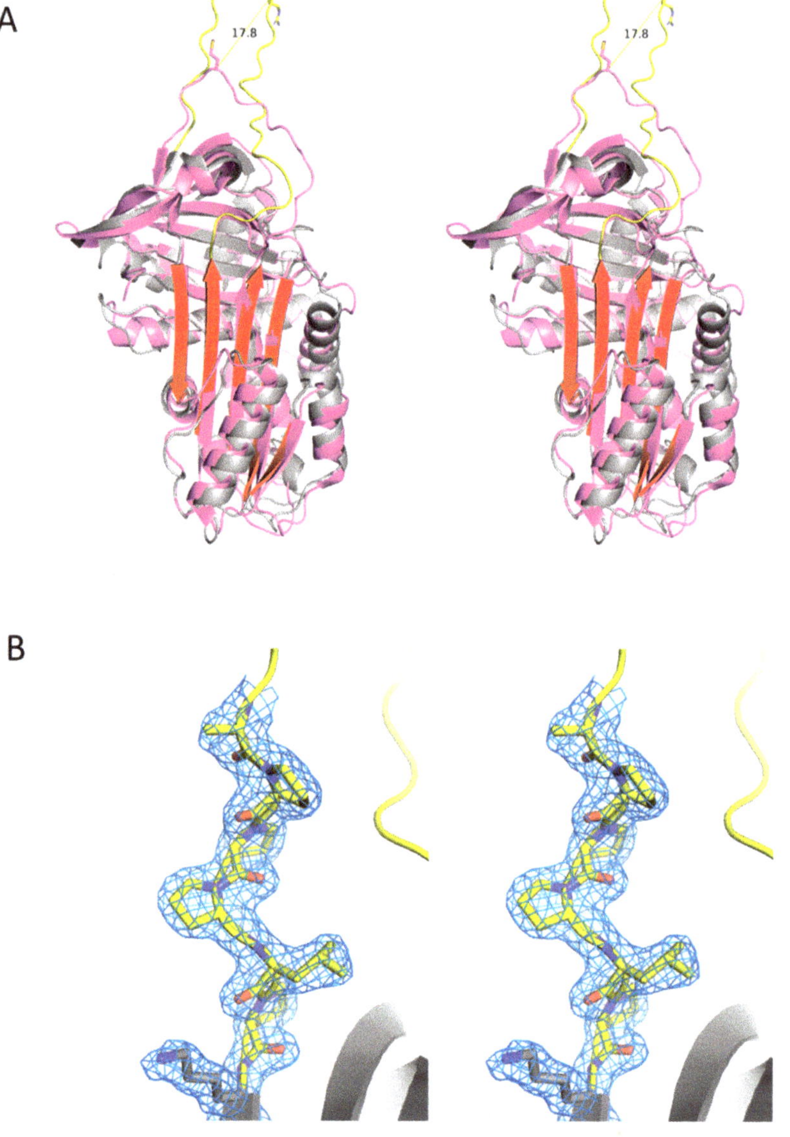

Figure 5. Crystal structure of native Iripin-8. (**A**) Stereo view of a ribbon diagram of Iripin-8 (gray with yellow RCL and red beta sheet A) superimposed with alpha-1-antitrypsin (PDB code 3ne4). The P1 side chains of both molecules are represented as sticks, and the distance between their Cα atoms is shown. (**B**) Stereo view of a close-up of the P′ region with surrounding electron density (contoured at 1 times the RMSD of the map), forming a rigid type II polyproline helix.

3. Discussion

Similar to other characterized tick salivary serpins [13], we found that Iripin-8 can modulate host complement and coagulation cascades to facilitate tick feeding [46].

Structurally, Iripin-8 has an unusually long, exposed, and rigid RCL, with an Arg in its P1 position. This potentially enables it to inhibit a range of proteases, as the RCL can interact independently from the body of the serpin molecule. We characterized Iripin-8 as an in vitro inhibitor of at least 10 serine proteases. The interference with the coagulation cascade through inhibition of kallikrein, thrombin, fVIIa, fIXa, fXa, fXIa, and fXIIa in vivo would be beneficial for tick feeding [3,47].

Iripin-8 also inhibited trypsin and kallikrein. Trypsin has a role in meal digestion and has also been linked to skin inflammation [48,49]. Potentially, trypsin inhibition in the host skin could be another mechanism by which the tick impairs the host immune response. Kallikrein has a role in the development of inflammation and pain. It is an activator of the nociceptive mediator bradykinin in the kinin–kallikrein system [50]. Through its inhibition, a deleterious inflammatory response could be altered to the tick's advantage.

Iripin-8 showed the greatest inhibition of plasmin, a protease involved in fibrin degradation and clot removal [51]. This was surprising, as clot removal should be beneficial for ticks. On the other hand, it is not fully understood whether fibrin clot formation occurs at a tick feeding site in the presence of tick anticoagulant molecules [52]. Apart from fibrinolysis, plasmin also modulates several immunological processes, interacting with leukocytes, endothelial cells, extracellular matrix components, and immune system factors [51,53,54]. Excessive plasmin generation can even lead to pathophysiological inflammatory processes [54]. Considering the proinflammatory role of plasmin, its inhibition by tick salivary serpin could be more relevant to the tick than unimpaired fibrinolysis. Although we did not see any effect of Iripin-8 in two immune assays, we cannot exclude the possibility that Iripin-8 exerts an immunomodulatory effect.

The anticomplement activity of tick saliva or its protein components has been known for decades and is described in numerous publications [14,15,55–57]. Although the active molecules originate from either unique tick protein families [58–62] or lipocalins [16], anticomplement activity has only recently been reported for a tick salivary serpin [44] as the only tick protease inhibitor with such activity. Since complement products might directly damage the tick hypostome or initiate a stronger immune response [11], we propose that the role of Iripin-8 is to attenuate these mechanisms. At the same time, an impaired complement system cannot effectively fight pathogens entering the wound at the same time as tick saliva [14]. In this context, we wanted to test a potential effect of Iripin-8 transcriptional downregulation on *Borrelia* transmission from ticks to the host. Although we saw some effect of RNA interference (RNAi) on tick fitness, it had no effect on the amount of *Borrelia* in host tissues. Such a result can be explained by a redundancy in tick salivary molecules, as ticks secrete a variety of effectors against the same host defense mechanism and knockdown of one molecule can be substituted by the activity of others [63].

The increased tick mortality after Iripin-8 knockdown might be due to a potential role for Iripin-8 within the tick body. As an anticoagulant, Iripin-8 can help to keep ingested blood in the tick midgut in an unclotted state for later intracellular digestion [64–66]. A similar principle has previously been suggested for other midgut serpins in various tick species [67]. Other potential functions of Iripin-8 include a role in hemolymph clotting [68,69] or in reproduction and egg development [70,71].

The broad inhibitory specificity combined with a conserved, long, and rigid RCL implies that Iripin-8's role does not necessarily have to only be the modulation of host defense mechanisms. The function of a protruded RCL can be adapted to fit the active site of an unknown protease of tick origin, independently of the serpin body, thus regulating physiological processes in the tick itself, such as melanization and immune processes, which are also regulated by serpins in arthropods.

Interestingly, several PEG molecules from the crystallization buffer bind to Iripin-8, including one in a deep cavity, perhaps indicating the presence of a small-molecule binding site. Considering that serpins can act as transport proteins independently of their inhibitory properties [72,73], the binding properties of Iripin-8 could have physiological relevance in ticks or their tick–host interactions.

By comparing Iripin-8 with other members of the tick serpin group with an identical RCL, we confirmed that the anticoagulant features have also been reported for AAS19 [33] and RmS-15 [35]. RNAi knockdown of Iripin-8 reduced feeding success, while RNAi of AAS19 decreased the blood intake and morphological deformation of ticks [74], and RNAi of RHS8 had an effect on body weight, feeding time, and vitellogenesis [34]. However, these findings are difficult to correlate due to the use of different tick species and life stages. Although Iripin-8 was detected in tick saliva and therefore most likely plays a role in the regulation of host defense mechanisms, further experiments to define Iripin-8 functions in tick tissues would be of interest. Similar to AAS19 [74], Iripin-8 might also regulate hemolymph clotting in the tick body, which is naturally regulated by serpins [67]. Iripin-8 could also contribute to maintaining ingested blood in the tick midgut in an unclotted state to preserve availability for intracellular digestion [64,66].

Although the concentration of Iripin-8 in tick saliva is not known, we can expect it to be lower than most of the concentrations used in our assays. Tick saliva, as a complex mixture, contains an abundance of bioactive molecules that are redundant in their activities and contribute to the inhibition of host defense mechanisms [63]. Therefore, despite the fact that the concentrations of Iripin-8 used in our experiments do not reflect a physiological situation, they can reflect the overall concentration of functionally redundant salivary proteins.

We conclude that the tick serpin Iripin-8 is secreted into the host as a component of *I. ricinus* saliva. Based on its inhibitory activity, mainly of proteases of the coagulation cascade [47], we suggest that its main role as a salivary protein is in the modulation of host blood coagulation and complement activity, with possible function in regulating the immune response. As such, Iripin-8 alters host defense mechanisms and most likely facilitates tick feeding on hosts.

Nevertheless, a more detailed comparative study of tick serpins with conserved RCLs might shed some light on the role of this particular subgroup in different tick species. The conservation of Iripin-8 among tick species suggests a potential for targeting this serpin as a tick control strategy.

4. Materials and Methods

4.1. Ticks and Laboratory Animals

All animal experiments were carried out in accordance with the Animal Protection Law of the Czech Republic No. 246/1992 Coll., ethics approval No. MSMT-19085/2015-3, and protocols approved by the responsible committee of the Institute of Parasitology, Biology Center of the Czech Academy of Sciences (IP BC CAS). Male and female adult *I. ricinus* ticks were collected by flagging in a forest near České Budějovice in the Czech Republic and kept in 95% humidity chambers under a 12 h light/dark cycle at laboratory temperature. Tick nymphs were obtained from the tick rearing facility of the IP BC CAS. C3H/HeN mice were purchased from Velaz s.r.o. (Prague, Czech Republic). Mice were housed in individually ventilated cages under a 12 h light/dark cycle and used at 6–12 weeks. Laboratory rabbits were purchased from Velaz and housed individually in cages in the animal facility of the Institute of Parasitology. Guinea pigs were bred and housed in cages in the animal facility of the Institute of Parasitology. All mammals were fed a standard pellet diet and given water ad libitum.

4.2. Gene Expression Profiling

I. ricinus nymphs were fed on C3H/HeN mice for 1 day, 2 days, and until full engorgement (3–4 days); *I. ricinus* females were fed on guinea pigs for 1, 2, 3, 4, 6, and 8 days. Adult salivary glands, midguts, and ovaries, as well as whole nymph bodies, were dissected under RNase-free conditions, and total RNA was isolated using TRI Reagent solution

(MRC, Cincinnati, OH, USA). cDNA was prepared using 1 µg of total RNA from pools of ticks fed on three different guinea pigs using the Transcriptor First Strand cDNA Synthesis kit (Roche, Basel, Switzerland) according to the manufacturer's instructions. The cDNA was subsequently used for the analysis of *Iripin-8* expression by qPCR in a Rotor-Gene 6000 cycler (Qiagen, Hilden, Germany) using FastStart Universal SYBR® Green Master Mix (Roche). *Iripin-8* expression profiles were calculated using the Livak and Schmittgen mathematical model [75] and normalized to *I. ricinus* elongation factor 1α (ef1; GenBank No. GU074829.1) [76,77]. Primer sequences are shown in Supplementary Table S1.

4.3. RNA Silencing and Borrelia Transmission

Borrelia afzelii-infected *I. ricinus* nymphs were prepared as described previously [78,79]. A fragment of the *Iripin-8* gene was amplified from *I. ricinus* cDNA using primers containing restriction sites for ApaI and XbaI (Supplementary Table S1; Iripin-8 RNAi) and cloned into the pll10 vector with two T7 promoters in reverse orientations [80]. Double-stranded RNA (dsRNA) of *Iripin-8* and dsRNA of green fluorescent protein (*gfp*) used for control were synthesized using the MEGAscript T7 transcription kit (Ambion, Austin, TX, USA), as described previously [81]. The dsRNA (32 nl; 3 µg/µL) was injected into the hemocoel of sterile or infected nymphs using a Nanoject II instrument (Drummond Scientific, Broomall, PA). After 3 days of rest in a humid chamber at laboratory temperature, ticks were fed on C3H/HeN mice (15–20 nymphs per mouse) until full engorgement. Two weeks later, mice were sacrificed, and the numbers of *Borrelia* spirochetes in the earlobe, urinary bladder, heart tissue, and ankle joint were estimated by qPCR [82] and normalized to the number of mouse genomes [83] (primer and probe sequences in Supplementary Table S1). The level of gene knockdown was checked by qPCR in an independent experiment.

4.4. Cloning, Expression, and Purification of Iripin-8

The full cDNA sequence of the gene encoding Iripin-8 was amplified with the primers presented in Supplementary Table S1 using cDNA prepared from the salivary glands of female *I. ricinus* ticks fed for 3 and 6 days on rabbits as a template. The Iripin-8 gene without a signal peptide was cloned into a linearized Champion™ pET SUMO expression vector (Life Technologies, Carlsbad, CA, USA) using NEBuilder® HiFi DNA Assembly Master Mix (New England Biolabs, Ipswich, MA, USA) and transformed into *Escherichia coli* strain Rosetta 2(DE3)pLysS (Novagen, Merck Life Science, Darmstadt, Germany) for expression. Bacterial cultures were fermented in autoinduction TB medium supplemented with 50 mg/L kanamycin at 25 °C for 24 h.

SUMO-tagged Iripin-8 was purified from clarified cell lysate using a HisTrap FF column (GE Healthcare, Chicago, IL, USA) and eluted with 200 mM imidazole. After the first purification, His and SUMO tags were cleaved using a SUMO protease (1:100 *w/w*) overnight at laboratory temperature. Samples were then reapplied to the HisTrap column to separate tags from the native serpin. This step was followed by ion exchange chromatography using a HiTrap Q HP column (GE Healthcare) and by size exclusion chromatography using a HiLoad 16/60 Superdex 75 column (GE Healthcare) to ensure sufficient protein purity.

4.5. SDS-PAGE of Complex Formation

Iripin-8 and proteases were incubated at 1 µM final concentrations in a buffer corresponding to each protease (please see below) for 1 h at laboratory temperature. For the assay with fVIIa, we added 1 µM tissue factor (TF). Covalent complex formation was then analyzed in a reducing SDS-PAGE using 4–12% and 12% NuPAGE gels, followed by silver staining.

4.6. Determination of Inhibition Constants

Second-order rate constants of protease inhibition were measured by a discontinuous method under pseudo first-order conditions, using at least a 20-fold molar excess of serpin over protease. Reactions were incubated at laboratory temperature and were stopped at each time point by the addition of the chromogenic/fluorogenic substrate appropriate for the protease used. The slope of the linear part of absorbance/fluorescence increase over time gave the residual protease activity at each time point. The apparent (observed) first-order rate constant k_{obs} was calculated from the slope of a plot of the natural log of residual protease activity over time. k_{obs} was measured for 5–6 different serpin concentrations, each of them consisting of 8 different time points and plotted against serpin concentration. The slope of this linear plot gave the second-order rate constant k_2. For each determination, the standard error of the mean is given.

The assay buffer was 20 mM Tris, 150 mM NaCl, 5 mM $CaCl_2$, 0.2% BSA, 0.1% PEG 8000, pH 7.4 for thrombin, fXa, and fXIa; 20 mM Tris, 150 mM NaCl, 5 mM $CaCl_2$, 0.1% PEG 6000, 0.01% Triton X-100, pH 7.5 for activated protein C (APC), fVIIa, fIXa, fXIIa, plasmin, and chymotrypsin; 20 mM Tris, 150 mM NaCl, 0.02% Triton X-100, pH 8.5 for kallikrein and trypsin.

Substrates were: 400 µM S-2238 (Diapharma, Chester, OH, USA) for thrombin; 400 µM S-2222 for fXa (Diapharma); 400 µM S-2366 (Diapharma) for fXIa; 250 µM Boc-QAR-AMC for fVIIa; 250 µM D-CHA-GR-AMC for fXIIa; 250 µM Boc-VPR-AMC for kallikrein, trypsin, and APC; 250 µM D-VLK-AMC for plasmin; and 250 µM Boc-G(OBzl)GR-AMC for fIXa.

Final concentrations and origin of human proteases were as follows: 2 nM thrombin (Haematologic Technologies, Essex Junction, VT, USA), 20 nM fVIIa (Haematologic Technologies), 20 nM TF (BioLegend), 200 nM fIXa (Haematologic Technologies), 5 nM fXa (Haematologic Technologies), 2 nM fXIa (Haematologic Technologies), 10 nM fXIIa (Molecular Innovations, Novi, MI), 8 nM plasma kallikrein (Sigma-Aldrich, St Louis, MO, USA), 1.25 nM plasmin (Haematologic Technologies), 15 nM APC (Haematologic Technologies), 20 pM trypsin (RnD); 10 nM chymotrypsin (Merck).

4.7. Anti-Iripin-8 Serum Production and Western Blotting

Serum with antibodies against Iripin-8 was produced by immunization of a rabbit with pure recombinant protein as described previously [84]. Tick saliva was collected from ticks fed for 6 days on guinea pigs by pilocarpine induction as described previously [85]. Tick saliva was separated by reducing electrophoresis using NuPAGE™ 4–12% Bis-Tris gels. Proteins were either visualized using Coomassie staining or transferred onto PVDF membranes (Thermo Fisher Scientific). Subsequently, membranes were blocked in 5% skimmed milk in Tris-buffered saline (TBS) with 0.1% Tween 20 (TBS T) for 1 h at laboratory temperature. Membranes were then incubated with rabbit anti-Iripin-8 serum diluted in 5% skimmed milk in TBS-T (1:100) overnight at 4 °C. After washing in TBS-T, the membranes were incubated with secondary antibody (goat anti-rabbit) conjugated with horseradish peroxidase (Cell Signaling Technology; Danvers, MA, USA; 1:2000). Proteins were visualized using the enhanced chemiluminescent substrate WesternBright™ Quantum (Advansta, San Jose, CA, USA) and detected using a CCD imaging system (Uvitec, Cambridge, UK).

4.8. Coagulation Assays

All assays were performed at 37 °C using preheated reagents (Technoclone, Vienna, Austria). Normal human plasma (Coagulation Control N) was preincubated with Iripin-8 for 10 min prior to coagulation initiation. All assays were analyzed using the Ceveron four coagulometer (Technoclone).

For prothrombin time (PT) estimation, 100 µL plasma was preincubated with 6 µM Iripin-8, followed by the addition of 200 µL Technoplastin® HIS solution and estimation of fibrin clot formation time. For activated partial thromboplastin time (aPTT), 100 µL plasma was preincubated with various concentrations of Iripin-8 (94 nM–6 µM), followed by the addition of 100 µL of Dapttin® TC and incubation for 2 min. Coagulation was triggered by the addition of 100 µL 25 mM $CaCl_2$ solution. For thrombin time (TT), 200 µL of thrombin reagent was incubated with various concentrations of Iripin-8 for 10 min and subsequently added to 200 µL of plasma to initiate clot formation.

4.9. Crystal Structure Determination

Iripin-8 was concentrated to 6.5 mg/mL and dialyzed into 20 mM Tris pH 7.4, 20 mM NaCl. Crystals were obtained from the PGA screen [86] (Molecular Dimensions, Maumee, OH) in 0.1 M Tris pH 7.8, 5% PGA-LM, 30% *v/v* PEG 550 MME. Crystals were flash-frozen in liquid nitrogen straight from the well condition without additional cryoprotection. Data were collected at the Diamond Light Source (Didcot) on a beamline I04-1 and processed using the CCP4 suite [87] as follows: integration by Mosflm [88] and scaling and merging with Aimless [89]. The structure was solved by molecular replacement with Phaser [90]. The template for molecular replacement was generated from the structure of conserpin (PDB ID 5CDX [91]), which was truncated to remove flexible regions and mutated using Chainsaw [92] based on a sequence alignment to Iripin-8 using Expresso [93]. The structure was refined with Refmac [94]. Model quality was assessed by MolProbity [95,96], and figures were generated using PyMOL [97].

4.10. Complement Assay

Fresh rabbit erythrocytes were collected in Alsever's solution from the rabbit marginal ear artery, washed three times in excess PBS buffer, and finally diluted to a 2% suspension (*v/v*). Fresh human serum was obtained from three healthy individuals. The assay was performed in a 96-well round-bottomed microtiter plate (Nunc, Thermo Fisher Scientific). Each well contained 100 μL 50% human serum in PBS premixed with different concentrations of Iripin-8 (315 nM–10 μM). After 10 min incubation at laboratory temperature, 100 μL of erythrocyte suspension was added (i.e., 25% final serum concentration after the addition of erythrocyte suspension to a final 1%). Reaction wells were observed individually under a stereomicroscope using oblique illumination and an aluminum pad, and the time needed for erythrocyte lysis was measured. When full lysis was achieved, the reaction mixture turned from opaque to transparent. Negative controls did not contain either Iripin-8 or human serum. Additional controls were performed with heat-inactivated serum (56 °C, 30 min). The assay was evaluated in technical and biological triplicates.

4.11. Immunological Assays

Both the CD4$^+$ T cell proliferation assay and neutrophil migration assay were performed following the protocols described by Kotál et al. [45]. Briefly, for the CD4$^+$ T cell proliferation assay, splenocytes were isolated from OT-II mice, fluorescently labeled, pre-incubated with serpin for 2 h, and their proliferation stimulated by the addition of OVA peptide. After 72 h, cells were labelled with anti-CD4 antibody and analyzed by flow cytometry. For the migration assay, neutrophils were isolated from mouse bone marrow by immunomagnetic separation and preincubated with serpin for 1 h. Cells were then seeded in the inserts of 5 μm pore Corning® Transwell® chambers (Corning, Corning, NY, USA) and were allowed to migrate towards an fMLP (Sigma-Aldrich) gradient for 1 h. The migration rate was determined by cell counting using the Neubauer chamber.

4.12. Statistical Analysis

All experiments were performed as three biological replicates. Data are presented as mean ± standard error of mean (SEM) in all graphs. Student's *t*-test or one-way ANOVA was used to calculate statistical differences between two or more groups, respectively. For RT-PCR, data for nymphs, salivary glands, midgut, and ovaries were analyzed separately using one-way ANOVA, followed by Dunnett's post hoc test. Statistically significant results are marked: * $p \leq 0.05$; ** $p \leq 0.01$; *** $p \leq 0.001$; n.s., not significant.

Supplementary Materials: The following are available online at https://www.mdpi.com/article/10.3390/ijms22179480/s1: Supplementary Table S1: List of primers; Supplementary Table S2: Data processing, refinement, and model; phylogenetic analysis of Iripin-8 group between tick species; Supplementary Figure S1: Alignment and phylogenetic analysis of Iripin-8; Supplementary Figure S2: Analysis of Iripin-8 purity by SDS-PAGE; circular dichroism (CD) spectroscopy; Supplementary Figure S3: CD spectrogram of Iripin-8; Supplementary Figure S4: Effect of Iripin-8 on T cell proliferation and neutrophil migration; Supplementary Figure S5: A ribbon diagram of two Iripin-8 symmetry-related molecules; Supplementary Figure S6: Ribbon diagram of Iripin-8 with highlighted molecules of PEG; Supplementary methods: Evolutionary analysis by the maximum likelihood method.

Author Contributions: J.K. designed and performed experiments, performed the analyses, and wrote the manuscript; H.L., M.E., L.A.M., Z.B., A.C., and O.H. designed and performed experiments; S.G.I.P. and J.A.H. solved, refined, and analyzed the structure; J.C. designed experiments, performed analyses, and edited the manuscript; M.K. revised the manuscript. All authors have read and agreed to the published version of the manuscript.

Funding: This work was supported by the Grant Agency of the Czech Republic (grant 19-14704Y to J.C.) and by ERD Funds and and Ministry of Education, Youth, and Sport, Czech Republic (MEYS), project "Centre for Research of Pathogenicity and Virulence of Parasites" (No. 384 CZ.02.1.01/0.0/0.0/16_019/0000759 to M.K. and O.H.).

Institutional Review Board Statement: All animal experiments were carried out in accordance with the Animal Protection Law of the Czech Republic No. 246/1992 Coll., ethics approval no. MSMT-19085/2015-3, and protocols approved by the responsible committee of the Institute of Parasitology, Biology Center of the Czech Academy of Sciences.

Informed Consent Statement: Not applicable.

Data Availability Statement: All data are either contained within the manuscript and supporting information or available from the corresponding author on reasonable request.

Acknowledgments: We thank Jan Erhart and Miroslava Čenková from the tick rearing facility of the Institute of Parasitology, Biology Center, for providing the ticks and assisting during the experiments.

Conflicts of Interest: The authors declare no conflict of interest. The funders had no role in the design of the study; in the collection, analyses, or interpretation of data; in the writing of the manuscript; or in the decision to publish the results.

References

1. Lindgren, E.; Talleklint, L.; Polfeldt, T. Impact of climatic change on the northern latitude limit and population density of the disease-transmitting European tick Ixodes ricinus. *Environ. Health Perspect.* **2000**, *108*, 119–123. [CrossRef]
2. Sonenshine, D.E. *Biology of Ticks*, 2nd ed.; Oxford University Press: Oxford, UK, 2014.
3. Chmelar, J.; Calvo, E.; Pedra, J.H.; Francischetti, I.M.; Kotsyfakis, M. Tick salivary secretion as a source of antihemostatics. *J Proteom.* **2012**, *75*, 3842–3854. [CrossRef]
4. Kotal, J.; Langhansova, H.; Lieskovska, J.; Andersen, J.F.; Francischetti, I.M.; Chavakis, T.; Kopecky, J.; Pedra, J.H.; Kotsyfakis, M.; Chmelar, J. Modulation of host immunity by tick saliva. *J. Proteom.* **2015**, *128*, 58–68. [CrossRef]
5. Kazimirova, M.; Stibraniova, I. Tick salivary compounds: Their role in modulation of host defences and pathogen transmission. *Front. Cell Infect. Microbiol.* **2013**, *3*, 43. [CrossRef]
6. Simo, L.; Kazimirova, M.; Richardson, J.; Bonnet, S.I. The Essential Role of Tick Salivary Glands and Saliva in Tick Feeding and Pathogen Transmission. *Front. Cell Infect. Microbiol.* **2017**, *7*, 281. [CrossRef]
7. Stibraniova, I.; Bartikova, P.; Holikova, V.; Kazimirova, M. Deciphering Biological Processes at the Tick-Host Interface Opens New Strategies for Treatment of Human Diseases. *Front. Physiol.* **2019**, *10*, 830. [CrossRef]
8. Krober, T.; Guerin, P.M. An in vitro feeding assay to test acaricides for control of hard ticks. *Pest Manag. Sci.* **2007**, *63*, 17–22. [CrossRef]
9. Bajda, S.; Dermauw, W.; Panteleri, R.; Sugimoto, N.; Douris, V.; Tirry, L.; Osakabe, M.; Vontas, J.; Van Leeuwen, T. A mutation in the PSST homologue of complex I (NADH:ubiquinone oxidoreductase) from Tetranychus urticae is associated with resistance to METI acaricides. *Insect Biochem. Mol. Biol.* **2017**, *80*, 79–90. [CrossRef] [PubMed]
10. Amara, U.; Rittirsch, D.; Flierl, M.; Bruckner, U.; Klos, A.; Gebhard, F.; Lambris, J.D.; Huber-Lang, M. Interaction between the coagulation and complement system. *Adv. Exp. Med. Biol.* **2008**, *632*, 71–79. [PubMed]
11. Manning, J.E.; Cantaert, T. Time to Micromanage the Pathogen-Host-Vector Interface: Considerations for Vaccine Development. *Vaccines* **2019**, *7*, 10. [CrossRef] [PubMed]
12. Pingen, M.; Schmid, M.A.; Harris, E.; McKimmie, C.S. Mosquito Biting Modulates Skin Response to Virus Infection. *Trends Parasitol.* **2017**, *33*, 645–657. [CrossRef]
13. Chmelar, J.; Kotal, J.; Langhansova, H.; Kotsyfakis, M. Protease Inhibitors in Tick Saliva: The Role of Serpins and Cystatins in Tick-host-Pathogen Interaction. *Front. Cell Infect. Microbiol.* **2017**, *7*, 216. [CrossRef]
14. Wikel, S. Ticks and tick-borne pathogens at the cutaneous interface: Host defenses, tick countermeasures, and a suitable environment for pathogen establishment. *Front. Microbiol.* **2013**, *4*, 337. [CrossRef]

15. Schroeder, H.; Skelly, P.J.; Zipfel, P.F.; Losson, B.; Vanderplasschen, A. Subversion of complement by hematophagous parasites. *Dev. Comp. Immunol.* **2009**, *33*, 5–13. [CrossRef]
16. Nunn, M.A.; Sharma, A.; Paesen, G.C.; Adamson, S.; Lissina, O.; Willis, A.C.; Nuttall, P.A. Complement inhibitor of C5 activation from the soft tick Ornithodoros moubata. *J. Immunol.* **2005**, *174*, 2084–2091. [CrossRef]
17. Huntington, J.A. Serpin structure, function and dysfunction. *J. Thromb. Haemost. JTH* **2011**, *9* (Suppl. 1), 26–34. [CrossRef]
18. Law, R.H.; Zhang, Q.; McGowan, S.; Buckle, A.M.; Silverman, G.A.; Wong, W.; Rosado, C.J.; Langendorf, C.G.; Pike, R.N.; Bird, P.I.; et al. An overview of the serpin superfamily. *Genome Biol.* **2006**, *7*, 216. [CrossRef]
19. Silverman, G.A.; Whisstock, J.C.; Bottomley, S.P.; Huntington, J.A.; Kaiserman, D.; Luke, C.J.; Pak, S.C.; Reichhart, J.M.; Bird, P.I. Serpins flex their muscle: I. Putting the clamps on proteolysis in diverse biological systems. *J. Biol. Chem.* **2010**, *285*, 24299–24305. [CrossRef]
20. Whisstock, J.C.; Silverman, G.A.; Bird, P.I.; Bottomley, S.P.; Kaiserman, D.; Luke, C.J.; Pak, S.C.; Reichhart, J.M.; Huntington, J.A. Serpins flex their muscle: II. Structural insights into target peptidase recognition, polymerization, and transport functions. *J. Biol. Chem.* **2010**, *285*, 24307–24312. [CrossRef]
21. Rawlings, N.D.; Barrett, A.J.; Thomas, P.D.; Huang, X.; Bateman, A.; Finn, R.D. The MEROPS database of proteolytic enzymes, their substrates and inhibitors in 2017 and a comparison with peptidases in the PANTHER database. *Nucleic Acids Res.* **2018**, *46*, D624–D632. [CrossRef]
22. Whisstock, J.C.; Bottomley, S.P. Molecular gymnastics: Serpin structure, folding and misfolding. *Curr. Opin. Struct. Biol.* **2006**, *16*, 761–768. [CrossRef]
23. Meekins, D.A.; Kanost, M.R.; Michel, K. Serpins in arthropod biology. *Semin. Cell Dev. Biol.* **2017**, *62*, 105–119. [CrossRef]
24. Porter, L.; Radulovic, Z.; Kim, T.; Braz, G.R.; Da Silva Vaz, I., Jr.; Mulenga, A. Bioinformatic analyses of male and female Amblyomma americanum tick expressed serine protease inhibitors (serpins). *Ticks Tick-Borne Dis.* **2015**, *6*, 16–30. [CrossRef]
25. Karim, S.; Ribeiro, J.M. An Insight into the Sialome of the Lone Star Tick, Amblyomma americanum, with a Glimpse on Its Time Dependent Gene Expression. *PLoS ONE* **2015**, *10*, e0131292. [CrossRef]
26. Gaj, T.; Gersbach, C.A.; Barbas, C.F., 3rd. ZFN, TALEN, and CRISPR/Cas-based methods for genome engineering. *Trends Biotechnol.* **2013**, *31*, 397–405. [CrossRef] [PubMed]
27. Tirloni, L.; Kim, T.K.; Pinto, A.F.M.; Yates, J.R., 3rd; da Silva Vaz, I., Jr.; Mulenga, A. Tick-Host Range Adaptation: Changes in Protein Profiles in Unfed Adult Ixodes scapularis and Amblyomma americanum Saliva Stimulated to Feed on Different Hosts. *Front. Cell Infect. Microbiol.* **2017**, *7*, 517. [CrossRef]
28. Leboulle, G.; Crippa, M.; Decrem, Y.; Mejri, N.; Brossard, M.; Bollen, A.; Godfroid, E. Characterization of a novel salivary immunosuppressive protein from Ixodes ricinus ticks. *J. Biol. Chem.* **2002**, *277*, 10083–10089. [CrossRef] [PubMed]
29. Prevot, P.P.; Adam, B.; Boudjeltia, K.Z.; Brossard, M.; Lins, L.; Cauchie, P.; Brasseur, R.; Vanhaeverbeek, M.; Vanhamme, L.; Godfroid, E. Anti-hemostatic effects of a serpin from the saliva of the tick Ixodes ricinus. *J. Biol. Chem.* **2006**, *281*, 26361–26369. [CrossRef]
30. Chmelar, J.; Oliveira, C.J.; Rezacova, P.; Francischetti, I.M.; Kovarova, Z.; Pejler, G.; Kopacek, P.; Ribeiro, J.M.; Mares, M.; Kopecky, J.; et al. A tick salivary protein targets cathepsin G and chymase and inhibits host inflammation and platelet aggregation. *Blood* **2011**, *117*, 736–744. [CrossRef]
31. Palenikova, J.; Lieskovska, J.; Langhansova, H.; Kotsyfakis, M.; Chmelar, J.; Kopecky, J. Ixodes ricinus salivary serpin IRS-2 affects Th17 differentiation via inhibition of the interleukin-6/STAT-3 signaling pathway. *Infect. Immun.* **2015**, *83*, 1949–1956. [CrossRef]
32. Chlastakova, A.; Kotal, J.; Berankova, Z.; Kascakova, B.; Martins, L.A.; Langhansova, H.; Prudnikova, T.; Ederova, M.; Kuta Smatanova, I.; Kotsyfakis, M.; et al. Iripin-3, a New Salivary Protein Isolated From Ixodes ricinus Ticks, Displays Immunomodulatory and Anti-Hemostatic Properties In Vitro. *Front. Immunol.* **2021**, *12*, 626200. [CrossRef]
33. Kim, T.K.; Tirloni, L.; Radulovic, Z.; Lewis, L.; Bakshi, M.; Hill, C.; da Silva Vaz, I., Jr.; Logullo, C.; Termignoni, C.; Mulenga, A. Conserved Amblyomma americanum tick Serpin19, an inhibitor of blood clotting factors Xa and XIa, trypsin and plasmin, has anti-haemostatic functions. *Int. J. Parasitol.* **2015**, *45*, 613–627. [CrossRef]
34. Xu, Z.; Yan, Y.; Zhang, H.; Cao, J.; Zhou, Y.; Xu, Q.; Zhou, J. A serpin from the tick Rhipicephalus haemaphysaloides: Involvement in vitellogenesis. *Vet. Parasitol.* **2020**, *279*, 109064. [CrossRef]
35. Xu, T.; Lew-Tabor, A.; Rodriguez-Valle, M. Effective inhibition of thrombin by Rhipicephalus microplus serpin-15 (RmS-15) obtained in the yeast Pichia pastoris. *Ticks Tick-Borne Dis.* **2016**, *7*, 180–187. [CrossRef]
36. Kovarova, Z.; Chmelar, J.; Sanda, M.; Brynda, J.; Mares, M.; Rezacova, P. Crystallization and diffraction analysis of the serpin IRS-2 from the hard tick Ixodes ricinus. *Acta Crystallogr. Sect. F Struct. Biol. Cryst. Commun.* **2010**, *66 Pt 11*, 1453–1457. [CrossRef]
37. Ellisdon, A.M.; Zhang, Q.; Henstridge, M.A.; Johnson, T.K.; Warr, C.G.; Law, R.H.; Whisstock, J.C. High resolution structure of cleaved Serpin 42 Da from Drosophila melanogaster. *BMC Struct. Biol.* **2014**, *14*, 14. [CrossRef]
38. Schreuder, H.A.; de Boer, B.; Dijkema, R.; Mulders, J.; Theunissen, H.J.; Grootenhuis, P.D.; Hol, W.G. The intact and cleaved human antithrombin III complex as a model for serpin-proteinase interactions. *Nat. Struct. Biol.* **1994**, *1*, 48–54. [CrossRef]
39. Akazawa, T.; Ogawa, M.; Hayakawa, S.; Hirata, M.; Niwa, T. Structural change of ovalbumin-related protein X by alkali treatment. *Poult. Sci.* **2018**, *97*, 1730–1737. [CrossRef]
40. Yang, L.; Irving, J.A.; Dai, W.; Aguilar, M.I.; Bottomley, S.P. Probing the folding pathway of a consensus serpin using single tryptophan mutants. *Sci. Rep.* **2018**, *8*, 2121. [CrossRef]

41. Pereira, M.H.; Souza, M.E.; Vargas, A.P.; Martins, M.S.; Penido, C.M.; Diotaiuti, L. Anticoagulant activity of Triatoma infestans and Panstrongylus megistus saliva (Hemiptera/Triatominae). *Acta Trop.* **1996**, *61*, 255–261. [CrossRef]
42. Lestinova, T.; Rohousova, I.; Sima, M.; de Oliveira, C.I.; Volf, P. Insights into the sand fly saliva: Blood-feeding and immune interactions between sand flies, hosts, and Leishmania. *PLoS Negl. Trop. Dis.* **2017**, *11*, e0005600. [CrossRef]
43. Hoffman, M. Remodeling the blood coagulation cascade. *J. Thromb. Thrombolysis* **2003**, *16*, 17–20. [CrossRef]
44. Kascakova, B.; Kotal, J.; Martins, L.A.; Berankova, Z.; Langhansova, H.; Calvo, E.; Crossley, J.A.; Havlickova, P.; Dycka, F.; Prudnikova, T.; et al. Structural and biochemical characterization of the novel serpin Iripin-5 from Ixodes ricinus. *Acta Crystallogr. Sect. D* **2021**, *77*. [CrossRef]
45. Kotal, J.; Stergiou, N.; Busa, M.; Chlastakova, A.; Berankova, Z.; Rezacova, P.; Langhansova, H.; Schwarz, A.; Calvo, E.; Kopecky, J.; et al. The structure and function of Iristatin, a novel immunosuppressive tick salivary cystatin. *Cell. Mol. Life Sci. CMLS* **2019**, *76*, 2003–2013. [CrossRef] [PubMed]
46. Ricklin, D.; Hajishengallis, G.; Yang, K.; Lambris, J.D. Complement: A key system for immune surveillance and homeostasis. *Nat. Immunol.* **2010**, *11*, 785–797. [CrossRef] [PubMed]
47. Oliveira, F.; de Carvalho, A.M.; de Oliveira, C.I. Sand-fly saliva-leishmania-man: The trigger trio. *Front. Immunol.* **2013**, *4*, 375. [CrossRef]
48. Awuoche, E.O. Tsetse fly saliva: Could it be useful in fly infection when feeding in chronically aparasitemic mammalian hosts. *Open Vet. J.* **2012**, *2*, 95–105.
49. Caljon, G.; Van Den Abbeele, J.; Sternberg, J.M.; Coosemans, M.; De Baetselier, P.; Magez, S. Tsetse fly saliva biases the immune response to Th2 and induces anti-vector antibodies that are a useful tool for exposure assessment. *Int. J. Parasitol.* **2006**, *36*, 1025–1035. [CrossRef]
50. Kashuba, E.; Bailey, J.; Allsup, D.; Cawkwell, L. The kinin-kallikrein system: Physiological roles, pathophysiology and its relationship to cancer biomarkers. *Biomark. Biochem. Indic. Expo. Response Susceptibility Chem.* **2013**, *18*, 279–296. [CrossRef]
51. Caljon, G.; De Ridder, K.; De Baetselier, P.; Coosemans, M.; Van Den Abbeele, J. Identification of a tsetse fly salivary protein with dual inhibitory action on human platelet aggregation. *PLoS ONE* **2010**, *5*, e9671. [CrossRef]
52. Mans, B.J. Chemical Equilibrium at the Tick-Host Feeding Interface:A Critical Examination of Biological Relevance in Hematophagous Behavior. *Front. Physiol.* **2019**, *10*, 530. [CrossRef]
53. Calvo, E.; Dao, A.; Pham, V.M.; Ribeiro, J.M. An insight into the sialome of Anopheles funestus reveals an emerging pattern in anopheline salivary protein families. *Insect Biochem. Mol. Biol.* **2007**, *37*, 164–175. [CrossRef]
54. Draxler, D.F.; Sashindranath, M.; Medcalf, R.L. Plasmin: A Modulator of Immune Function. *Semin. Thromb. Hemost.* **2017**, *43*, 143–153. [CrossRef]
55. Cong, L.; Ran, F.A.; Cox, D.; Lin, S.; Barretto, R.; Habib, N.; Hsu, P.D.; Wu, X.; Jiang, W.; Marraffini, L.A.; et al. Multiplex genome engineering using CRISPR/Cas systems. *Science* **2013**, *339*, 819–823. [CrossRef]
56. Lawrie, C.H.; Sim, R.B.; Nuttall, P.A. Investigation of the mechanisms of anti-complement activity in Ixodes ricinus ticks. *Mol. Immunol.* **2005**, *42*, 31–38. [CrossRef]
57. Miller, J.C.; Tan, S.; Qiao, G.; Barlow, K.A.; Wang, J.; Xia, D.F.; Meng, X.; Paschon, D.E.; Leung, E.; Hinkley, S.J.; et al. A TALE nuclease architecture for efficient genome editing. *Nat. Biotechnol.* **2011**, *29*, 143–148. [CrossRef]
58. Valenzuela, J.G.; Charlab, R.; Mather, T.N.; Ribeiro, J.M. Purification, cloning, and expression of a novel salivary anticomplement protein from the tick, Ixodes scapularis. *J. Biol. Chem.* **2000**, *275*, 18717–18723. [CrossRef]
59. Tyson, K.; Elkins, C.; Patterson, H.; Fikrig, E.; de Silva, A. Biochemical and functional characterization of Salp20, an Ixodes scapularis tick salivary protein that inhibits the complement pathway. *Insect Mol. Biol.* **2007**, *16*, 469–479. [CrossRef]
60. Tyson, K.R.; Elkins, C.; de Silva, A.M. A novel mechanism of complement inhibition unmasked by a tick salivary protein that binds to properdin. *J. Immunol.* **2008**, *180*, 3964–3968. [CrossRef] [PubMed]
61. Daix, V.; Schroeder, H.; Praet, N.; Georgin, J.P.; Chiappino, I.; Gillet, L.; de Fays, K.; Decrem, Y.; Leboulle, G.; Godfroid, E.; et al. Ixodes ticks belonging to the Ixodes ricinus complex encode a family of anticomplement proteins. *Insect Mol. Biol.* **2007**, *16*, 155–166. [CrossRef]
62. Schuijt, T.J.; Coumou, J.; Narasimhan, S.; Dai, J.; Deponte, K.; Wouters, D.; Brouwer, M.; Oei, A.; Roelofs, J.J.; van Dam, A.P.; et al. A tick mannose-binding lectin inhibitor interferes with the vertebrate complement cascade to enhance transmission of the lyme disease agent. *Cell Host Microbe* **2011**, *10*, 136–146. [CrossRef]
63. Chmelar, J.; Kotal, J.; Kopecky, J.; Pedra, J.H.; Kotsyfakis, M. All For One and One For All on the Tick-Host Battlefield. *Trends Parasitol.* **2016**, *32*, 368–377. [CrossRef]
64. Sojka, D.; Franta, Z.; Horn, M.; Caffrey, C.R.; Mares, M.; Kopacek, P. New insights into the machinery of blood digestion by ticks. *Trends Parasitol.* **2013**, *29*, 276–285. [CrossRef]
65. Franta, Z.; Frantova, H.; Konvickova, J.; Horn, M.; Sojka, D.; Mares, M.; Kopacek, P. Dynamics of digestive proteolytic system during blood feeding of the hard tick Ixodes ricinus. *Parasites Vectors* **2010**, *3*, 119. [CrossRef] [PubMed]
66. Lara, F.A.; Lins, U.; Bechara, G.H.; Oliveira, P.L. Tracing heme in a living cell: Hemoglobin degradation and heme traffic in digest cells of the cattle tick Boophilus microplus. *J. Exp. Biol.* **2005**, *208 Pt 16*, 3093–3101. [CrossRef]
67. Blisnick, A.A.; Foulon, T.; Bonnet, S.I. Serine Protease Inhibitors in Ticks: An Overview of Their Role in Tick Biology and Tick-Borne Pathogen Transmission. *Front. Cell Infect. Microbiol.* **2017**, *7*, 199. [CrossRef]

68. Mulenga, A.; Tsuda, A.; Onuma, M.; Sugimoto, C. Four serine proteinase inhibitors (serpin) from the brown ear tick, Rhiphicephalus appendiculatus; cDNA cloning and preliminary characterization. *Insect Biochem. Mol. Biol.* **2003**, *33*, 267–276. [CrossRef]
69. Imamura, S.; da Silva Vaz Junior, I.; Sugino, M.; Ohashi, K.; Onuma, M. A serine protease inhibitor (serpin) from Haemaphysalis longicornis as an anti-tick vaccine. *Vaccine* **2005**, *23*, 1301–1311. [CrossRef]
70. Rodriguez-Valle, M.; Vance, M.; Moolhuijzen, P.M.; Tao, X.; Lew-Tabor, A.E. Differential recognition by tick-resistant cattle of the recombinantly expressed Rhipicephalus microplus serine protease inhibitor-3 (RMS-3). *Ticks Tick-Borne Dis.* **2012**, *3*, 159–169. [CrossRef] [PubMed]
71. Rodriguez-Valle, M.; Xu, T.; Kurscheid, S.; Lew-Tabor, A.E. Rhipicephalus microplus serine protease inhibitor family: Annotation, expression and functional characterisation assessment. *Parasites Vectors* **2015**, *8*, 7. [CrossRef]
72. Schussler, G.C. The thyroxine-binding proteins. *Thyroid* **2000**, *10*, 141–149. [CrossRef]
73. Lin, H.Y.; Muller, Y.A.; Hammond, G.L. Molecular and structural basis of steroid hormone binding and release from corticosteroid-binding globulin. *Mol. Cell Endocrinol.* **2010**, *316*, 3–12. [CrossRef]
74. Kim, T.K.; Radulovic, Z.; Mulenga, A. Target validation of highly conserved Amblyomma americanum tick saliva serine protease inhibitor 19. *Ticks Tick-Borne Dis.* **2016**, *7*, 405–414. [CrossRef] [PubMed]
75. Mans, B.J. Tick histamine-binding proteins and related lipocalins: Potential as therapeutic agents. *Curr. Opin. Investig. Drugs* **2005**, *6*, 1131–1135.
76. Mans, B.J.; Louw, A.I.; Neitz, A.W. Evolution of hematophagy in ticks: Common origins for blood coagulation and platelet aggregation inhibitors from soft ticks of the genus Ornithodoros. *Mol. Biol. Evol.* **2002**, *19*, 1695–1705. [CrossRef] [PubMed]
77. Maraganore, J.M.; Chao, B.; Joseph, M.L.; Jablonski, J.; Ramachandran, K.L. Anticoagulant activity of synthetic hirudin peptides. *J. Biol. Chem.* **1989**, *264*, 8692–8698. [CrossRef]
78. Stepanova-Tresova, G.; Kopecky, J.; Kuthejlova, M. Identification of Borrelia burgdorferi sensu stricto, Borrelia garinii and Borrelia afzelii in Ixodes ricinus ticks from southern Bohemia using monoclonal antibodies. *Zent. Bakteriol. Int. J. Med. Microbiol.* **2000**, *289*, 797–806.
79. Pospisilova, T.; Urbanova, V.; Hes, O.; Kopacek, P.; Hajdusek, O.; Sima, R. Tracking of Borrelia afzelii transmission from infected Ixodes ricinus nymphs to mice. *Infect. Immun.* **2019**, *87*, e00896-18. [CrossRef]
80. Fong, S.W.; Kini, R.M.; Ng, L.F.P. Mosquito Saliva Reshapes Alphavirus Infection and Immunopathogenesis. *J. Virol.* **2018**, *92*, e01004-17. [CrossRef]
81. Stibraniova, I.; Lahova, M.; Bartikova, P. Immunomodulators in tick saliva and their benefits. *Acta Virol.* **2013**, *57*, 200–216. [CrossRef]
82. Guerrero, D.; Cantaert, T.; Misse, D. Aedes Mosquito Salivary Components and Their Effect on the Immune Response to Arboviruses. *Front. Cell Infect. Microbiol.* **2020**, *10*, 407. [CrossRef] [PubMed]
83. Dai, J.; Wang, P.; Adusumilli, S.; Booth, C.J.; Narasimhan, S.; Anguita, J.; Fikrig, E. Antibodies against a tick protein, Salp15, protect mice from the Lyme disease agent. *Cell Host Microbe* **2009**, *6*, 482–492. [CrossRef] [PubMed]
84. Kopacek, P.; Zdychova, J.; Yoshiga, T.; Weise, C.; Rudenko, N.; Law, J.H. Molecular cloning, expression and isolation of ferritins from two tick species–Ornithodoros moubata and Ixodes ricinus. *Insect Biochem. Mol. Biol.* **2003**, *33*, 103–113. [CrossRef]
85. Kyckova, K.; Kopecky, J. Effect of tick saliva on mechanisms of innate immune response against Borrelia afzelii. *J. Med. Entomol.* **2006**, *43*, 1208–1214. [CrossRef] [PubMed]
86. Ribeiro, J.M.; Arca, B.; Lombardo, F.; Calvo, E.; Phan, V.M.; Chandra, P.K.; Wikel, S.K. An annotated catalogue of salivary gland transcripts in the adult female mosquito, Aedes aegypti. *BMC Genom.* **2007**, *8*, 6. [CrossRef]
87. Friedrich, T.; Kroger, B.; Bialojan, S.; Lemaire, H.G.; Hoffken, H.W.; Reuschenbach, P.; Otte, M.; Dodt, J. A Kazal-type inhibitor with thrombin specificity from Rhodnius prolixus. *J. Biol. Chem.* **1993**, *268*, 16216–16222. [CrossRef]
88. Lange, U.; Keilholz, W.; Schaub, G.A.; Landmann, H.; Markwardt, F.; Nowak, G. Biochemical characterization of a thrombin inhibitor from the bloodsucking bug Dipetalogaster maximus. *Haemostasis* **1999**, *29*, 204–211. [CrossRef]
89. Bussacos, A.C.; Nakayasu, E.S.; Hecht, M.M.; Parente, J.A.; Soares, C.M.; Teixeira, A.R.; Almeida, I.C. Diversity of anti-haemostatic proteins in the salivary glands of Rhodnius species transmitters of Chagas disease in the greater Amazon. *J. Proteom.* **2011**, *74*, 1664–1672. [CrossRef]
90. Cappello, M.; Bergum, P.W.; Vlasuk, G.P.; Furmidge, B.A.; Pritchard, D.I.; Aksoy, S. Isolation and characterization of the tsetse thrombin inhibitor: A potent antithrombotic peptide from the saliva of Glossina morsitans morsitans. *Am. J. Trop. Med. Hyg.* **1996**, *54*, 475–480. [CrossRef]
91. Cappello, M.; Li, S.; Chen, X.; Li, C.B.; Harrison, L.; Narashimhan, S.; Beard, C.B.; Aksoy, S. Tsetse thrombin inhibitor: Bloodmeal-induced expression of an anticoagulant in salivary glands and gut tissue of Glossina morsitans morsitans. *Proc. Natl. Acad. Sci. USA* **1998**, *95*, 14290–14295. [CrossRef]
92. Chagas, A.C.; Oliveira, F.; Debrabant, A.; Valenzuela, J.G.; Ribeiro, J.M.; Calvo, E. Lundep, a sand fly salivary endonuclease increases Leishmania parasite survival in neutrophils and inhibits XIIa contact activation in human plasma. *PLoS Pathog.* **2014**, *10*, e1003923. [CrossRef]
93. Kato, H.; Gomez, E.A.; Fujita, M.; Ishimaru, Y.; Uezato, H.; Mimori, T.; Iwata, H.; Hashiguchi, Y. Ayadualin, a novel RGD peptide with dual antihemostatic activities from the sand fly Lutzomyia ayacuchensis, a vector of Andean-type cutaneous leishmaniasis. *Biochimie* **2015**, *112*, 49–56. [CrossRef] [PubMed]

94. Maraganore, J.M.; Bourdon, P.; Jablonski, J.; Ramachandran, K.L.; Fenton, J.W. 2nd, Design and characterization of hirulogs: A novel class of bivalent peptide inhibitors of thrombin. *Biochemistry* **1990**, *29*, 7095–7101. [CrossRef]
95. USAN Council. List No. 374. New names. Bivalirudin. Desirudin. *Clin. Pharmacol. Ther.* **1995**, *58*, 241. [CrossRef]
96. Poulin, R. *Evolutionary Ecology of Parasites*, 2nd ed.; Princeton University Press: Princeton, NJ, USA, 2007.
97. Schrodinger, LLC. *The PyMOL Molecular Graphics System*; Version 1.8; Schrodinger, LLC: New York, NY, USA, 2015.

International Journal of
Molecular Sciences

Addendum

Addendum: Kotál et al. *Ixodes ricinus* Salivary Serpin Iripin-8 Inhibits the Intrinsic Pathway of Coagulation and Complement. *Int. J. Mol. Sci.* 2021, 22, 9480

Jan Kotál [1,2], Stéphanie G. I. Polderdijk [3], Helena Langhansová [1], Monika Ederová [1], Larissa A. Martins [2], Zuzana Beránková [1], Adéla Chlastáková [1], Ondřej Hajdušek [4], Michail Kotsyfakis [1,2], James A. Huntington [3] and Jindřich Chmelař [1,*]

[1] Department of Medical Biology, Faculty of Science, University of South Bohemia in České Budějovice, Branišovská 1760c, 37005 České Budějovice, Czech Republic; jan.kotal@nih.gov (J.K.); hlanghansova@prf.jcu.cz (H.L.); ederom01@prf.jcu.cz (M.E.); beranz02@prf.jcu.cz (Z.B.); chlasa00@prf.jcu.cz (A.C.); kotsyfakis@paru.cas.cz (M.K.)
[2] Laboratory of Genomics and Proteomics of Disease Vectors, Institute of Parasitology, Biology Center CAS, Branišovská 1160/31, 37005 České Budějovice, Czech Republic; larissa.martins@nih.gov
[3] Cambridge Institute for Medical Research, Department of Haematology, University of Cambridge, The Keith Peters Building, Hills Road, Cambridge CB2 0XY, UK; stephanie.polderdijk@cantab.net (S.G.I.P.); jah52@cam.ac.uk (J.A.H.)
[4] Laboratory of Vector Immunology, Institute of Parasitology, Biology Center CAS, Branišovská 1160/31, 37005 České Budějovice, Czech Republic; hajdus@paru.cas.cz
* Correspondence: chmelar@prf.jcu.cz

Citation: Kotál, J.; Polderdijk, S.G.I.; Langhansová, H.; Ederová, M.; Martins, L.A.; Beránková, Z.; Chlastáková, A.; Hajdušek, O.; Kotsyfakis, M.; Huntington, J.A.; et al. Addendum: Kotál et al. *Ixodes ricinus* Salivary Serpin Iripin-8 Inhibits the Intrinsic Pathway of Coagulation and Complement. *Int. J. Mol. Sci.* 2021, 22, 9480. *Int. J. Mol. Sci.* **2021**, 22, 11271. https://doi.org/10.3390/ijms222011271

Received: 18 September 2021
Accepted: 22 September 2021
Published: 19 October 2021

Publisher's Note: MDPI stays neutral with regard to jurisdictional claims in published maps and institutional affiliations.

The accession code will be added in Section 2.8. Structural Features of Iripin-8: The coordinates and structure factors of Iripin-8 are deposited in the Protein Data Bank under accession code 7PMU.

This addendum does not cause any changes to the results or conclusions in the original published paper [1].

Reference

1. Kotál, J.; Polderdijk, S.G.I.; Langhansová, H.; Ederová, M.; Martins, L.A.; Beránková, Z.; Chlastáková, A.; Hajdušek, O.; Kotsyfakis, M.; Huntington, J.A.; et al. *Ixodes ricinus* Salivary Serpin Iripin-8 Inhibits the Intrinsic Pathway of Coagulation and Complement. *Int. J. Mol. Sci.* **2021**, 22, 9480. [CrossRef] [PubMed]

International Journal of
Molecular Sciences

Article

Mialostatin, a Novel Midgut Cystatin from *Ixodes ricinus* Ticks: Crystal Structure and Regulation of Host Blood Digestion

Jan Kotál [1,2,†], Michal Buša [3,4,†], Veronika Urbanová [1], Pavlína Řezáčová [3], Jindřich Chmelař [2], Helena Langhansová [2], Daniel Sojka [1], Michael Mareš [3,*] and Michail Kotsyfakis [1,*]

1. Institute of Parasitology, Biology Centre, Academy of Sciences of the Czech Republic, Branišovská 1160/31, 37005 České Budějovice, Czech Republic; jankotal@gmail.com (J.K.); veronika@paru.cas.cz (V.U.); sojka@paru.cas.cz (D.S.)
2. Department of Medical Biology, Faculty of Science, University of South Bohemia in České Budějovice, Branišovská 1760c, 37005 České Budějovice, Czech Republic; chmelar@prf.jcu.cz (J.C.); hlanghansova@prf.jcu.cz (H.L.)
3. Institute of Organic Chemistry and Biochemistry, Academy of Sciences of the Czech Republic, Flemingovo n. 2, 16610 Praha, Czech Republic; michal.busa@uochb.cas.cz (M.B.); rezacova@img.cas.cz (P.Ř.)
4. Department of Biochemistry, Faculty of Science, Charles University, Hlavova 2030/8, 12800 Prague, Czech Republic
* Correspondence: mares@uochb.cas.cz (M.M.); mich_kotsyfakis@yahoo.com (M.K.)
† These authors contributed equally to this work.

Citation: Kotál, J.; Buša, M.; Urbanová, V.; Řezáčová, P.; Chmelař, J.; Langhansová, H.; Sojka, D.; Mareš, M.; Kotsyfakis, M. Mialostatin, a Novel Midgut Cystatin from *Ixodes ricinus* Ticks: Crystal Structure and Regulation of Host Blood Digestion. *Int. J. Mol. Sci.* **2021**, 22, 5371. https://doi.org/10.3390/ijms22105371

Academic Editor: Andrea Battistoni

Received: 18 April 2021
Accepted: 17 May 2021
Published: 20 May 2021

Publisher's Note: MDPI stays neutral with regard to jurisdictional claims in published maps and institutional affiliations.

Abstract: The hard tick *Ixodes ricinus* is a vector of Lyme disease and tick-borne encephalitis. Host blood protein digestion, essential for tick development and reproduction, occurs in tick midgut digestive cells driven by cathepsin proteases. Little is known about the regulation of the digestive proteolytic machinery of *I. ricinus*. Here we characterize a novel cystatin-type protease inhibitor, mialostatin, from the *I. ricinus* midgut. Blood feeding rapidly induced mialostatin expression in the gut, which continued after tick detachment. Recombinant mialostatin inhibited a number of *I. ricinus* digestive cysteine cathepsins, with the greatest potency observed against cathepsin L isoforms, with which it co-localized in midgut digestive cells. The crystal structure of mialostatin was determined at 1.55 Å to explain its unique inhibitory specificity. Finally, mialostatin effectively blocked in vitro proteolysis of blood proteins by midgut cysteine cathepsins. Mialostatin is likely to be involved in the regulation of gut-associated proteolytic pathways, making midgut cystatins promising targets for tick control strategies.

Keywords: cathepsin; crystal structure; cysteine protease; digestion; *Ixodes ricinus*; midgut; parasite

1. Introduction

Ticks are globally distributed ectoparasitic arthropods that strictly feed on host blood. While soft ticks (family Argasidae) feed only for a few hours, hard ticks (family Ixodidae) usually attach to their hosts for several days to fully engorge and proceed to their next developmental stage. The hard tick *Ixodes ricinus* is found mainly in Europe but also in neighboring parts of Africa and the Middle East, where it is a major vector of pathogens such as Lyme disease spirochetes (*Borrelia burgdorferi sensu lato*), tick-borne encephalitis virus [1] or *Babesia* spp. [2]. Adult *I. ricinus* females feed for 6–9 days on a vertebrate host to enlarge over 100 times in weight [3].

Since blood is a highly specific and sole source of nutrients for these ticks, they have adapted to efficiently process large amounts of host blood. Blood degrades in the acidic endolysosomes of digestive cells of the tick midgut. Gut lumen uptake of the two main blood constituents, albumin and hemoglobin, is facilitated by two different mechanisms [4]: albumin is taken up non-specifically by fluid-phase endocytosis, while hemoglobin is recognized by specific receptor-mediated endocytosis. Subsequently, albumin is directed to small acidic vesicles and hemoglobin to a population of large digestive vesicles [4]. Despite these differences, both albumin and hemoglobin are cleaved and processed to single amino acids and short peptides by the same proteolytic system [5,6]. The degradation pathway for hemoglobin is described in detail elsewhere [4]. Briefly, the initial phase is catalyzed by three *I. ricinus* digestive endopeptidases at low pH (3.5 to 4.5) including a cysteine protease legumain (IrAE) [7] and aspartic protease cathepsin D (IrCD1) [8]. Two cysteine protease cathepsin L isoforms, IrCL1 [9] (GenBank: EF428205) and IrCL3 (GenBank: QBK51063), complement the initial phase: IrCL1 expression in tick gut cells peaks at the end of tick feeding [9], while its ortholog, IrCL3, is present in the tick midgut predominantly after feeding, where it complements the activity of IrCL1 (D. Sojka, personal communication, December 2020). Cysteine proteases with exopeptidase activity, cathepsins B and C (IrCB and IrCC), continue hemoglobin degradation to dipeptides at an optimal pH of 5.5–6.0 in digestive cells [6,10,11]. Digestion to single amino acids is facilitated by carboxypeptidase and leucine aminopeptidase [6]. Blood processing by ticks and the roles of individual proteases are reviewed in detail elsewhere [12,13].

Under physiological conditions, cysteine protease activity is regulated by proteinaceous inhibitors, including those in the cystatin family [14,15]. Cystatins are tight binding, reversible inhibitors of legumain and papain-like cysteine proteases [16]. According to MEROPS nomenclature, cystatins are subdivided into three subfamilies: I25A (type 1, stefins), I25B (type 2 and type 3, kininogens), and I25C (type 4, fetuins) [17]. Only type 1 and 2 cystatins have so far been identified in ticks [18]. Cystatins are mostly associated with the regulation of proteases involved in blood digestion and heme detoxification in the tick midgut [18] and with the modulation of the host immune system as components of tick saliva [19,20], although they have also been detected in other tick tissues [21,22].

In soft ticks, only two midgut cystatins have been functionally characterized: Om-cystatins 1 and 2 from *Ornithodoros moubata* [23]. While Om-cystatin 1 is exclusively expressed in the midgut, Om-cystatin 2 can be found in all tissues and has immunomodulatory properties when secreted into the host [24,25]. Both inhibit cathepsins B, C, and H and are involved in blood processing [23]. Gut-associated cystatins from only two *Ixodes* species have been reported to date: a gut-secreted cystatin JpIocys2a from *Ixodes ovatus* was shown to inhibit cathepsins B, C, and L [26], while the expression of three *Ixodes persulcatus* cystatins, JpIpcys2a, b, and c, was demonstrated in almost all tissues and instars [27].

Despite the relatively good characterization of the digestive proteases present in the *I. ricinus* midgut [13], there has been little functional characterization of their inhibitors and regulatory mechanisms. Here we report a novel cystatin from the *I. ricinus* midgut, mialostatin, and present its crystal structure, inhibitory specificity, tissue localization, and role in the regulation of blood digestion.

2. Results

2.1. Mialostatin Transcript Predominantly Accumulates in the Tick Midgut

In order to clone mialostatin, we used primers based on available cystatin sequences identified in *Ixodes scapularis* tick genome. To obtain the longest possible reads, we also focused on the 5′ UTRs and 3′ UTRs regions. In the course of our study, an *I. ricinus* transcriptome was published with a transcript of an identical sequence to mialostatin (Genbank accession number GFVZ01041806.1) [28]. However, since this particular transcript was obtained from whole body tick sequencing, we used BLAST to search for highly similar sequences in other transcriptomic studies to specifically localize its expression (https://blast.ncbi.nlm.nih.gov/Blast.cgi, accessed on 20 May 2021). As a result we found a highly similar transcript SigP-158801 upregulated mainly in the tick midgut [29]; similarly to another transcript GCJO01026918.1 identified in a study focusing on the tick gut [30]. To verify the localization of mialostatin, we examined its expression in different tick tissues and feeding stages and confirmed its predominantly midgut expression.

Increased transcription of mialostatin over the feeding course implies an important role in tick metabolism [18]. Figure 1 shows the expression of mialostatin in the tick midgut, ovaries, and salivary glands before, during, and after feeding. Mialostatin transcript was predominantly present in the tick midgut, where expression oscillated throughout feeding and after detachment but at consistently higher expression than the other examined tissues. The presence and upregulation of mialostatin transcript in tick salivary glands and ovaries were low, peaking in fully fed ticks and at the early phase of detachment at maximum levels of only 10–20% of midgut expression (Figure 1).

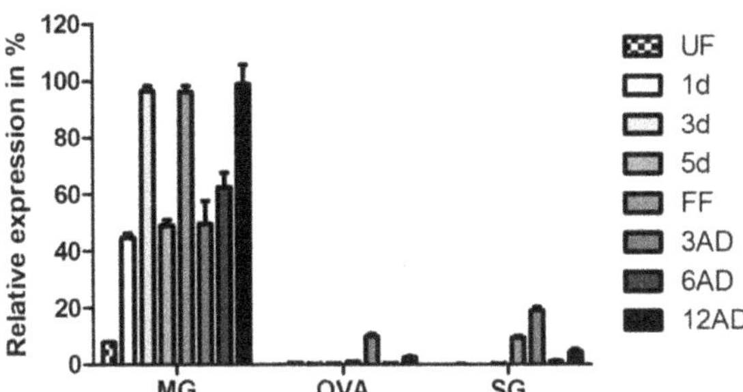

Figure 1. Mialostatin is predominantly produced in the tick midgut and its expression is upregulated by tick feeding. Expression maxima are prior to rapid engorgement, in fully fed females, and at two weeks post tick detachment from the host. Mialostatin expression was determined by quantitative PCR using cDNA templates prepared from a pool of three tissues from female ticks (MG—midgut, OVA—ovaries, SG—salivary glands). The qPCR output was normalized to the *I. ricinus* elongation factor 1 gene and compared across all values with the highest expression set to 100%. Data show an average of three biological replicates ± SEM. Categories: UF—unfed ticks; 1d, 3d, 5d—ticks after 1, 3, or 5 days of feeding; FF—fully fed ticks after 7–8 days of feeding; 3AD, 6AD, 12AD—ticks 3, 6, or 12 days after detachment.

2.2. Mialostatin Is a Broad-Spectrum Inhibitor of Cysteine Cathepsins and Is Highly Effective against Cathepsin L

Purified recombinant mialostatin was screened in vitro for its inhibitory potential against major endogenous digestive proteases present in the *I. ricinus* gut [6]. These proteases were tested in the form of recombinant enzymes or proteolytic activities in the gut tissue extract (Table 1, left and middle panels). The strongest inhibition was found for recombinant *I. ricinus* cathepsins L1 and L3 (IC_{50}s of 0.071 and 0.39 nM, respectively), which are papain-type cysteine proteases and consistent with sub-nanomolar inhibition of cathepsin L-like activity by the extract (IC_{50} of 0.18 nM). The cathepsin B-like and cathepsin C-like activities of the extract were inhibited with lower potency, with IC_{50} values in the double-digit nanomolar range (12.1 and 91.7 nM, respectively). Mialostatin did not inhibit *I. ricinus* digestive proteases out of the papain family, including the aspartic protease cathepsin D1 and the clan CD cysteine protease legumain (asparaginyl endopeptidase).

We next expanded a spectrum of papain-type cysteine proteases and screened mialostatin against a representative panel of human cathepsins selected to cover a wide range of endo- and exopeptidase activities (Table 1, right panel) including endopeptidases cathepsins L, K, and S and exopeptidases cathepsin B (a peptidyl dipeptidase and endopeptidase), cathepsin C (a dipeptidyl peptidase), and cathepsin H (an aminopeptidase). Human cathepsin L was inhibited at subnanomolar concentrations (IC_{50} 0.38 nM), similar to its *I. ricinus* homologs, and all other human cysteine cathepsins were inhibited with IC_{50} values in a narrow range from 2.2 to 24 nM.

In conclusion, mialostatin displays an unusually broad inhibitory specificity against cysteine cathepsins, with a particularly strong interaction with cathepsin L isoforms. The high affinity for cysteine cathepsins with endopeptidase and exopeptidase activities clearly distinguishes mialostatin from other described *Ixodes* cystatins, which display weak or no inhibition of these exopeptidases (see below).

Table 1. Inhibitory effect of mialostatin on the activity of tick and human proteases. The inhibitory potency of mialostatin was determined against: (i) native cysteine cathepsins present in *I. ricinus* gut tissue extract using protease-specific assays (left panel); (ii) selected digestive proteases of *I. ricinus* prepared as recombinant proteins (middle panel); and (iii) a representative set of human cysteine cathepsins (right panel). The IC_{50} values (mean values $\pm$ SE) were measured by kinetic activity assays using specific fluorogenic peptide substrates (for details, see Methods).

Inhibition of *I. ricinus* Midgut Homogenate		Inhibition of Recombinant Digestive *I. ricinus* Proteases		Inhibition of Human Cysteine Cathepsins	
Targeted Activity	IC_{50} (nM)	Proteases	IC_{50} (nM)	Protease	IC_{50} (nM)
Cathepsin L	0.18 ± 0.02	*Ir*-cathepsin L1 (IrCL1)	0.071 ± 0.01	*Hs*-cathepsin L	0.38 ± 0.03
Cathepsins L and B	3.1 ± 0.4	*Ir*-cathepsin L3 (IrCL3)	0.39 ± 0.18	*Hs*-cathepsin C	2.1 ± 0.8
Cathepsin B	12.1 ± 1.5	*Ir*-legumain (IrAE)	n.i.	*Hs*-cathepsin S	2.2 ± 0.4
Cathepsin C	91.7 ± 5.5	*Ir*-cathepsin D1 (IrCD1)	n.i.	*Hs*-cathepsin B	9.0 ± 0.3
				Hs-cathepsin K	9.7 ± 1.3
				Hs-cathepsin H	24.0 ± 3.5

Abbreviation: n.i.—no significant inhibition at 10 μM mialostatin concentration.

2.3. Mialostatin Is Present in the Tick Gut Wall and Lumen

We further investigated mialostatin's distribution within the tick midgut. Gut epithelia and lumina were collected from fully fed *I. ricinus* adult females and subjected to proteomic analysis to directly determine the presence or absence of mialostatin. The LC-MS/MS strategy was based on the enzymatic digestion of a complex protein mixture and MS/MS peptide sequencing. This analysis provided 11–71% peptide coverage of the mialostatin sequence and a mass accuracy of <5 p.p.m. (Table S1), allowing us to conclude that mialostatin is present in both the gut tissue and luminal contents of *I. ricinus* ticks.

Immunolabeling was used to evaluate the potential biological selectivity of mialostatin towards different papain-like enzymes present in tick gut tissue. Localization of mialostatin with IrCB, IrCL1, and IrCL3 was examined using multicolor immunohistochemistry (Figure 2), with the sample collection and section preparation time points selected based on qPCR-determined dynamic expression profiles of individual proteases (sixth day of feeding for IrCLB and IrCL1; eleventh day post feeding for IrCL3) to establish the availability of these proteases for co-localization with mialostatin at these timepoints. IrCL3 was the most probable target protease for mialostatin, as co-localization signals at the surface of large vesicles in tick gut cells (specific ring patterns) was nearly complete. However, there was also some co-localization of mialostatin with IrCL1 but not the cysteine protease cathepsin B (IrCB).

Immunoblot analyses of tick gut tissue were performed to (i) confirm mialostatin selectivity for IrCL3 and further evaluate potential interactions with IrCL1; and (ii) evaluate potential secretion of mialostatin and the cathepsin-L-like tick proteases into the gut lumen. The latter could not be observed by immunohistochemical labeling due to the rapid dilution of gut epithelial cell secretions with the large amount of imbibed host blood in the lumen. However, mialostatin was detected in the gut wall (Figure 3A) at all collected time-points during feeding. Gut tissue originating from ticks membrane-fed on pure bovine blood serum (without erythrocytes) [31] was used to avoid interference between mialostatin-specific signals and host hemoglobin proteins of identical molecular weight. IrCL1 and IrCL3 signals were also detected in both the gut epithelium (cell wall) and the gut lumen (Figure 3B). The multiple IrCL1 bands corresponded to the proenzyme and mature enzyme forms [9].

2.4. Mialostatin Inhibits Blood-Protein Digestion Catalyzed by Tick Gut Cysteine Cathepsins

In tick gut tissue, cysteine cathepsins play a critical role in the acidic degradation of the two most abundant host blood proteins, hemoglobin and serum albumin. In particular, cathepsin L is involved in the initial phase of the degradation pathway, which is continued by the action of cathepsins B and C [5,6]. We evaluated the effect of mialostatin on the in vitro degradation of hemoglobin and serum albumin by the proteolytically active extract of *I. ricinus* gut tissue (limited to cysteine proteases by treatment with inhibitors of other protease classes). Both blood proteins were digested at optimal acidic pH, and SDS-PAGE analysis demonstrated highly efficient degradation of these substrates (Figure 4A,B). These processes were effectively blocked by mialostatin in a dose-dependent manner, with complete inhibition at a low nanomolar concentration of mialostatin. A similar effect was achieved by adding E-64, a small molecule general inhibitor of cysteine cathepsins. Further, we tested the stability of mialostatin exposed to the complex proteolytic environment of the gut tissue extract (Figure 4C), which revealed that mialostatin was generally stable and showed only partial degradation over long-term treatment. In summary, the blood protein digestion catalyzed by cysteine cathepsins of *I. ricinus* can be effectively controlled by mialostatin under native-like conditions.

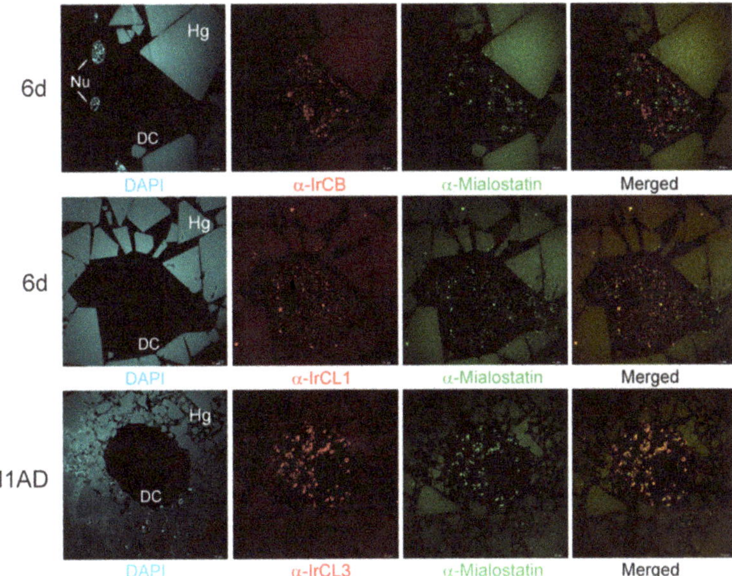

Figure 2. Mialostatin co-localizes with IrCL3 inside gut digestive cells of female *I. ricinus* ticks. Multicolored confocal immunofluorescence indicates variable colocalization of mialostatin (green signal) with the cathepsin-type proteases IrCB, IrCL1, and IrCL3 (red signal) in female tick gut sections at the sixth day of feeding (6d) and eleventh day post detachment from the host (11AD). Mialostatin and IrCL3 show the greatest co-localization (yellow signal in merged images), thus IrCL3 represents the most probable target protease. DAPI counterstaining is shown in cyan. DC—digestive cells; Nu—nucleus; Hg—hemoglobin crystals in gut lumen, scale bar-10 μm.

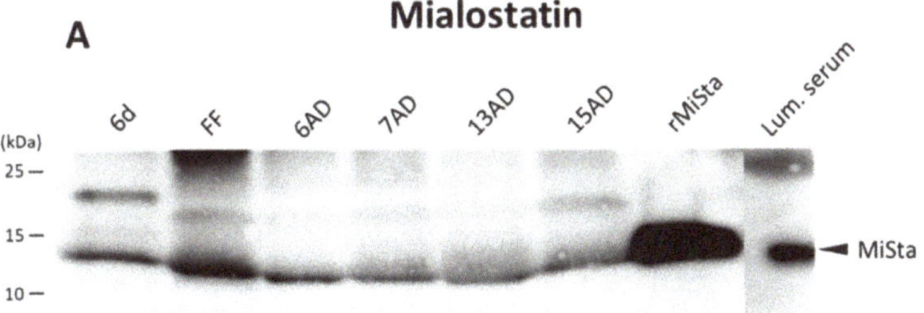

Figure 3. *Cont.*

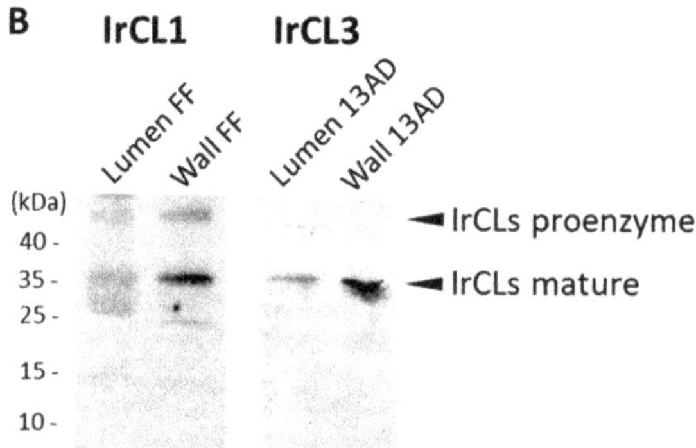

Figure 3. Western blot analysis of mialostatin, IrCL1, and IrCL3 in the tick midgut. (**A**) Tick midgut wall tissue extracts from various stages of tick feeding and midgut lumina from fully fed ticks were analyzed by SDS-PAGE and Western blotting. Mialostatin was labeled with a mouse monoclonal antibody and its signal detected using a fluorescently labeled secondary antibody. (**B**) Tick midgut wall and midgut lumen homogenates from fully fed ticks were analyzed by SDS-PAGE and immunoblotting with α-IrCL1 and α-IrCL3 rabbit polyclonal antibodies. Goat α-rabbit IgG Alexa 488 fluorescent secondary antibody was used to visualize protein bands using ChemiDoc MP imager. 6d—sixth day of feeding; FF—fully fed; 6, 7, 13, 15 AD—days post detachment from the host. Lum. serum—luminal fluid from ticks fed on erythrocyte-free serum; rMista—recombinant mialostatin. Full view of presented Western blots can be found in the Supplementary Materials, Figures S2–S4.

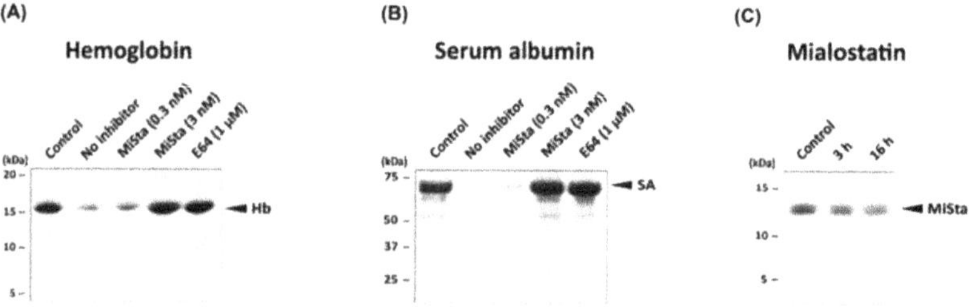

Figure 4. Blood protein digestion with tick gut proteases is inhibited by mialostatin. Hemoglobin (**A**) and serum albumin (**B**) were digested in vitro with *I. ricinus* gut tissue extracts in the presence and absence of mialostatin. Blood protein substrate (5 µg of hemoglobin or 10 µg of serum albumin) was incubated with 0.4 µg gut tissue extract of cysteine proteases at pH 3.6 for 16 h. The extract was pre-incubated with mialostatin (MiSta) or the general cysteine protease inhibitor E-64 at the indicated concentrations prior to initiation of digestion. The digests were subjected to Tricine-SDS-PAGE (**A**) or Laemmli-SDS-PAGE (**B**) and visualized by protein staining. The hemoglobin (Hb) and serum albumin (SA) substrates are marked; the non-digested control is indicated. (**C**) Proteolytic stability of mialostatin in the gut tissue extract. Mialostatin (5 µg) was incubated with 0.4 µg gut extract protein under the same conditions as in (**A**,**B**), subjected to Tricine-SDS-PAGE, and visualized by protein staining. Mialostatin (MiSta) is marked; the non-digested control is indicated.

2.5. Phylogenetic Analysis and Three-Dimensional Structure of Mialostatin and Its Reactive Site

Phylogenetic analysis clearly demonstrated that mialostatin belongs to the cystatin superfamily. According to the maximum likelihood method, the tick cystatin phylogenetic tree contained three separate prostriate clades (Figure 5A and Figure S1). As shown in the simplified tree in Figure 5A, mialostatin fell into a strongly supported group with four other cystatins from the genus *Ixodes*. This clade was distant from other clades, including the recently described iristatin [32] and previously characterized sialostatins L and L2 [33,34]. In general, tick cystatins cluster into several clades specific for either prostriate, metastriate, or argasid tick species, suggesting fast evolution of cystatin genes in ticks. The full phylogenetic tree presented in Figure S1 shows strong bootstrap support for smaller clades, but the topology is less clear closer to the root of the tree. The analysis shows mialostatin as a new distinct *Ixodes* cystatin consistent with its presumed major role in the midgut, as most previously characterized cystatins from *Ixodes* spp. are of salivary origin.

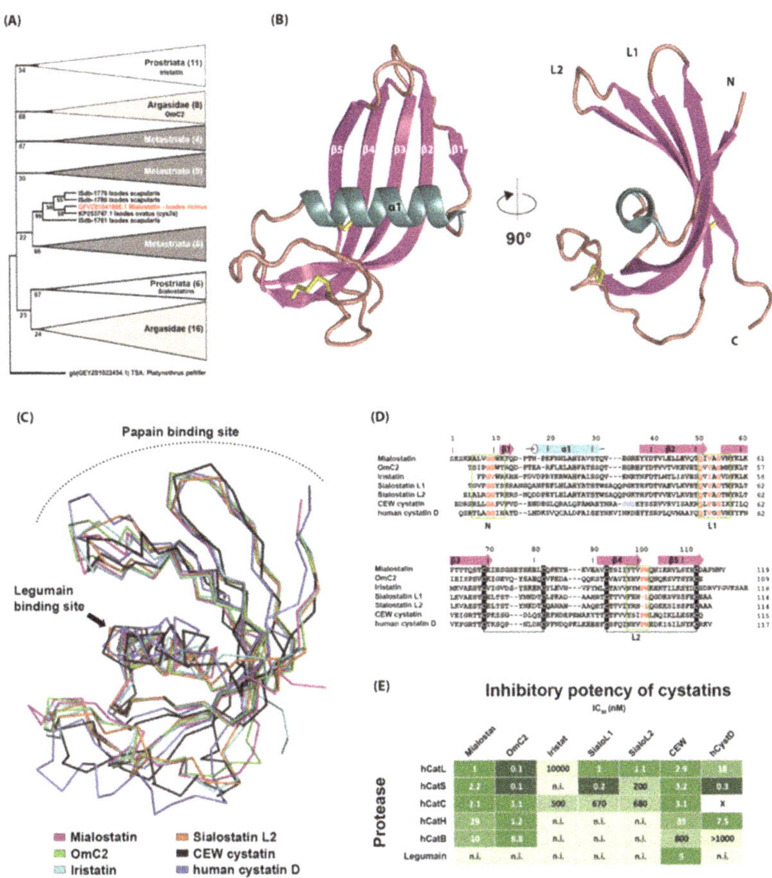

Figure 5. Crystal structure of mialostatin and its comparison with other family 2 cystatins. (**A**) Molecular phylogenetic analysis (maximum likelihood model) of secreted tick cystatins. Simplified consensus tree based on the maximum likelihood method with 1000-repeat bootstrap support. All clades except the one with mialostatin (highlighted in red) are condensed. Cystatin from the mite *Platynothrus peltifer* was used as an outgroup. The tree with the highest log-likelihood (−4870, 9711) is shown. Branches corresponding to partitions reproduced in less than 20% bootstrap replicates are collapsed. Numbers next to branches represent percentage of trees, in which the associated taxa clustered together during bootstrap analysis. For full tree see Figure S1. (**B**) The three-dimensional structure of mialostatin (PDB code 6ZTK) is shown as a cartoon representation colored by secondary structural elements (α1—cyan; β1-5—magenta). The N- and C-termini and two disulfide bridges,

Cys69–Cys82 and Cys93–Cys113 (yellow sticks), are indicated. The hairpin loops L1 and L2 and the N-terminus of cystatins are involved in the binding of papain-type cysteine proteases. (**C**) Superposition of Cα traces of the mialostatin structure with five other cystatin structures including OmC2 from the soft tick *O. moubata* (PDB code 3L0R), iristatin from the hard tick *I. ricinus* (5O46), sialostatin L2 from the hard tick *I. scapularis* (3LH4), and representative vertebrate members of family 2 cystatins: chicken egg white (CEW) cystatin (1CEW) and human cystatin D (1RN7). The orientation of mialostatin is as in (**B**). Color coding of the structures and positions of the binding sites for papain-type cysteine proteases and legumains are indicated. (**D**) Structure-based sequence alignment of mialostatin with OmC2, iristatin, sialostatins L1 and L2, CEW cystatin, and human cystatin D. Residues identical to those of mialostatin are shaded grey. The secondary structural elements of mialostatin are depicted in magenta for β-strands and cyan for α-helices. The conserved disulfide bridges are indicated by the connecting black lines. Three regions involved in the interaction between cystatins and papain-type cysteine proteases are boxed in green and labeled (the region size was selected based on the predominant binding residues in the available complex structures); the consensus core residues are highlighted in red. The legumain binding site in CEW cystatin is highlighted in blue. Mature protein sequences were used in the alignment; residue numbering is according to mialostatin. (**E**) A comparison of the inhibitory potency of mialostatin with the other family 2 cystatins (shown in **D**) against various cysteine proteases including human papain-type cathepsins L to B (hCatL to hCatB) and mammalian legumains. IC_{50} values are presented [23,32,33,35–37] and displayed as a heat map (green scale); n.i.—not inhibited; x—no literature data are available.

The crystal structure of mialostatin was determined by molecular replacement using the structure of the tick cystatin OmC2 as a search model and refined using data to 1.55 Å resolution (Table S2). The hexagonal prism crystal form contained two molecules in the asymmetric unit with a solvent content of about 57%. All protein residues could be modeled into a well-defined electron density map with the exception of the first nine residues, which formed a flexible N-terminus (Ser1 to Gly9), and the last two C-terminal residues (Asn118, Val119) of chain A. The final model consisted of two mialostatin molecules, chains A and B, containing 108 and 110 residues, respectively. The root-mean-square deviation (RMSD) for superposition of the Cα atoms of the two chains was 0.14 Å, a low value within the range observed for different crystal structures of identical proteins.

Figure 5B shows the overall structure of mialostatin. The molecule adopts a typical cystatin fold (so called 'hot dog' fold [38]) characterized by a five-stranded twisted antiparallel β-sheet wrapped around a central α-helix. Mialostatin contains two conserved disulfide bridges connecting Cys69 with Cys82 and Cys93 with Cys113. Structural comparison and sequence alignment with other known cystatin structures clearly demonstrated that mialostatin belongs to family 2 of the cystatin superfamily (Figure 5C,D). The closest structural homolog of mialostatin was the salivary/gut cystatin OmC2 from the soft tick *O. moubata* [24] with the highest sequence identity (53%) and lowest RMSD for Cα (0.86 Å), followed by salivary homologs iristatin (41% identity, 1.56 Å RMSD) from the hard tick *I. ricinus* and sialostatins L1 (42%, 2.09 Å) and L2 (40%, 2.50 Å) from the hard tick *I. scapularis* [32,33]. Lower structural similarity was found with vertebrate members, namely human cystatin D (35%, 4.80 Å) and chicken egg white (CEW) cystatin (23%, 3.80 Å) (Figure 5D) [39,40].

The interaction between family 2 cystatins and papain-type cysteine proteases is mediated by three regions, the N-terminal segment and two hairpin loops L1 and L2, which form a tripartite wedge-shaped edge that binds to the enzyme active site cleft (Figure 5B,C) [40–42]. In mialostatin, the first part of the binding site is formed by the N-terminal segment around Gly10, which is the first visible residue in the electron density map. The conserved pair of glycines (Gly9, Gly10) provide conformational flexibility to the N-terminal segment to adopt an optimal conformation for target binding. The L1 loop (between β1 and β2) of mialostatin exposes the segment Gln51-Ile52-Val53-Ala54-Gly55 corresponding to the critical binding motif Gln-Xaa-Val-Xaa-Gly conserved in cystatins (Figure 5B). The L2 loop (between β3 and β4) is characterized in mialostatin and other cystatins, except sialostatins, by the presence of a conservative Pro101-Trp102 segment. To conclude, the structural analysis of mialostatin demonstrated a functionally competent reactive site against papain-type cysteine proteases. The binding motif for legumain-type cysteine proteases, which has been characterized in several cystatins (e.g., CEW cystatin), was absent in mialostatin (Figure 5C,D), consistent with the fact that mialostatin and other tick cystatins do not suppress legumain activity (Table 1, Figure 5E).

The inhibitory selectivity of the structurally analyzed cystatins is illustrated in Figure 5E. Mialostatin and OmC2 represent broad-spectrum inhibitors of papain-type cathepsins and are the most versatile in terms of their interactions of the analyzed tick cystatins. However, the other tick homologs displayed a distinct selectivity profile limited to effective inhibition of only some cathepsins. This may reflect structural changes in the conserved motifs on the L1 and L2 loops of iristatin and sialostatins, respectively, and in their N-terminal sequence potentially clashing with the partially occluded active sites of exopeptidases such as cathepsins B or H. Conversely, binding events to, for example, cathepsins B and C, can be supported by the electrostatic interactions formed by a positively charged basic patch (residues 12, 20, 106, 107) located at the reactive site of mialostatin and OmC2.

3. Discussion

Ixodes ricinus has previously been used as a model tick species to investigate and describe the complex intestinal digestive proteolytic mechanisms occurring in hematophagous arthropods. Blood proteins have been shown to be processed intracellularly by a multienzyme network of cysteine and aspartic proteases, with major involvement of cysteine cathepsin-type proteases from the CA clan [6]. However, previous studies have not investigated the regulation of digestive proteolysis, including the control mechanisms that protect the gut epithelium from excessive proteolysis and potential cell damage. Cystatins, naturally occurring cysteine protease inhibitors, are among the primary molecules of interest in the *I. ricinus* anti-proteolytic system, as they have been previously proposed to interact with digestive proteases in several other tick species [26,43,44].

In this study, we identified mialostatin as the first gut-associated cystatin to be identified in *I. ricinus* and present its comprehensive functional and structural characterization. Mialostatin was a potent inhibitor of *I. ricinus* digestive cysteine proteases of clan CA, covering both exopeptidases cathepsins B and C and endopeptidases cathepsins L1 and L3 (named IrCB, IrCC, IrCL1, and IrCL3, respectively). Its broad inhibitory selectivity is in clear contrast with *Ixodes* salivary cystatins such as sialostatins L1, L2, and iristatin, which have much narrower selectivity and mainly target endopeptidases [32–34]. On the other hand, similar broad anti-protease activity has been reported for OmC2 and partially also for OmC1 [23], cystatins present in the midgut of *O. moubata* soft ticks, or rBrBmcys2b from *Rhipicephalus microplus* [26] hard ticks. The 3D structural analysis identified mialostatin as a close homolog of OmC2 and provided a structural explanation for its binding selectivity through comparison of the architecture of the reactive site of mialostatin with other publicly available tick cystatin structures. Specifically, we highlight a combination of structural changes in three segments forming a tripartite wedge on mialostatin and OmC2 that slots into the cathepsin active site cleft. Based on structure-activity relationships and phylogenetic data, we propose that well-characterized mialostatin and OmC2 represent a new evolutionary subgroup of tick gut-associated cystatins that differ from salivary cystatins modulating host immune responses. Functional diversification of the cystatin superfamily is described in vertebrates [45]. It is likely that similar process occurs in ticks due to fast evolution of secreted proteins, therefore the phylogenetic tree reflects both localization and function of the cystatins. It is interesting to note that OmC2 also exhibits immunomodulatory properties, which correlate with its dual expression pattern in both the salivary glands and gut of *O. moubata* ticks, while OmC1 and mialostatin are expressed predominantly in tick midguts [23].

The biological role(s) of mialostatin in the tick gut can be inferred in the context of tick feeding behavior and associated physiological processes. Adult *I. ricinus* females engorge an enormous amount of host blood that exceeds the weight of the unfed tick more than a hundred-fold. The current model of the multienzyme digestive protease network responsible for blood protein processing is based mainly on investigations of the well-developed digestive midgut cells occurring in partially engorged *I. ricinus* females at the end of the slow feeding period at day 6–7 [13]. This period is followed by a rapid engorgement phase lasting 12–24 h, which accounts for about two-thirds of the total blood volume ingested before detachment from the host. Most blood proteins are used for vitellogenesis and massive egg production during several weeks off-host [5,46]. The molecular mechanisms underlying the associated protein turnover and long-term blood meal storage in the tick gut lumen remain unexplored, mainly due to technical limitations in studying fully fed females. Nevertheless, advances in the field and initial results led to the hypothesis that off-host digestion may include extracellular proteolysis of blood proteins in the gut lumen, which supports or replaces intracellular digestive proteolysis in the gut epithelium [47]. Despite the broad biochemical selectivity of mialostatin, its biological selectivity is limited due to compartmentalization in tick midgut cells. Our immunohistochemistry results demonstrated that mialostatin is localized to the same population of intracellular vesicles as IrCL3 on the 11th day post detachment, suggesting that mialostatin predominantly targets IrCL3. Mialostatin is stored in these vesicles in some cells even during tick feeding. Forming an inhibitory complex between mialostatin and IrCL3 might be relevant for intracellular trafficking of enzymatically inactive IrCL3 in tick gut cells. The localization of the mialostatin-IrCL3 complex to the surface of the large dense granules two weeks post detachment is probably associated with an excretion/secretion mechanism allowing translocation of the complex to the gut lumen. We hypothesize that IrCL3 might partially restore its proteolytic activity in the diluted contents of the gut lumen, where mialostatin can competitively interact with other secreted cysteine cathepsins including IrCL1 as the strongest mialostatin binder. This would enable cathepsin-mediated luminal proteolysis of blood proteins or the generation of antimicrobial peptides under general mialostatin control [48]. Luminal IrCL3 might also act as an anti-coagulation factor, as recently reported for a related *R. microplus* cathepsin L [49].

In conclusion, mialostatin is the first gut-associated cystatin characterized from *I. ricinus* at the functional and structural levels. Mialostatin localized to both digestive cells and the gut lumen, where it targets cathepsin L isoforms and regulates their activity during trafficking and processing of host blood proteins. As components of gut-associated proteolytic pathways, mialostatin and homologous cystatins in other tick species represent potential vaccination antigens for novel anti-tick interventions targeting tick reproduction. The vaccination efficacy of proteins derived from the tick gut ("concealed" antigens) in controlling tick infestations has already been successfully demonstrated [50], and new candidate antigens are increasingly in demand to combat tick infestations and to limit the global spread of tick-borne diseases.

4. Materials and Methods

4.1. Ticks and Laboratory Animals

All animal experiments were carried out in accordance with the Animal Protection Law of the Czech Republic No. 246/1992 Sb., ethics approval No. 34/2018, and protocols approved by the responsible committee of the Institute of Parasitology, Biology Centre of the Czech Academy of Sciences. Male and female adult *I. ricinus* ticks were collected by flagging in a forest near České Budějovice in the Czech Republic and then kept in 95% humidity chambers under a 12 h light/dark cycle at room temperature. Female BALB/c mice were purchased from Velaz (Prague, Czech Republic). Mice were housed in individually ventilated cages maintained under a 12 h light/dark cycle. Mice were used at 8–12 weeks of age. Laboratory rabbits were purchased from RABBIT CZ a. s. (Trhový Štěpánov, Czech Republic) and housed individually in cages in the animal facility of the Institute of Parasitology. Guinea pigs were bred and housed in cages in the animal facility of the Institute of Parasitology. All mammals were fed a standard pellet diet and provided with water ad libitum.

4.2. Quantitative Real-Time PCR

Female *I. ricinus* ticks were fed on rabbits and allowed to mate with male ticks. Salivary glands, midguts, and ovaries from five ticks per time point were dissected on a petri dish under a drop of ice-cold DEPC-treated PBS. Total RNA was isolated from dissected tissue using the NucleoSpin RNA kit (Macherey-Nagel, Düren, Germany) and its quality checked by agarose gel electrophoresis before storing the RNA at −80 °C. cDNA was prepared from 500 ng of total RNA from independent biological triplicates using the Transcriptor High-Fidelity cDNA Synthesis Kit (Roche Applied Science, Penzberg, Germany). The cDNAs served as templates for subsequent quantitative expression analyses of mialostatin transcription by qRT-PCR. Samples were analysed with a LightCycler 480 (Roche Applied Science, Penzberg, Germany) using FastStart Universal SYBR Green Master Mix (Roche Applied Science, Penzberg, Germany). Reaction conditions over 50 cycles were as follows: denaturation, 95 °C/10 s; annealing, 60 °C/10 s; extension, 72 °C/10 s. Relative expression values were standardized to a reference gene, *I. ricinus* elongation factor 1 (*ef1*; GenBank: GU074828) [51–53], and normalized to the sample with the highest level of expression. The primers sequences for mialostatin and *ef1* RT-PCR are shown in Table S3.

4.3. Mialostatin Cloning, Expression, Refolding, and Purification and Antibody Production

The full cDNA sequence of the gene encoding mialostatin was amplified using primers designed based on the GFVZ01041806.1 [28] transcript from NCBI GenBank. The primer sequences used for the final cloning of mialostatin are presented in Table S3. A pool of *I. ricinus* cDNA prepared from the salivary glands of female ticks fed for three and six days on rabbits was used as a template. The 372 base pair DNA fragment encoding mialostatin without a signal peptide and with an inserted ATG codon was cloned into a pET-17b vector (Novagen, Darmstadt, Germany) and transformed into *Escherichia coli* strain BL21(DE3)pLysS (Novagen) for expression. Bacterial cultures were grown in LB medium with 100 µg/mL ampicillin and 34 µg/mL chloramphenicol to an OD600 of 0.8, when protein expression was induced by the addition of isopropyl 1-thio-β-D-galactopyranoside to a final concentration of 1 mM. Cultures were harvested after 2 h of incubation at 37 °C at 200 rpm shaking speed. Isolated inclusion bodies were dissolved in 6 M guanidine hydrochloride, 20 mM Tris, and 10 mM DTT, pH 8 for 1 h followed by centrifugation (10 min, 10,000× *g*) to remove undissolved impurities. Refolding was performed by rapid dilution in 160 × excess of 20 mM Tris and 300 mM NaCl, pH 8.5. The resulting refolded protein was purified by HiLoad Superdex 200 26/60 gel filtration chromatography and HiPrep Q FF 16/10 ion exchange chromatography. Endotoxin was removed using a detergent-based method. Purified recombinant mialostatin was used to raise antibodies in a mouse and rabbit as described previously [54,55]. The immunoglobulin (Ig) fraction of rabbit serum was obtained by caprylic acid precipitation of serum proteins as described previously [56]. Hybridoma cells were raised by fusing splenocytes from immunized mice and mouse myeloma SP 2/0-Ag14 cells. Monoclonal antibodies were produced in cell culture following the previously described protocol [55].

4.4. Preparation of Tick Gut Samples

I. ricinus midguts were dissected from female *I. ricinus* fed on laboratory guinea pigs (samples for proteolysis analysis and Western blotting) or from females' membrane fed on erythrocyte-depleted blood serum (samples for mass spectrometry analysis) [5]. The gut contents were carefully removed without disrupting the epithelium, and the gut tissue was washed in phosphate buffered saline (PBS). For mass spectrometry analysis, the gut contents were processed as described previously [6]. Gut tissue extract (150 mg protein/mL) was prepared by homogenization of the pooled gut tissue in 0.1 M Na acetate pH 4.5, 1% CHAPS on ice. The extract was cleared by centrifugation (16,000× *g*, 10 min, 4 °C), filtered through Ultrafree MC 0.22 µm (Millipore, Bedford, MA, USA), and stored at −80 °C.

4.5. Protease Inhibition Assays

Inhibition measurements were performed in triplicate in 96-well microplates (100 μL assay volume) at 37 °C. Recombinant mialostatin was preincubated with protease for 10 min followed by the addition of specific fluorogenic substrate (see Sections 4.5.1–4.5.3). The kinetics of product release were continuously monitored using an Infinite M1000 (Tecan, Männedorf, Switzerland) microplate fluorescence reader at 360 nm excitation and 465 nm emission wavelengths (for AMC-containing substrates) or at 320 nm excitation and 420 nm emission wavelengths (for Abz-containing substrate). IC_{50} values were determined from residual velocities using dose-response plots; nonlinear regression was fitted using GraFit software (Erithacus, East Grinstead, UK).

4.5.1. Inhibition of Proteases in Tick Gut Homogenates

To prevent interference of non-target proteases, homogenates (80 ng) were treated with specific low molecular weight inhibitors (final assay concentrations are indicated) including 1 μM pepstatin and 1 mM EDTA (against aspartic proteases and metalloproteases; all assays), 1 μM E-64 (against cathepsins L/B; cathepsin C assay), 1 μM CA-074 (against cathepsin B; cathepsin L assay), and 1 μM Z–Phe–Phe–DMK (against cathepsin L; cathepsin B assay) [6]. The assay substrates and buffers were as follows: 20 μM Z–Phe–Arg–AMC substrate and 0.1 M Na acetate pH 4.5 or 5.0 in the cathepsin L and L/B assays, respectively; 20 μM Z–Arg–Arg–AMC substrate and 0.1 M MES pH 6.5 in the cathepsin B assay; 20 μM Gly–Arg–AMC substrate in 0.1 M Na acetate pH 5.5, 25 mM NaCl in the cathepsin C assay; all assay buffers contained 2.5 mM DTT and 0.1% PEG 1500.

4.5.2. Inhibition Assays of Recombinant Tick Proteases

The assay conditions for individual proteases were as follows: 1.2 nM IrCD1 and 20 μM Abz–Lys–Pro–Ala–Glu–Phe–Nph–Ala–Leu substrate in 0.1 M Na acetate pH 4.0; 1.25 nM IrAE and 20 μM Z–Ala–Ala–Asn–AMC substrate in 0.1 M MES pH 5.0, 2.5 mM DTT, 1 μM E-64; 0.1 nM IrCL1 or 20 pM IrCL3 and 20 μM Z–Phe–Arg–AMC substrate in 0.1 M Na acetate pH 4.5, 2.5 mM DTT; all assay buffers contained 0.1% PEG 1500. The tick proteases were prepared as described elsewhere [7,9,10,57,58].

4.5.3. Inhibition Assays of Human Proteases

Inhibition assays were performed following the same protocol used in our previous publications [24,32]. The assay conditions for individual proteases were as follows: 35 pM cathepsin B or 33 pM cathepsin L or 5 nM cathepsin K and 250 μM Z–Leu–Arg–AMC substrate in 0.1 M Na acetate pH 5.5, 0.1 M NaCl; 350 pM cathepsin S and 250 μM Z–Val–Val–Arg–AMC substrate in the same buffer; 0.5 nM cathepsin C and 250 μM Gly-Arg-AMC substrate in the same buffer; 20 nM cathepsin H and 40 μM Z–Leu–Arg–AMC substrate in 0.1 M Na/K phosphate pH 6.8; all assay buffers contained 1 mM EDTA, 2.5 mM DTT, and 0.01% Triton X-100. The human proteases were purchased from Merck (Kenilworth, NJ, USA) and Biomol (Hamburg, Germany).

4.6. Protein Digestion Assay

Digestion of 10 μg human serum albumin (Sigma Aldrich, St Louis, MO, USA), 5 μg bovine hemoglobin (Sigma Aldrich, St Louis, MO, USA), and 5 μg of mialostatin was performed with the tick gut tissue homogenate (0.4 μg protein) in 50 mM Na citrate pH 3.6, 2.5 mM DTT, in a total volume of 100 μL for 16 h at 26 °C. In the albumin and hemoglobin digestion assays, the homogenate was preincubated (15 min) in the same buffer with non-cysteine protease inhibitors: 1 μM pepstatin, 100 μM Pefablock, and 1 mM EDTA. The albumin digest was resolved with Laemmli SDS-PAGE gels (15%) and the hemoglobin and mialostatin digests by Tricine-SDS-PAGE gels (16% T/6% C) containing 6 M urea [59]. Electrophoresis was performed under reducing conditions, and protein was stained with Coomassie Blue G250.

4.7. Reducing SDS-PAGE and Western Blotting

Tick tissue homogenates were separated by reducing SDS-PAGE using 4–20% Mini-PROTEAN® TGX™ Precast Protein Gels (Bio-Rad Laboratories, Hercules, CA, USA). Separated protein loads were visualized using TGX stain-free chemistry in the ChemiDoc MP imager (Bio-Rad, Hercules, CA, USA). After protein load documentation, separated proteins were electro-transferred from the gel onto an Immun-Blot® LF PVDF membrane using the Trans-Blot Turbo system (Bio-Rad, Hercules, CA, USA). Prior to Western blot analyses, membranes were blocked with 3% non-fat milk in PBS with 0.05% Tween 20 (PBS-Tween) for 1 h at room temperature. Blocked membranes were incubated with the rabbit Ig fraction of α-IrCL1 or α-IrCL3 polyclonal sera diluted 1:1000 in PBS-Tween containing 1% milk. Goat anti-rabbit IgG Alexa 488-labeled antibody (1:1000, Thermo Fisher Scientific, Waltham, MA, USA) was used as a secondary antibody. For mialostatin detection, α-mialostatin monoclonal antibody (1:30) diluted in in PBS-Tween containing 1% milk and the goat anti-mouse Alexa 546-labeled antibody (1:1000, Thermo Fisher Scientific, Waltham, MA, USA) were used. In between individual steps of the whole procedure, membranes were washed 3×5 min in PBS-Tween on a rotating shaker platform at room temperature. Labeling with primary antibodies was performed on a rotating shaker platform at 4 °C overnight. Labeling with secondary antibodies was performed on a rotating shaker platform at room temperature for 1 h. Fluorescent signals were again visualized using the ChemiDoc MP imager and analyzed using Image Lab Software (Bio-Rad, Hercules, CA, USA).

4.8. Immunohistochemistry

Samples of *I. ricinus* gut tissues were prepared as described previously [11]. Briefly, the gut was dissected from adult females at specific days of feeding on the host and days post-attachment and fixed in 4% formaldehyde and 0.1% glutaraldehyde solution, washed with PBS, dehydrated using ascending ethanol dilutions, then infiltrated with LR White resin (London Resin Company, Stansted, UK) and polymerized. Semi-thin sections (0.5 μm) were blocked with 1% BSA and 1% milk in PBS-Tween (0.3% (*v/v*) Tween 20) for 45 min. Immunohistochemical double-staining was performed gradually, with the initial antibody labeling of the respective intestinal protease (*I. ricinus* cathepsin L1 IrCL1 [9]; cathepsin L3 IrCL3; cathepsin B IrCB [11]) subsequently followed with immunolabeling of mialostatin. First, semi-thin tick gut tissue sections were blocked with blocking solution (1% BSA, 1% milk solution in PBS-Tween) for 45 min at room temperature. For protease immunostaining, sections were first labeled (4 °C overnight) with primary antibodies: (i) rabbit α-IrCL1 affinity-purified polyclonal serum diluted 1:5 in PBS-Tween; (ii) rabbit α-IrCB affinity-purified polyclonal serum diluted 1:5 in PBS-Tween; (iii) isolated Ig fraction of α-IrCL3 polyclonal serum diluted 1:5 (IrCL1) in PBS-Tween. After washing 3×5 min in PBS-Tween, sections were subsequently labeled with Alexa Fluor® 647 goat α-rabbit secondary antibody (diluted 1:500 in PBS-Tween; Thermo Fisher Scientific, Waltham, MA, USA). Sections were subsequently used for mialostatin immunolabeling: sections were once again washed 3×5 min in PBS-Tween and incubated with mouse α-mialostatin monoclonal antibody diluted 1:50 in PBS-Tween. Incubation was performed in a humid chamber at room temperature for 90 min. Sections were once again washed (3×5 min in PBS-Tween) and incubated with secondary goat α-mouse Alexa Fluor® 488 (Thermo Fisher Scientific, Waltham, MA, USA) diluted 1:500 in PBS-Tween for 1 h at room temperature. Finally, all sections were washed in PBS-Tween and counterstained with DAPI (4′,6′-diamidino-2-phenylindole; 2.5 μg/mL; Sigma Aldrich, St Louis, MO, USA) for 7 min, washed again with PBS-Tween, mounted in Fluoromount medium (Sigma Aldrich, St Louis, MO, USA), and examined with the IX83 confocal microscope (Olympus, Tokyo, Japan). Images were processed with FluoView FV3000 software (Olympus, Tokyo, Japan).

4.9. Evolutionary Analysis by the Maximum Likelihood Method

The evolutionary history was inferred using the maximum likelihood method and JTT matrix-based model [60]. The bootstrap consensus tree inferred from 1000 replicates was taken to represent the evolutionary history of the taxa analyzed [61]. Branches corresponding to partitions reproduced in less than 20% bootstrap replicates were collapsed. The percentage of replicate trees in which the associated taxa clustered together in the bootstrap test (1000 replicates) are shown next to the branches [61]. Initial tree(s) for the heuristic search were obtained automatically by applying Neighbor-Join and BioNJ algorithms to a matrix of pairwise distances estimated using the JTT model and then selecting the topology with the superior log-likelihood value. This analysis involved 71 amino acid sequences. There were 108 positions in the final dataset. Evolutionary analyses were conducted in MEGA X [62].

4.10. Crystallization and Data Collection

Screening for crystallization conditions was performed using the JCSG-plus kit (Molecular Dimensions, Sheffield, UK) by the sitting drop vapor diffusion technique. Preliminary crystals of mialostatin were obtained in 0.1 M citric acid pH 3.5, 0.8 M ammonium sulfate. Optimal crystals were prepared at 18 °C using the hanging drop vapor diffusion technique in 15-well NeXtal plates (Qiagen, Hilden, Germany). The crystallization drop consisted of 2 μL of the mialostatin protein solution (12.5 mg/mL in 10 mM Tris buffer, pH 8.0) and 1 μL of the precipitant solution equilibrated over a reservoir containing 300 μL precipitant solution (0.1 M citric acid pH 4.0, 0.8 M ammonium sulfate). Crystals shaped as hexagonal prisms reached their final size of $0.6 \times 0.3 \times 0.3$ mm within 1 month. For data collection, crystals were soaked in reservoir solution supplemented with 20% glycerol and flash cooled in liquid nitrogen. Diffraction data at 100 K were collected using a BL14.1 beamline operated by the Helmholtz-Zentrum Berlin (HZB) at the BESSY II electron storage ring (Berlin-Adlershof, Germany) [63] and processed using the XDS suite of programs [64]. Crystals exhibited the symmetry of space group P6222 and contained two molecules in the asymmetric unit. Crystal parameters and data collection statistics are shown in Table S2.

4.11. Structure Determination

The phase problem was solved by molecular replacement using Molrep [65] from the CCP4 package [66]. The search model was derived from the structure of cystatin OmC2 (PDB code 3L0R) [24] sharing 53% sequence identity with mialostatin. Model refinement was carried out using REFMAC 5.8 [66] from the CCP4 package with 5% of the reflections reserved for cross-validation. Manual building and addition of water molecules was performed using Coot [67]. The quality of the final model was validated with Molprobity [68]. Final refinement statistics are given in Supporting Information Table S2. Figures showing structural representations were prepared with the PyMOL Molecular Graphics System (Schrödinger, New York, NY, USA). Atomic coordinates and structure factors were deposited in the PDB under accession code 6ZTK.

4.12. Statistical Analysis

All experiments were performed in biological triplicate. Data are presented as mean ± standard error of mean (SEM) in all graphs. Student's t-test or one-way ANOVA were used to calculate statistical differences between two or more groups, respectively. Statistically significant results are marked: $* \, p \leq 0.05$; $** \, p \leq 0.01$; $*** \, p \leq 0.001$; $**** \, p \leq 0.0001$.

Supplementary Materials: The following are available online at https://www.mdpi.com/article/10.3390/ijms22105371/s1, Figure S1: The phylogenetic tree of 71 cystatins from both Ixodidae and Argasidae tick species, Figure S2: Detection of mialostatin in midgut of blood fed ticks, Figure S3: Detection of mialostatin in midgut of serum fed ticks, Figure S4: Detection of *I. ricinus* cathepsins L in midgut of blood fed ticks, Table S1: Identification of mialostatin by mass spectrometry, Table S2: X-ray data collection and refinement statistics, Table S3: Primer sequences.

Author Contributions: J.K., M.B., J.C., D.S., M.M. and M.K. designed and performed the experiments, performed the analyses, and wrote the manuscript; V.U., P.Ř. and H.L. designed and performed the experiments and revised the manuscript. All authors have read and agreed to the published version of the manuscript.

Funding: This work was supported by the Grant Agency of the Czech Republic (No. 19-382 07247S to M.K.), the project CePaViP OPVVV (No. CZ.02.1.01/0.0/0.0/16_019/0000759 to M.K.) from the European Regional Development Fund and institutional project RVO 60077344 to M.K., M.B., P.Ř., and M.M. were supported by project ChemBioDrug CZ.02.1.01/0.0/0.0/16_019/0000729 from the European Regional Development Fund (OP RDE) and the institutional project RVO 61388963. M.M., D.S. and V.U. were supported by the Grant Agency of the Czech Republic (No. 21-08826S). J.C. was supported by the Grant Agency of the Czech Republic (No. 19-14704Y).

Institutional Review Board Statement: All animal experiments were carried out in accordance with the Animal Protection Law of the Czech Republic No. 246/1992 Sb., ethics approval No. 34/2018, and protocols approved by the responsible committee of the Institute of Parasitology, Biology Centre of the Czech Academy of Sciences.

Informed Consent Statement: Not applicable.

Data Availability Statement: All data are either contained within the manuscript and supporting information or available from the corresponding author on reasonable request.

Acknowledgments: The authors thank Jiří Brynda and Martin Hubálek for technical assistance with crystallography and mass spectrometry analysis, respectively. Diffraction data were collected on BL14.1 at the BESSY II electron storage ring operated by the Helmholtz-Zentrum Berlin. The funders had no role in the design, data collection and analysis, decision to publish, or preparation of the manuscript.

Conflicts of Interest: The authors declare no conflict of interest.

References

1. Lindgren, E.; Talleklint, L.; Polfeldt, T. Impact of climatic change on the northern latitude limit and population density of the disease-transmitting European tick Ixodes ricinus. *Environ. Health Perspect.* **2000**, *108*, 119–123. [CrossRef]
2. Yang, Y.; Christie, J.; Köster, L.; Du, A.; Yao, C. Emerging Human Babesiosis with "Ground Zero" in North America. *Microorganisms* **2021**, *9*, 440. [CrossRef]
3. Sonenshine, D.E.; Roe, R.M. *Biology of Ticks*, 2nd ed.; Oxford University Press: New York, NY, USA, 2014.
4. Lara, F.A.; Lins, U.; Bechara, G.H.; Oliveira, P.L. Tracing heme in a living cell: Hemoglobin degradation and heme traffic in digest cells of the cattle tick Boophilus microplus. *J. Exp. Biol.* **2005**, *208*, 3093–3101. [CrossRef] [PubMed]
5. Sojka, D.; Pytelková, J.; Perner, J.; Horn, M.; Konvičková, J.; Schrenková, J.; Mareš, M.; Kopáček, P. Multienzyme degradation of host serum albumin in ticks. *Ticks Tick-Borne Dis.* **2016**, *7*, 604–613. [CrossRef] [PubMed]
6. Horn, M.; Nussbaumerová, M.; Šanda, M.; Kovářová, Z.; Srba, J.; Franta, Z.; Sojka, D.; Bogyo, M.; Caffrey, C.R.; Kopáček, P.; et al. Hemoglobin Digestion in Blood-Feeding Ticks: Mapping a Multipeptidase Pathway by Functional Proteomics. *Chem. Biol.* **2009**, *16*, 1053–1063. [CrossRef] [PubMed]
7. Sojka, D.; Hajdušek, O.; Dvořák, J.; Sajid, M.; Franta, Z.; Schneider, E.L.; Craik, C.S.; Vancová, M.; Burešová, V.; Bogyo, M.; et al. IrAE—An asparaginyl endopeptidase (legumain) in the gut of the hard tick Ixodes ricinus. *Int. J. Parasitol.* **2007**, *37*, 713–724. [CrossRef]
8. Sojka, D.; Franta, Z.; Frantová, H.; Bartošová, P.; Horn, M.; Váchová, J.; O'Donoghue, A.J.; Eroy-Reveles, A.A.; Craik, C.S.; Knudsen, G.M.; et al. Characterization of Gut-associated Cathepsin D Hemoglobinase from Tick Ixodes ricinus (IrCD1). *J. Biol. Chem.* **2012**, *287*, 21152–21163. [CrossRef] [PubMed]
9. Franta, Z.; Sojka, D.; Frantova, H.; Dvorak, J.; Horn, M.; Srba, J.; Talacko, P.; Mares, M.; Schneider, E.; Craik, C.S.; et al. IrCL1—The haemoglobinolytic cathepsin L of the hard tick, Ixodes ricinus. *Int. J. Parasitol.* **2011**, *41*, 1253–1262. [CrossRef] [PubMed]
10. Sojka, D.; Franta, Z.; Horn, M.; Hajdušek, O.; Caffrey, C.R.; Mares, M.; Kopáček, P. Profiling of proteolytic enzymes in the gut of the tick Ixodes ricinus reveals an evolutionarily conserved network of aspartic and cysteine peptidases. *Parasites Vectors* **2008**, *1*, 7. [CrossRef]
11. Franta, Z.; Frantová, H.; Konvičková, J.; Horn, M.; Sojka, D.; Mareš, M.; Kopáček, P. Dynamics of digestive proteolytic system during blood feeding of the hard tick Ixodes ricinus. *Parasites Vectors* **2010**, *3*, 119. [CrossRef]
12. Sojka, D.; Francischetti, I.M.B.; Calvo, E.; Kotsyfakis, M. Cysteine Proteases from Bloodfeeding Arthropod Ectoparasites. *Adv. Exp. Med. Biol.* **2011**, *712*, 177–191. [CrossRef]

13. Sojka, D.; Franta, Z.; Horn, M.; Caffrey, C.R.; Mareš, M.; Kopáček, P. New insights into the machinery of blood digestion by ticks. *Trends Parasitol.* **2013**, *29*, 276–285. [CrossRef] [PubMed]

14. Caffrey, C.R.; Goupil, L.; Rebello, K.M.; Dalton, J.P.; Smith, D. Cysteine proteases as digestive enzymes in parasitic helminths. *PLoS Negl. Trop. Dis.* **2018**, *12*, e0005840. [CrossRef] [PubMed]

15. Novinec, M.; Lenarčič, B.; Turk, B. Cysteine Cathepsin Activity Regulation by Glycosaminoglycans. *BioMed Res. Int.* **2014**, *2014*, 1–9. [CrossRef]

16. Turk, V.; Bode, W. The cystatins: Protein inhibitors of cysteine proteinases. *FEBS Lett.* **1991**, *285*, 213–219. [CrossRef]

17. Rawlings, N.D.; Waller, M.; Barrett, A.J.; Bateman, A. MEROPS: The database of proteolytic enzymes, their substrates and inhibitors. *Nucleic Acids Res.* **2014**, *42*, D503–D509. [CrossRef]

18. Schwarz, A.; Valdés, J.J.; Kotsyfakis, M. The role of cystatins in tick physiology and blood feeding. *Ticks Tick-Borne Dis.* **2012**, *3*, 117–127. [CrossRef] [PubMed]

19. Chmelař, J.; Kotál, J.; Langhansová, H.; Kotsyfakis, M. Protease Inhibitors in Tick Saliva: The Role of Serpins and Cystatins in Tick-host-Pathogen Interaction. *Front. Cell. Infect. Microbiol.* **2017**, *7*, 216. [CrossRef] [PubMed]

20. Martins, L.A.; Kotál, J.; Bensaoud, C.; Chmelař, J.; Kotsyfakis, M. Small protease inhibitors in tick saliva and salivary glands and their role in tick-host-pathogen interactions. *Biochim. Biophys. Acta BBA Proteins Proteom.* **2020**, *1868*, 140336. [CrossRef] [PubMed]

21. Lima, C.A.; Sasaki, S.D.; Tanaka, A.S. Bmcystatin, a cysteine proteinase inhibitor characterized from the tick Boophilus microplus. *Biochem. Biophys. Res. Commun.* **2006**, *347*, 44–50. [CrossRef] [PubMed]

22. Parizi, L.F.; Githaka, N.W.; Acevedo, C.; Benavides, U.; Seixas, A.; Logullo, C.; Konnai, S.; Ohashi, K.; Masuda, A.; Vaz, I.D.S. Sequence characterization and immunogenicity of cystatins from the cattle tick Rhipicephalus (Boophilus) microplus. *Ticks Tick-Borne Dis.* **2013**, *4*, 492–499. [CrossRef] [PubMed]

23. Grunclová, L.; Horn, M.; Vancová, M.; Sojka, D.; Franta, Z.; Mares, M.; Kopáček, P. Two secreted cystatins of the soft tick Ornithodoros moubata: Differential expression pattern and inhibitory specificity. *Biol. Chem.* **2006**, *387*, 1635–1644. [CrossRef]

24. Salát, J.; Paesen, G.C.; Řezáčová, P.; Kotsyfakis, M.; Kovářová, Z.; Šanda, M.; Majtán, J.; Grunclová, L.; Horká, H.; Andersen, J.F.; et al. Crystal structure and functional characterization of an immunomodulatory salivary cystatin from the soft tick Ornithodoros moubata. *Biochem. J.* **2010**, *429*, 103–112. [CrossRef] [PubMed]

25. Zavašnik-Bergant, T.; Vidmar, R.; Sekirnik, A.; Fonović, M.; Salát, J.; Grunclová, L.; Kopáček, P.; Turk, B. Salivary Tick Cystatin OmC2 Targets Lysosomal Cathepsins S and C in Human Dendritic Cells. *Front. Cell. Infect. Microbiol.* **2017**, *7*, 288. [CrossRef] [PubMed]

26. Parizi, L.F.; Sabadin, G.A.; Alzugaray, M.F.; Seixas, A.; Logullo, C.; Konnai, S.; Ohashi, K.; Masuda, A.; da Sliva Vaz, I., Jr. Rhipicephalus microplus and Ixodes ovatus cystatins in tick blood digestion and evasion of host immune response. *Parasites Vectors* **2015**, *8*, 122. [CrossRef]

27. Rangel, C.K.; Parizi, L.F.; Sabadin, G.A.; Costa, E.P.; Romeiro, N.C.; Isezaki, M.; Githaka, N.W.; Seixas, A.; Logullo, C.; Konnai, S.; et al. Molecular and structural characterization of novel cystatins from the taiga tick Ixodes persulcatus. *Ticks Tick-Borne Dis.* **2017**, *8*, 432–441. [CrossRef] [PubMed]

28. Charrier, N.P.; Couton, M.; Voordouw, M.J.; Rais, O.; Durand-Hermouet, A.; Hervet, C.; Plantard, O.; Rispe, C. Whole body transcriptomes and new insights into the biology of the tick Ixodes ricinus. *Parasites Vectors* **2018**, *11*, 1–15. [CrossRef]

29. Kotsyfakis, M.; Schwarz, A.; Erhart, J.; Ribeiro, J.M.C. Tissue- and time-dependent transcription in Ixodes ricinus salivary glands and midguts when blood feeding on the vertebrate host. *Sci. Rep.* **2015**, *5*, srep09103. [CrossRef]

30. Cramaro, W.J.; Revets, D.; Hunewald, O.E.; Sinner, R.; Reye, A.L.; Muller, C.P. Integration of Ixodes ricinus genome sequencing with transcriptome and proteome annotation of the naïve midgut. *BMC Genom.* **2015**, *16*, 871. [CrossRef]

31. Perner, J.; Provazník, J.; Schrenková, J.; Urbanová, V.; Ribeiro, J.M.C.; Kopáček, P. RNA-seq analyses of the midgut from blood- and serum-fed Ixodes ricinus ticks. *Sci. Rep.* **2016**, *6*, 36695. [CrossRef]

32. Kotál, J.; Stergiou, N.; Buša, M.; Chlastáková, A.; Beránková, Z.; Řezáčová, P.; Langhansová, H.; Schwarz, A.; Calvo, E.; Kopecký, J.; et al. The structure and function of Iristatin, a novel immunosuppressive tick salivary cystatin. *Cell. Mol. Life Sci.* **2019**, *76*, 2003–2013. [CrossRef] [PubMed]

33. Kotsyfakis, M.; Horka, H.; Salat, J.; Andersen, J.F. The crystal structures of two salivary cystatins from the tick Ixodes scapularis and the effect of these inhibitors on the establishment of Borrelia burgdorferi infection in a murine model. *Mol. Microbiol.* **2010**, *77*, 456–470. [CrossRef] [PubMed]

34. Kotsyfakis, M.; Sá-Nunes, A.; Francischetti, I.M.B.; Mather, T.N.; Andersen, J.F.; Ribeiro, J.M.C. Antiinflammatory and Immunosuppressive Activity of Sialostatin L, a Salivary Cystatin from the Tick Ixodes scapularis. *J. Biol. Chem.* **2006**, *281*, 26298–26307. [CrossRef]

35. Balbin, M.; Hall, A.; Grubb, A.; Mason, R.W.; Lopez-Otin, C.; Abrahamson, M. Structural and functional characterization of two allelic variants of human cystatin D sharing a characteristic inhibition spectrum against mammalian cysteine proteinases. *J. Biol. Chem.* **1994**, *269*, 23156–23162. [CrossRef]

36. Anastasi, A.; Brown, M.A.; Kembhavi, A.A.; Nicklin, M.J.H.; Sayers, C.A.; Sunter, D.C.; Barrett, A.J. Cystatin, a protein inhibitor of cysteine proteinases. Improved purification from egg white, characterization, and detection in chicken serum. *Biochem. J.* **1983**, *211*, 129–138. [CrossRef] [PubMed]

37. Vasiljeva, O.; Dolinar, M.; Turk, V.; Turk, B. Recombinant Human Cathepsin H Lacking the Mini Chain Is an Endopeptidase. *Biochemistry* **2003**, *42*, 13522–13528. [CrossRef] [PubMed]
38. Pidugu, L.S.; Maity, K.; Ramaswamy, K.; Surolia, N.; Suguna, K. Analysis of proteins with the 'Hot dog' fold: Prediction of function and identification of catalytic residues of hypothetical proteins. *BMC Struct. Biol.* **2009**, *9*, 37. [CrossRef]
39. Alvarez-Fernandez, M.; Liang, Y.-H.; Abrahamson, M.; Su, X.-D. Crystal Structure of Human Cystatin D, a Cysteine Peptidase Inhibitor with Restricted Inhibition Profile. *J. Biol. Chem.* **2005**, *280*, 18221–18228. [CrossRef]
40. Bode, W.; Engh, R.; Musil, D.; Thiele, U.; Huber, R.; Karshikov, A.; Brzin, J.; Kos, J.; Turk, V. The 2.0 A X-ray crystal structure of chicken egg white cystatin and its possible mode of interaction with cysteine proteinases. *EMBO J.* **1988**, *7*, 2593–2599. [CrossRef]
41. Turk, V.; Stoka, V.; Vasiljeva, O.; Renko, M.; Sun, T.; Turk, B.; Turk, D. Cysteine cathepsins: From structure, function and regulation to new frontiers. *Biochim. Biophys. Acta BBA Proteins Proteom.* **2012**, *1824*, 68–88. [CrossRef]
42. Nandy, S.K.; Seal, A. Structural Dynamics Investigation of Human Family 1 & 2 Cystatin-Cathepsin L1 Interaction: A Comparison of Binding Modes. *PLoS ONE* **2016**, *11*, e0164970. [CrossRef]
43. Wang, Y.; Yu, X.; Cao, J.; Zhou, Y.; Gong, H.; Zhang, H.; Li, X.; Zhou, J. Characterization of a secreted cystatin from the tick Rhipicephalus haemaphysaloides. *Exp. Appl. Acarol.* **2015**, *67*, 289–298. [CrossRef] [PubMed]
44. Lu, S.; da Rocha, L.A.; Torquato, R.J.; Junior, I.D.S.V.; Florin-Christensen, M.; Tanaka, A.S. A novel type 1 cystatin involved in the regulation of Rhipicephalus microplus midgut cysteine proteases. *Ticks Tick-Borne Dis.* **2020**, *11*, 101374. [CrossRef] [PubMed]
45. Kordiš, D.; Turk, V. Phylogenomic analysis of the cystatin superfamily in eukaryotes and prokaryotes. *BMC Evol. Biol.* **2009**, *9*, 266. [CrossRef]
46. Kopáček, P.; Perner, J.; Sojka, D.; Šíma, R.; Hajdušek, O. Molecular Targets to Impair Blood Meal Processing in Ticks. In *Ectoparasites*; Wiley: New York, NY, USA, 2018; Volume 8, pp. 139–165.
47. Reyes, J.; Ayala-Chavez, C.; Sharma, A.; Pham, M.; Nuss, A.B.; Gulia-Nuss, M. Blood Digestion by Trypsin-Like Serine Proteases in the Replete Lyme Disease Vector Tick, Ixodes scapularis. *Insects* **2020**, *11*, 201. [CrossRef] [PubMed]
48. Cruz, C.E.; Fogaça, A.C.; Nakayasu, E.S.; Angeli, C.B.; Belmonte, R.; Almeida, I.C.; Miranda, A.; Miranda, M.T.M.; Tanaka, A.S.; Braz, G.R.; et al. Characterization of proteinases from the midgut of Rhipicephalus (Boophilus) microplus involved in the generation of antimicrobial peptides. *Parasites Vectors* **2010**, *3*, 63. [CrossRef]
49. Xavier, M.A.; Tirloni, L.; Torquato, R.; Tanaka, A.; Pinto, A.F.M.; Diedrich, J.K.; Yates, J.R., 3rd; da Silva Vaz, I., Jr.; Seixas, A.; Termignoni, C. Blood anticlotting activity of a Rhipicephalus microplus cathepsin L-like enzyme. *Biochimie* **2019**, *163*, 12–20. [CrossRef]
50. Rodríguez-Mallon, A. Developing Anti-tick Vaccines. *Methods Mol. Biol.* **2016**, *1404*, 243–259. [CrossRef]
51. Nijhof, A.M.; Balk, J.A.; Postigo, M.; Jongejan, F. Selection of reference genes for quantitative RT-PCR studies in Rhipicephalus (Boophilus) microplus and Rhipicephalus appendiculatus ticks and determination of the expression profile of Bm86. *BMC Mol. Biol.* **2009**, *10*, 112. [CrossRef]
52. Urbanová, V.; Hartmann, D.; Grunclová, L.; Šíma, R.; Flemming, T.; Hajdušek, O.; Kopáček, P. IrFC—An Ixodes ricinus injury-responsive molecule related to Limulus Factor C. *Dev. Comp. Immunol.* **2014**, *46*, 439–447. [CrossRef]
53. Vechtova, P.; Fussy, Z.; Cegan, R.; Sterba, J.; Erhart, J.; Benes, V.; Grubhoffer, L. Catalogue of stage-specific transcripts in Ixodes ricinus and their potential functions during the tick life-cycle. *Parasites Vectors* **2020**, *13*, 1–19. [CrossRef]
54. Kopácek, P.; Zdychová, J.; Yoshiga, T.; Weise, C.; Rudenko, N.; Law, J.H. Molecular cloning, expression and isolation of ferritins from two tick species–Ornithodoros moubata and Ixodes ricinus. *Insect Biochem. Mol. Biol.* **2003**, *33*, 103–113. [CrossRef]
55. Hurrell, J.G.R. *Monoclonal Hybridoma Antibodies: Techniques and Applications*; CRC Press: Boca Raton, FL, USA, 2017; p. 239.
56. Russo, C.; Callegaro, L.; Lanza, E.; Ferrone, S. Purification of IgG monoclonal antibody by caprylic acid precipitation. *J. Immunol. Methods* **1983**, *65*, 269–271. [CrossRef]
57. Hartmann, D.; Šíma, R.; Konvičková, J.; Perner, J.; Kopáček, P.; Sojka, D. Multiple legumain isoenzymes in ticks. *Int. J. Parasitol.* **2018**, *48*, 167–178. [CrossRef]
58. Hánová, I.; Brynda, J.; Houštecká, R.; Alam, N.; Sojka, D.; Kopáček, P.; Marešová, L.; Vondrášek, J.; Horn, M.; Schueler-Furman, O.; et al. Novel Structural Mechanism of Allosteric Regulation of Aspartic Peptidases via an Evolutionarily Conserved Exosite. *Cell Chem. Biol.* **2018**, *25*, 318–329.e4. [CrossRef]
59. Schagger, H. Tricine-SDS-PAGE. *Nat. Protoc.* **2006**, *1*, 16–22. [CrossRef] [PubMed]
60. Jones, D.T.; Taylor, W.R.; Thornton, J.M. The rapid generation of mutation data matrices from protein sequences. *Bioinformatics* **1992**, *8*, 275–282. [CrossRef] [PubMed]
61. Felsenstein, J. Confidence Limits on Phylogenies: An Approach Using the Bootstrap. *Evol. Int. J. Org. Evol.* **1985**, *39*, 783–791. [CrossRef]
62. Kumar, S.; Stecher, G.; Li, M.; Knyaz, C.; Tamura, K. MEGA X: Molecular evolutionary genetics analysis across computing platforms. *Mol. Biol. Evol.* **2018**, *35*, 1547–1549. [CrossRef]
63. Mueller, U.; Darowski, N.; Fuchs, M.R.; Förster, R.; Hellmig, M.; Paithankar, K.S.; Pühringer, S.; Steffien, M.; Zocher, G.; Weiss, M.S. Facilities for macromolecular crystallography at the Helmholtz-Zentrum Berlin. *J. Synchrotron Radiat.* **2012**, *19*, 442–449. [CrossRef] [PubMed]
64. Kabsch, W. Integration, scaling, space-group assignment and post-refinement. *Acta Cryst. Sect. D Biol. Cryst.* **2010**, *66*, 133–144. [CrossRef] [PubMed]

65. Vagin, A.; Teplyakov, A. Molecular replacement with MOLREP. *Int. Tables Crystallogr.* **2012**, *66*, 364–366. [CrossRef]
66. Winn, M.D.; Ballard, C.C.; Cowtan, K.D.; Dodson, E.J.; Emsley, P.; Evans, P.R.; Keegan, R.M.; Krissinel, E.B.; Leslie, A.G.W.; McCoy, A.; et al. Overview of theCCP4 suite and current developments. *Acta Cryst. Sect. D Biol. Cryst.* **2011**, *67*, 235–242. [CrossRef] [PubMed]
67. Emsley, P.; Lohkamp, B.; Scott, W.G.; Cowtan, K. Features and development ofCoot. *Acta Cryst. Sect. D Biol. Cryst.* **2010**, *66*, 486–501. [CrossRef] [PubMed]
68. Chen, V.B.; Arendall, W.B., 3rd; Headd, J.J.; Keedy, D.A.; Immormino, R.M.; Kapral, G.J.; Murray, L.W.; Richardson, J.S.; Richardson, D.C. MolProbity: All-atom structure validation for macromolecular crystallography. *Acta Cryst. D Biol. Cryst.* **2010**, *66*, 12–21. [CrossRef] [PubMed]

International Journal of
Molecular Sciences

Review

Arboviruses: How Saliva Impacts the Journey from Vector to Host

Christine A. Schneider [1], Eric Calvo [2] and Karin E. Peterson [1,*]

[1] Laboratory of Persistent Viral Diseases, Rocky Mountain Laboratories, National Institute of Allergy and Infectious Diseases, National Institutes of Health, Hamilton, MT 59840, USA; christine.schneiderlewis@nih.gov

[2] Laboratory of Malaria and Vector Research, National Institute of Allergy and Infectious Diseases, National Institutes of Health, Rockville, MD 20852, USA; ecalvo@od.nih.gov

* Correspondence: petersonka@niaid.nih.gov

Abstract: Arthropod-borne viruses, referred to collectively as arboviruses, infect millions of people worldwide each year and have the potential to cause severe disease. They are predominately transmitted to humans through blood-feeding behavior of three main groups of biting arthropods: ticks, mosquitoes, and sandflies. The pathogens harbored by these blood-feeding arthropods (BFA) are transferred to animal hosts through deposition of virus-rich saliva into the skin. Sometimes these infections become systemic and can lead to neuro-invasion and life-threatening viral encephalitis. Factors intrinsic to the arboviral vectors can greatly influence the pathogenicity and virulence of infections, with mounting evidence that BFA saliva and salivary proteins can shift the trajectory of viral infection in the host. This review provides an overview of arbovirus infection and ways in which vectors influence viral pathogenesis. In particular, we focus on how saliva and salivary gland extracts from the three dominant arbovirus vectors impact the trajectory of the cellular immune response to arbovirus infection in the skin.

Keywords: viral infection; skin; immune enhancement; mosquito

Citation: Schneider, C.A.; Calvo, E.; Peterson, K.E. Arboviruses: How Saliva Impacts the Journey from Vector to Host. *Int. J. Mol. Sci.* **2021**, 22, 9173. https://doi.org/10.3390/ijms22179173

Academic Editor: József Tőzsér

Received: 31 July 2021
Accepted: 22 August 2021
Published: 25 August 2021

Publisher's Note: MDPI stays neutral with regard to jurisdictional claims in published maps and institutional affiliations.

1. Introduction

Arbovirus infection involves complex interactions between the virus, the vector, and the host. In extracting a blood meal, the vector induces skin damage at the bite site, initiating a tissue repair response from the host. Simultaneous deposition of an arbovirus into the bite site sets off a battle of viral immune evasion and host immune defense. Blood-feeding arthropods (BFAs) further complicate these interactions because their saliva contains highly bioactive salivary proteins that can block, enhance, or alter each of these interactions. This review will examine these relationships (virus, vector, and host) with a focus on BFA saliva and salivary proteins.

2. Arboviruses

Arthropod-borne viruses (arboviruses) can be found in multiple virus families, although the majority cluster within three: Bunyaviridae, Flaviviridae, and Togaviridae (Figure 1). Arboviruses within the Bunyaviridae family have extensive genus and vector diversity, with viruses from this family transmitted to humans by sandflies, mosquitoes, and ticks [1]. The severity of bunyavirus-induced disease in humans also fluctuates from asymptomatic to severe disease, including seizures, coma, and in rare cases, death. The Flaviviridae family contains one genus of arboviruses, all of which are transmitted to human hosts by mosquitoes or ticks [2]. As a genus, flaviviruses are the most prevalent and widely studied arboviruses, with extensive recent focus placed on Dengue virus (DENV) and Zika Virus (ZIKV). ZIKV is capable of vertical and sexual transmission in addition to being vector-borne, although these are not the dominant transmission mechanisms [3].

Flaviviruses are globally distributed and can cause widespread morbidity and mortality. Togaviridae contains a single genus, alphaviruses, which are exclusively mosquito-borne, with *Aedes* and *Culex* mosquitoes serving as important vectors [4,5]. These alphaviruses have less world-wide distribution and cause fewer cases of human disease overall than flaviviruses [3,6]. However, a recent increase in cases of severe febrile illness has been observed with the alphavirus, Chikungunya virus (CHIKV), as mosquito vector populations have shifted due to changing climates and infection of different mosquito species [7,8]. Additional arboviruses have been identified outside of these families and include both tick- and sandfly-borne viruses (Figure 1). Overwhelmingly, individual arboviruses are transmitted by a single type of vector (i.e., mosquitoes, ticks, or sandflies), with the exception of Vesicular Stomatitis virus (VSV; Rhabdoviridae), which is carried and transmitted readily by both sandflies and mosquitoes [9,10]. It is unclear why most viruses are not carried by multiple vectors simultaneously, but this may be related to host biology or a function of geographical location. Indeed, some viruses are vector-restricted even within an arthropod family. La Crosse virus (LACV; Bunyaviridae), for example, can infect both *Aedes triseriatus* and *Aedes hendersoni* mosquitoes and is passed transovarially to mosquito offspring by both, but is only readily saliva-transmitted to hosts by *A. triseriatus* [11]. Despite being part of diverse virus families, arboviruses share many functional similarities in how they interact with the mammalian host and arthropod vector.

Virus Family	Viral Genus	Virus	Vector Species
	Orthobunyavirus	California serogroup viruses	Mosquito (Aedes sp.)
	Phlebovirus	Rift Valley Fever virus	Mosquito (various)
	Phlebovirus	Toscana virus	Sandfly (Phelbotomus sp.)
Bunyaviridae	Phlebovirus	Phlebotomus fever virus	Sandfly (phelbotomus)
	Phlebovirus	Sandfly Fever Naples virus	Sandfly (phelbotomus)
	Phlebovirus	Sandfly Fever Sicilian virus	Sandfly (phelbotomus)
	Phlebovirus	Heartland virus	Tick (A. americanum)
	Phlebovirus	Severe fever with thrombocytopenia syndrome virus	Tick (H. longicornis)
	Nairovirus	Crimean Hemorrhagic Fever virus	Tick (Hyalomma sp.)
	Flavivirus	Dengue Virus	Mosquito (Aedes sp.)
	Flavivirus	Zika virus	Mosquito (Aedes sp.)
	Flavivirus	Yellow fever virus	Mosquito (Aedes sp.)
	Flavivirus	West Nile Virus	Mosquito (Culex sp.)
	Flavivirus	St. Louis Encephalitis virus	Mosquito (Culex sp.)
Flaviviridae	Flavivirus	Japanese encephalitis virus	Mosquito (Culex sp.)
	Flavivirus	Murray Valley encephalitis virus	Mosquito (Culex sp.)
	Flavivirus	Usutu	Mosquito (various)
	Flavivirus	Omsk Hemorrhagic fever virus	Tick (dermacentor)
	Flavivirus	Kyasanur Forest Diesease virus	Tick (Haemaphysalis sp.)
	Flavivirus	Tick-borne encephalitis virus	Tick (Ixodes and Haemaphysalis sp.)
	Flavivirus	Powassan virus	Tick (Ixodes sp.)
Orthomyxoviridae	Thogotovirus	Bourbon virus	Tick (A. americanum)
Reoviridae	Coltivirus	Colorado tick fever	Tick (dermacentor)
Rhabdoviridae	Vesiculovirus	Vesicular Stomatitis (New Jersey) virus	Sandflies (Lutz. Sp) | Mosquitos (various)
	Vesiculovirus	Chandipura	Sandify (Phlebotomus Sp.)
	Alphavirus	Barmah Forest Virus	Mosquito (Aedes and Culex sp.)
	Alphavirus	Chikungunya virus	Mosquito (Aedes sp.)
Togaviridae	Alphavirus	Venezuelan equine encephalitis virus	Mosquito (Culex sp.)
	Alphavirus	Sindbis virus	Mosquito (Culex sp.)
	Alphavirus	Equine encephalitis virus	Mosquito (Culex sp.)
	Alphavirus	Mayaro virus	Mosquito (Haemagogus sp.)

Figure 1. Virus families and common arthropod vectors of viral disease.

Arbovirus Infection of Vectors

Successful transfer of arboviruses to mammalian hosts begins first with infection of the arthropod vector. After taking a bloodmeal from an infected host, the virus replicates within the arthropod midgut and eventually reaches the salivary glands where it is transferred to the host during feeding. Although most research focus is placed on host pathology during mammalian infection, arboviruses can also impact the vector in both beneficial and harmful ways. Flaviviruses like DENV and WNV (Flaviviridae) reach the mosquito central nervous system (CNS), resulting in altered feeding behaviors that negatively impact successful bloodmeal acquisition [12–15]. Rift Valley Fever virus (RVFV; Bunyaviridae) similarly alters feeding behavior, but also reduces egg output and is ultimately lethal to mosquitoes [16]. Conversely, Sindbis virus (SINV; Togaviridae) is neuroinvasive within the mosquito, resulting in mosquitoes gaining resistance to the insect repellent compound DEET, reducing options available for preventing mosquito bites in humans [17,18]. Not all mosquito arboviruses alter feeding behaviors, however. Feeding disruption is only transiently observed with La Crosse virus (LACV; Bunyaviridae), a mosquito-borne virus in the same family as RVFV [16,19]. The virus' ability to induce pathology in the vector species may be a unique feature to mosquitoes. There are no currently reported benefits or detriments to ticks or sandflies from virus infection, although many of these questions are just beginning to be explored.

3. Arbovirus Infection of the Host

During feeding, BFAs inject protein-rich saliva as an aid in blood extraction, carrying arboviruses that are deposited into host skin. Vector saliva intercepts important coagulation pathways in the host but also can enhance viral dissemination and may increase pathogenesis of virus infections [20,21]. For example, neurological disease is more severe in experimental mouse models of Semliki Forest virus (SFV; Togaviridae) and West Nile virus (WNV; Flaviviridae) infection when salivary gland extract (SGE) is co-injected with the virus [22,23]. Dengue virus (DENV; Flaviviridae) replicates to higher viral titers in the skin in mice injected with *A. aegypti* SGE compared to DENV infection alone [24]. Similarly, Zika virus (ZIKV; Flaviviridae) titers are higher in many peripheral organs and the brain when inoculated into mice alongside isolated *A. aegypti* salivary protein LTRIN [25]. In some cases, salivary enhancement occurs even if virus and saliva are not injected at the same time. For WNV, mosquito saliva enhances viral titers even if injected 12 h after the virus, so long as it is spatially close to the initial virus injection site [26]. Interestingly, there appears to be a vector species-specific enhancement in some cases, with higher viral titers and more rapid death when RVFV (Bunyaviridae) is administered with *A. aegypti* SGE but not *C. pipens* SGE despite both being competent vectors for transmission [27–30]. Notably, salivary enhancement is also observed with tick-borne Powassan virus (POWV; Flaviviridae) and tick-borne encephalitis virus (TBEV; Flaviviridae), revealing that this is not a unique feature reserved solely to mosquitoes [31,32].

In addition to heightened viral pathology and titer, SGE has the potential to increase viral spread to new hosts. Naïve ticks fed on guinea pigs infected with Thogoto virus (THOGV; Orthomyxoviridae) become virus-positive more often when the host guinea-pigs are injected with tick SGE than virus injection alone [33]. Naïve ticks can become infected with THOGV after feeding on a naïve mouse if they co-feed alongside infected ticks—even in the absence of host seroconversion [33]. This spread in the absence of viremia in the host is of particular interest as ticks are frequently found in tightly-spaced clusters when feeding on natural virus reservoir species, suggesting localized transmission between ticks [34]. This is not a vector-specific observation as it also occurs in mouse models of WNV. If naïve mosquitoes feed on mice within 40 mm of an infected mosquito feeding site and no more than 45 min after the infected mosquito, naïve mosquitoes can become infected with virus [35,36].

Many theories have been put forth to explain why saliva and SGE increase local viral titer, including enhancing viral fusion to certain cell types, as has been shown with DENV in vitro [37]. Tick salivary proteins also prevent upregulation of interferon (IFN) β by infected dendritic cells (DCs), which leads to unchecked TBEV replication in vitro [38]. Evidence suggests viral dissemination in mammalian hosts may also be enhanced through immune modulation, positively influencing the mobilization and motility of immune cells from the skin to the lymph node (LN) and promoting viral spread [24,37]. Increased motility of infected immune cells may explain the spread of THOGV between co-feeding ticks despite a lack of detectable host viremia [33].

3.1. Feeding Differences of Vectors

Hematophagous arthropods rely on acquisition of blood meals as a rich source of proteins and heme for producing eggs and, in ticks, for progressing to the next stage of development [39]. In general, biting arthropods obtain blood from hosts by the same general mechanism, making use of sharp mouthparts to penetrate the host skin to reach blood vessels. During this process, protein-rich saliva is injected into the skin, facilitating blood extraction by intercepting the normal processes of hemostasis that prevent blood loss. Mosquitoes, ticks, and sandflies have therefore evolved mechanisms to prevent coagulation, limit vasoconstriction, and suppress pain receptors to facilitate feeding off mammalian hosts. The specific mechanisms used are, in part, tailored by the unique feeding style employed by each BFA.

3.1.1. Mosquitoes

The three largest mosquito genera, *Culex*, *Aedes*, and *Anopheles*, all serve as competent vectors for viral diseases. Mosquitoes have substantial genetic diversity between genera as well as the viruses they carry. *Aedes* mosquitoes are the dominant vector for mosquito-borne diseases, although *Culex* and *Anopheles* also carry pathogens that cause disease in humans [40,41]. Divergence within the mosquito genera has led to strains with substantially different feeding timetables and behaviors; however, all mosquitoes feed using the same physical mechanisms. Two thin serrated cutting projections called maxillae are used to penetrate through the thinner outer epidermis into the dermis layer below [42]. Sandwiched between the maxillae is a structure of two attached tubes, one for extracting blood and one which passes saliva into the host (Figure 2). The labrum resembles a needle with a beveled tip that has sensors for detecting blood vessels and ultimately for extracting blood once a vessel is located [42–44]. Laying on top of the labrum is the hypopharynx through which saliva is continually secreted [43]. The process of locating a vessel sometimes requires multiple penetrations into the skin for successful feeding to begin, but the penetration of such a small microneedle structure often elicits little pain [45]. Once a vessel is located and cannulated, most mosquitoes feed to repletion within 1–2 min [46]. Both male and female mosquitoes feed on plant juices for most of their nutrients, but female mosquitoes feed on mammals and birds when ready to produce eggs. The composition of the sialome (i.e., salivary proteome) changes to match this shift in food source [47]. Proteomic comparisons between sugar and blood-feeding female *Aedes aegypti* mosquitoes revealed increased expression of proteins intercepting host responses to biting, including vasoconstriction, pain sensation, and cytokine production, when females were fed on blood [47]. This represents the intensely dynamic nature of vector saliva, including the ability to tailor the composition to match shifting needs so that essential proteins are only expressed when most needed.

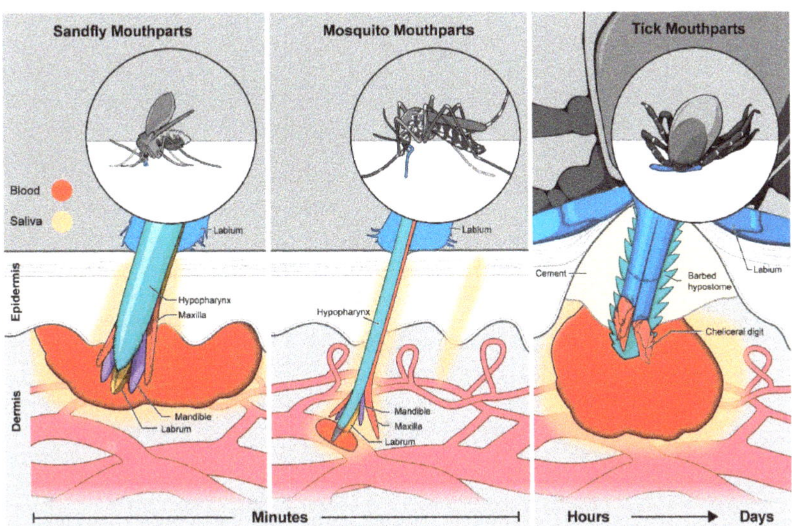

Figure 2. Sandfly, mosquito, and tick feeding methods and mouthparts. Each vector employs a feeding style influenced by the arrangement of their mouthparts. Saliva (shown in yellow) is secreted into the skin by each of the three arthropod vectors.

3.1.2. Ticks

Disease-transmitting ticks are subdivided into two main families based on their body composition—hard-bodied (Ixodidae) and soft-bodied (Argasidae). Members of both families are obligate blood-feeders, although Ixodidae ticks actively search out hosts for extended feedings (days to weeks) while Argasidae reside within host nests, allowing for repeated brief feedings (hours) from developing rodents and birds. As a result, Ixodidae have increased potential to transmit bacterial pathogens to hosts as pathogen transfer increases with longer duration of skin attachment [48]. Long duration feeding is not required for transfer of POWV, which reach hosts within the first 15 min of attachment to mouse skin [49]. Both sexes of ticks ingest blood to progress between developmental stages, after first emerging from eggs through nymph stages to adults [50]. Adult females require a blood meal as a rich source of heme for producing viable eggs, and many adult males require a blood meal to produce viable sperm [39,51]. In at least two tick species, males and females have different salivary protein profiles [52–54], although it has not been investigated if this impacts viral transmission.

Both hard and soft-bodied ticks have large cutting mouth parts called chelicerae to saw through the epidermis and deep into the dermis to access blood [55]. The back and forth sawing motion of the chelicerae forms a wound and results in the formation of a blood pool into which the barbed single-channel hypostome feeding tube is inserted [55] (Figure 2). Histology of mouse skin performed during active tick feeding revealed that the hypostome can even reach the skin fat cells on the border of the hypodermis [56]. Remarkably, ticks are able to attain the same depth of penetration across the nymph and adult stages, despite having vastly different body sizes [57]. However, the wound size and the amount of blood pooling into it is proportional to the size of the tick as well as the tick developmental stage, with adult ticks able to extract more whole blood than smaller nymphs [57]. Although nymphs have a smaller feeding volume, they are still able to acquire viral and bacterial infections from infected hosts so long as the feeding duration is sufficiently long for a given pathogen [48].

Often, a single skin penetration is sufficient to establish feeding for the duration of the attachment to the host—a period of a few hours (soft-bodied ticks) to weeks (hard-bodied ticks), depending on the species [58]. To enhance long-term attachment to the host, some species secrete a thick cement-like substance to surround the entire surface between their body and the host to prevent detachment [59]. This firm attachment facilitates tick engorgement, where female hard-bodied ticks may swell 30–100x their pre-attachment body weight in a feeding session [50]. During pauses in blood extraction, saliva is secreted back into the host, with the longer feeding hard-bodied ticks returning up to 74% of the water extracted from the total blood meal volume [60]. This saliva exchange uses the salivary glands to concentrate the removal of the most essential blood components, while also providing a mechanism for direct host manipulation through injection of salivary proteins that aid in feeding [60]. Tick-derived components in the saliva also help maintain the fluidity and integrity of the blood pool by interrupting clotting and vasoconstriction while also preventing immune cell repair to the wound area and suppressing pain [61]. As with mosquitoes, tick saliva is highly dynamic, but a shift in the sialome content occurs not because the food source changes, but instead by the duration of host attachment. In addition, compared to mosquitoes who feed only briefly, tick saliva is more complex, containing many more secreted proteins of uncharacterized function [62]. This is suggested to result from the need to interfere with numerous extensive processes involved with being long-term feeders, including interrupting tissue repair programs designed to close the wound [62]. Simo et al. have compiled a nice in-depth review of vasoactive and functionally active proteins identified to date in tick saliva [61].

3.1.3. Sandflies

Around 1000 species of sandflies have been identified with the arbovirus-relevant species found within two genera: *Phlebotomus* and *Lutzomyia* [63]. Although more commonly vectors for parasitic diseases like Leishmaniasis, infected female sandflies can transmit phleboviruses during blood feeding, and cases are often found clustered geographically during an outbreak [64]. The sandfly proboscis is structurally and functionally similar to the mosquito, with a two-channeled extraction—secretion mechanism surrounded by sharp maxillae [65]. However, the feeding style employed is much more similar to ticks. Female sandflies use the extensive sharp barbs on their maxillae and mandibles to cut a wound in the skin, creating a blood pool at the very surface of the dermis [65,66]. This wound formation is essential given that the labrum is too short to extend deeply into the dermis. This results in a comparatively shallow blood pool formed primarily in the epidermis, although saliva penetrates down into the dermis, inducing vascular leakage to supply the blood pool (Figure 2) [67]. In contrast to ticks, however, the blood meal is rapidly extracted over a period of minutes in a timescale similar to mosquitoes. This short duration makes maintenance of the blood pool less critical, although efficient interruption of hemostasis is still continuous and essential. Hematophagous female sandflies have saliva with 20x the protein concentration of non-blood feeding males, and work is ongoing to characterize the sialome of different sandfly species [68]. The very small body size of sandflies (<3 mm) reduces the likelihood of host detection during feeding, which is an important quality as the laceration formed by the bite does cause sharp pain in the host which is not inhibited by sandfly saliva [69].

3.2. Secretion of Saliva by Vectors

Despite differences in penetration depth and style of feeding, the injection of enzyme-rich, bioactive saliva into host skin is an essential component of arthropod blood feeding. For all three vectors, saliva injection begins from the moment of skin penetration and continues until the blood meal is finished [21,61,67,70,71]. Importantly, saliva is deposited differentially based on biting style—mosquitoes inject saliva both into the epidermis and dermis during initial skin penetration as well as adjacent to the blood stream once a vessel is successfully cannulated as feeding begins (Figure 2) [70,71]. Sandflies deposit saliva primarily into the epidermis due to their short mouthparts, but they are able to seep deeper into the epidermis and dermis as hyaluronidase enzymes and salivary proteins liquify the surrounding environment (Figure 2) [72–77]. Ticks, by sawing directly into the dermis through the epidermis, secrete saliva through both dermal layers and continue saliva injection for the entire long duration of feeding, resulting in substantial saliva coverage of the bite area (Figure 2) [21,71]. This contrast in saliva injection location has the potential to differentially influence the local immune response, given the carefully arranged cellular immune organization in different skin layers.

4. Vector Transmission Impacts on the Anti-Virus Host Response

4.1. The Skin Immune Environment

The skin serves as the first line of defense against insect-borne viruses, with specialized cells primed to react to both the cellular damage caused by the arthropod bite and to any infectious organism the bite delivers. The skin is divided into three structural layers—the thin epidermis is at the top, bridging the distance between the air interface and blood-vessel rich dermis below [78]. In the absence of infection, the epidermis contains only keratinocytes and Langerhans cells (LCs), both of which perform surveillance functions reminiscent of myeloid cells [78]. The highly vascularized dermis contains a much wider repertoire of immune cells including dermal macrophages, T and B cells, dendritic cells (DCs), and additional LCs [79,80]. The hypodermis is populated almost exclusively with adipose cells, although there are resident immune cells at low levels in the absence of inflammation [81].

During infection or injury, epidermal cells initiate the first reaction to foreign material through detection by pattern-recognition receptors (PRRs) expressed on host cells. These initiate cytokine production and the release of anti-microbial peptides [82,83]. Dermal innate and adaptive immune cells, along with DCs and resident macrophages, amplify inflammation during infection or wounding [80,84,85]. Chemokines recruit neutrophils, monocytes, and other peripheral immune cells to infiltrate into the highly vascularized dermis, followed by migration into the epidermis or hypodermis as necessary [83–85]. These processes work to reduce viral load and facilitate clearance of the infection. Epidermal and dermal LCs perform hybrid functions often ascribed to both macrophages and DCs, including migrating to the LN to present antigens to initiate adaptive immunity [83,86]. As the innate immune phase wanes and gives way to the adaptive response, T and B cells work to target virus-infected cells and produce antibodies specifically developed against the virus [79,80]. These processes all work together to resolve the infection and promote healing of the bite site.

4.2. Arboviruses Hijack Skin Immune Mechanisms

Arboviruses have evolved means of interrupting the development of a normal skin immune response by directly infecting and hijacking skin immune cells. Focusing first on the epidermis, Keratinocytes are permissive to infection with mosquito-borne flaviviruses, including DENV and WNV, both in vitro and in vivo, potentiating them as a replicative niche in the skin [87,88]. In addition, epidermal Langerhans cells are readily infected by mosquito-borne DENV, WNV, and CHIKV as well as tick-borne TBEV [20,89,90]. Because LCs are migratory, moving from the epidermis through the dermis and even to the draining LN, they may not only serve to amplify viral load, but also as a transport mechanism for the virus to the rest of the body. Indeed, infected migratory LCs are thought to contribute directly to the transfer of TBEV from infected ticks to naïve ones feeding on the same naïve host through their migration [32].

Ticks and mosquitoes, by virtue of their longer mouthparts, are capable of also depositing virions below the epidermis and deep into the dermis. Here, virus particles encounter a more diverse repertoire of immune cells, increasing the diversity of cells available for viral invasion. Tick-borne POWV is deposited deeply into the skin by infected ticks, and the virus is detectable, flanking the bite site in the dermis and hypodermis, co-localized with myeloid and T cell markers, suggesting these cells are actively infected [91].

Depending on the cells infected, virus deposited in the dermis is not restricted to remain there. DENV, Crimean—Congo Hemorrhagic Fever virus (CCHFV), and TBEV replicate readily within dermal DCs, which are capable of migration, potentiating their transport to the descending LN [90,92]. As the infection progresses, immune cells become activated and cytokine production recruits additional hematopoietic-origin monocytes from the blood. Monocytes are permissive to TBEV infection and are shown to shuttle virus to peripheral tissues and even into the CNS [93–95]. ZIKV preferentially targets monocytes in both adult and fetal infection, modulating adhesion molecule expression that may assist with viral dissemination [96–99]. Similarly, DENV targets monocytes for infection and may participate in inducing endothelial damage, precipitating vascular permeability [100]. Thus, arboviruses take advantage of the dermal immune mechanisms essential for anti-viral defense to facilitate their access to the rest of the host.

4.3. Salivary Gland Protein Suppression of Skin Host Immune Responses

Intrinsic viral mechanisms can influence immune cell activation and function in the skin, but with vector-borne viruses, this is only one part of the overall infection picture. In addition to alterations in viral load and dissemination, vector saliva and SGE also exhibit immunomodulatory properties that can influence the trajectory and pathogenesis of viral infections [101]. This is particularly true early in infection where the innate immune system is the primary response to viral infection [102]. This is accomplished in three main ways: (1) modulation of cellular activation at the infection site in the skin, leading to (2) alterations of cytokine production, both related to propagation of inflammatory signaling or cellular recruitment, and (3) manipulation of immune cell motility, both to the skin and away to the skin-draining LN. These three processes are intertwined, and disentangling which event occurs first is often challenging. Despite profound evolutionary divergence between ticks, sandflies, and mosquitoes, their saliva retains remarkable functional conservation in how it interacts with the mammalian immune system, discussed below. While a nice body of literature has been generated characterizing SGE with mosquito and tick viruses, there is a relative lack of studies on sandfly saliva during viral infection. Sandflies, while the primary vectors for several phleboviruses, are much more commonly studied for their dissemination of *Leishmania sp.*, an intracellular parasite and cause of leishmaniasis worldwide. Although parasitic infections are largely beyond the scope of this review, there are potentially important parallels to be drawn by discussing a few aspects of sandfly saliva during *Leishmania* infection.

4.3.1. Modulation of Innate Immune Activation

Across the three arthropod families, SGE widely suppress the initial post-bite innate immune response. This is accomplished at multiple levels, by dampening activation of skin-resident macrophages and DCs and by suppressing immune cell recruitment to the skin. Both maturation and activation of DCs by standard immune stimulants, such as poly I:C, CpG, or LPS, in the absence of infection are suppressed by tick SGE [103,104]. At least a portion of this suppressive function is due to salivary Prostaglandin E2, including suppressed cytokine production and reduced ability to stimulate T cells [105]. *Aedes aegypti* saliva also induces diminished DC—T cell stimulation but is the result of murine T-cell apoptosis in the absence of any alteration of DC maturation [106]. Macrophages in culture exposed to sandfly saliva have a blunted response to IFN-γ, producing less nitric oxide (NO), and leading to reduced ability to kill *L. major* parasites [107]. Tick SGE also lowers the capacity for re-activation with antigen when DCs are exposed to saliva during maturation, making them deficient at responding to additional exposures [103]. Finally, macrophages cultured with either WNV or SINV and treated with SGE from *A. aegypti* produced less IFN-β and iNOS compared to untreated infected cells [108], demonstrating that vector salivary components are capable of widespread immune suppression.

4.3.2. Alterations of Cytokine Production

SGE-mediated suppression of myeloid cell activation might help to explain how salivary proteins induce differential cytokine profiles during virus infection. Tick saliva reduces pro-inflammatory cytokine production in mice both if given alongside *Borrelia* spirochetes and if used to prime in advance of infection [109]. Mosquito SGE reduced viral spread in mice infected with Semliki Forest virus (SFV) by suppressing neutrophil-produced CCL2 responsible for recruiting pro-inflammatory monocytes [22]. One potent inhibitor of cytokines in mosquitoes is the salivary protein LTRIN, which acts by blocking of downstream activation of the lymphotoxin B receptor [25]. This inhibition prevents cytokine production downstream of NFκB activation, resulting in reduced neutrophil and macrophage recruitment to the blood during mouse models of ZIKV infection [25]. The reduction in chemokine expression also corresponds with decreased cellular recruitment to mouse skin and LN observed during viral infection in the presence of SGE [25].

Endothelial cells are sensitive to many pathogens and inflammatory environments [110]. Their activation during the early immune response can induce substantial cytokine production essential for facilitating immune cell recruitment and infiltration to sites of inflammation [110,111]. Isolated tick salivary protein Longistatin binds to an endothelial cell receptor used to detect foreign glycans, effectively preventing activation and cytokine produced by endothelial cells in the skin [112]. In contrast, however, sandfly salivary proteins LuloHya and Lundep enhance chemokine production from endothelial cells, contributing to vascular leakage and immune cell recruitment [73].

4.3.3. Prevention of Complement Activation

Activation of the complement cascade contributes to mobilization of many arms of the innate immune response, including increasing cytokine production, enhancing chemotaxis, and destroying viruses and infected cells for opsonization [113]. In arboviral infections, complement can be either protective against virus infection or contribute to viral pathology. Complement activation helps curb the spread of SINV and WNV by assembly of the membrane attack complex (MAC) that initiates lysis of viral particles or virus-infected cells [113–116]. Complement activation also results in enhanced antibody production in WNV and YFV, resulting in viral clearance [113,116–118]. However, in DENV infection, excessive cleavage of early complement proteins contributes to the development of vascular leakage and pathology in both animal models and patient samples [119–122].

BFA saliva contains proteins that can directly antagonize complement activation. At least three separate sandfly proteins have been identified that interfere with deposition of complement factors in both the alternative and classical pathways [123–125]. Similarly, *Anopheles sp.* mosquito saliva contains albicin, a protein directly antagonizing C3 cleavage and preventing full activation of the alternative pathway [53,126]. Perhaps most extensively studied, tick saliva contains nearly a dozen identified proteins that inhibit the classical or alternative complement pathways or block initial recognition by the lectin pathway [127]. Prevention of complement activation is thought to aid in BFA feeding by preventing inflammation and anaphylaxis that would be otherwise induced at the bite site. Whether these factors in BFA saliva enhance arbovirus transmission or augment host pathology is still under investigation and may vary depending on both host and vector.

4.3.4. Inhibition of Immune Cell Recruitment

Saliva and SGE may also be able to directly inhibit macrophage and monocyte motility. Rodriguez et al. noted decreased migration of human monocyte-derived DCs during recruitment by CCL19 when *H. marginatum* tick SGE was added to the cultures [92]. Similarly, mouse peritoneal exudate cells are prevented from migrating in transwell plates when cells are exposed to isolated tick salivary protein Longistatin [112]. RNAi-mediated blockade of Longistatin in ticks just before feeding on mice results in much higher immune cell recruitment to the bite site than in non-treated ticks [112]. Loss of Longistatin in tick saliva also reduces blood pool formation in mouse skin and prevents successful feeding, further emphasizing the importance of proper immune suppression to allow effective vector blood extraction [112]. Isolated tick salivary proteins directly inhibit recruitment of neutrophils during mouse skin infection while also dampening their effective use of reactive oxygen [128,129]. Tick saliva also contains a homologous protein to mammalian macrophage inhibitory factor (MIF), which blocks human macrophages and monocytes from being recruited to an inflammatory stimulus [130–133]. This is similar to reduction in mouse footpad swelling by injection of tick-derived Amregulin [134]. This direct motility inhibition has not been extensively discussed in the literature for mosquito or sandfly saliva but may represent a unique feature of tick feeding.

4.4. Sandfly SGE Enhances Chemotaxis to the Bite Site

In contrast to the above studies showing inhibition of innate immune responses, SGE from sandflies may instead promote activation of certain components of the immune response. For example, SGE from sandflies is chemotactic to mouse monocytes, influencing their motility towards the site of saliva injection through induced local expression of monocyte chemokine CCL2 [135,136]. This was suggested to increase the potential spread of the human parasite *Leishmania*, as the parasite preferentially replicates within macrophages, which can differentiate from infiltrating monocytes [135,136]. Indeed, injection of salivary gland lysates along with parasites results in mouse skin lesions that are larger and have higher parasite burdens than parasite injection alone [137]. Saliva exposure also induces more monocytes and NK cells to remain at the bite site up to a week post infection [138]. This represents an example of saliva-induced enhancement of host immunity not observed in the other two vector groups but suggests that sandfly evolution may be influenced in part by the needs of the predominant infection they carry. Additionally, mice are protected from infection by *L. major* by being bitten by uninfected *P. duboscqi* sandflies [138]. This protection is due to saliva exposure creating a primed environment for NK and T cell recruitment, including substantially greater activation and cytokine production [138]. Much of this function appears to be due to the protein Maxadilan, shown to be an effective vaccination strategy against *L. major* in vivo [139]. These results provide an interesting contrast to tick and mosquito phenotypes, but it remains to be established if these hold during viral infection in sandflies. Currently, there are no studies of sandfly sialomes during phlebovirus infection, but that work would provide an interesting potential for comparison with other viral vectors.

4.5. Modulations of Adaptive Immunity

Arthropod saliva can also affect the adaptive immune response. Tick saliva reduces T cell proliferation when co-incubated with DCs, thus preventing normal T cell mediated immune responses from developing [109]. This was mirrored by treatment with SGE from *A. aegypti*, with reduced mouse T cell proliferation and cytokine production observed with Concanavalin A activation [140]. This effect is mosquito-species-specific and not observed with *Culex* sp. SGE, which is toxic to murine T cells [140]. T cell recruitment is suppressed when mosquitoes are allowed to feed on mice near sites of WNV injection in the skin, reducing the effectiveness of skin-local adaptive immunity [108]. Finally, SGE from sandflies, mosquitoes, and ticks skews the development of naïve T cells from a normal antiviral Th1 response to Th2 both in vitro and in vivo, reducing their effectiveness in targeting pathogens [141,142]. Mosquito SGE alters cytokine production (including IL-4, IL-10, and IFN-g) from re-stimulated T cells ex vivo [143]. In vivo, both tick and sandfly salivary gland lysate injected into the skin upregulate IL-4 mRNA in the skin-draining LN [141]. Furthermore, tick cystatin proteins Iristatin and Sialostatin suppress normal cytokine production from T cells, resulting in reduced cellular recruitment and inflammation in mice in vivo [144,145]. Iristatin also prevents T cell activation during antigen recognition and blocks efficient T cell recruitment in mice [144]. This T cell phenomenon is not only observed in rodent models, as human cells also experience suppressed Th1 cytokines in favor of Th2 when exposed to sandfly SGE and isolated sandfly salivary protein Maxadilan [146].

The interruption of normal adaptive immunity by SGE and BFA saliva has the potential to widely influence arbovirus pathogenesis and resolution of infection.

5. Comparing Vectors—Overall Lessons to Be Learned and Looking Ahead

The three vectors discussed in this review employ different feeding styles to extract blood, harbor different viruses across different virus families, and are only distantly related. Despite this, there is profound functional conservation in anti-immune mechanisms employed by these vectors through their saliva. All three vectors withdraw blood after penetrating the skin but do so with varying degrees of tissue damage and from different places in the skin. Because the skin contains functionally distinct layers, this placement difference means biologically active salivary molecules and arboviruses will contact different resident and immune cells post deposition. The upper epidermis is full of innate immune sentinels poised to detect damage and pathogens while the lower dermis contains semi-resident populations of both innate and adaptive immune cells. Ticks, mosquitoes, and sandflies all inhibit T cell proliferation and production of key inflammatory cytokines using their saliva and salivary protein repertoire. It is essential for ticks to interrupt immune cell recruitment and tissue repair which would prevent their long-term feeding; however, short-duration feeding mosquitos and sandflies also alter cell chemotaxis and activation with their saliva. These immune-modulation functions result in similar viral outcomes—an overall enhancement of arboviral disease. In some cases, as with mosquito saliva and dengue virus, this occurs because of a direct interaction between salivary proteins and virus particles, resulting in improved cell entry and replication [24]. In other instances, it is indirect, resulting instead because of poor immune control at the site of virus deposition because saliva or salivary proteins have prevented immune activation. This work is still ongoing, and indeed, research on phleboviruses and sandflies is just beginning to be expanded. There is an expansion in research of the skin stages of arbovirus infection that precede the development of symptoms, which will further inform mechanisms that may be intercepted by salivary proteins.

Remarkably, the sialome among BFA shows remarkable functional similarity between species, despite a relative lack of structural or sequence homology. While it would be ill-advised to assume that all salivary enhancement functions found in one vector are mirrored in another, it may be worth looking more broadly at different mechanisms found amongst hematophagous arthropods to provide insights into viral enhancement shared between arboviral diseases.

Author Contributions: Conceptualization, C.A.S., E.C. and K.E.P.; writing—original draft preparation, C.A.S.; writing—review and editing, C.A.S., E.C. and K.E.P. All authors have read and agreed to the published version of the manuscript.

Funding: This work was supported by the Division of Intramural Research, National Institute of Allergy and Infectious Disease.

Institutional Review Board Statement: Not applicable.

Acknowledgments: The authors wish to thank Alyssa Evans, Paul Policastro, and Stephen Johnson for critical reading of the manuscript and Ryan Kissinger for figure illustrations.

Conflicts of Interest: The authors declare no conflict of interest.

References

1. Overturf, G.D. World arboviruses: The Bunyaviridae. *Pediatr. Infect. Dis. J.* **2009**, *28*, 1014–1015. [CrossRef]
2. Blitvich, B.J.; Firth, A.E. A review of flaviviruses that have no known arthropod vector. *Viruses* **2017**, *9*, 154. [CrossRef]
3. Pierson, T.C.; Diamond, M.S. The continued threat of emerging flaviviruses. *Nat. Microbiol.* **2020**, *5*, 796–812. [CrossRef] [PubMed]
4. Bhatt, S.; Gething, P.W.; Brady, O.J.; Messina, J.P.; Farlow, A.W.; Moyes, C.L.; Drake, J.M.; Brownstein, J.S.; Hoen, A.G.; Sankoh, O.; et al. The global distribution and burden of dengue. *Nature* **2013**, *496*, 504–507. [CrossRef] [PubMed]
5. Schmaljohn, A.L.; McClain, D.; McClain, D. Alphaviruses (*Togaviridae*) and Flaviviruses (*Flaviviridae*). In *Medical Microbiology*; The University of Texas Medical Branch at Galveston: Galveston, TX, USA, 1996.

6. Calisher, C.H.; Monath, T.P. Togaviridae and flaviviridae: The alphavirases and flaviviruses. In *Laboratory Diagnosis of Infectious Diseases Principles and Practice: VOLUME II Viral, Rickettsial, and Chlamydial Diseases*; Lennette, E.H., Halonen, P., Murphy, F.A., Balows, A., Hausler, W.J., Eds.; Springer New York: New York, NY, USA, 1988; pp. 414–434. [CrossRef]

7. Liu, Y.; Wu, B.; Paessler, S.; Walker, D.H.; Tesh, R.B.; Yu, X.J. The pathogenesis of severe fever with thrombocytopenia syndrome virus infection in alpha/beta interferon knockout mice: Insights into the pathologic mechanisms of a new viral hemorrhagic fever. *J. Virol.* **2014**, *88*, 1781–1786. [CrossRef] [PubMed]

8. Tsetsarkin, K.A.; Chen, R.; Yun, R.; Rossi, S.L.; Plante, K.S.; Guerbois, M.; Forrester, N.; Perng, G.C.; Sreekumar, E.; Leal, G.; et al. Multi-peaked adaptive landscape for chikungunya virus evolution predicts continued fitness optimization in Aedes albopictus mosquitoes. *Nat. Commun.* **2014**, *5*, 4084. [CrossRef] [PubMed]

9. Hajnicka, V.; Fuchsberger, N.; Slovak, M.; Kocakova, P.; Labuda, M.; Nuttall, P.A. Tick salivary gland extracts promote virus growth in vitro. *Parasitology* **1998**, *116 Pt 6*, 533–538. [CrossRef]

10. Limesand, K.H.; Higgs, S.; Pearson, L.D.; Beaty, B.J. Effect of mosquito salivary gland treatment on vesicular stomatitis New Jersey virus replication and interferon alpha/beta expression in vitro. *J. Med. Entomol.* **2003**, *40*, 199–205. [CrossRef]

11. Paulson, S.L.; Grimstad, P.R. Replication and dissemination of La Crosse virus in the competent vector Aedes triseriatus and the incompetent vector Aedes hendersoni and evidence for transovarial transmission by Aedes hendersoni (Diptera: Culicidae). *J. Med. Entomol.* **1989**, *26*, 602–609. [CrossRef]

12. Girard, Y.A.; Popov, V.; Wen, J.; Han, V.; Higgs, S. Ultrastructural study of West Nile virus pathogenesis in Culex pipiens quinquefasciatus (Diptera: Culicidae). *J. Med. Entomol.* **2005**, *42*, 429–444. [CrossRef]

13. Platt, K.B.; Linthicum, K.J.; Myint, K.S.; Innis, B.L.; Lerdthusnee, K.; Vaughn, D.W. Impact of dengue virus infection on feeding behavior of Aedes aegypti. *Am. J. Trop. Med. Hyg.* **1997**, *57*, 119–125. [CrossRef] [PubMed]

14. Salazar, M.I.; Richardson, J.H.; Sanchez-Vargas, I.; Olson, K.E.; Beaty, B.J. Dengue virus type 2: Replication and tropisms in orally infected Aedes aegypti mosquitoes. *BMC Microbiol.* **2007**, *7*, 9. [CrossRef] [PubMed]

15. Vogels, C.B.F.; Fros, J.J.; Pijlman, G.P.; van Loon, J.J.A.; Gort, G.; Koenraadt, C.J.M. Virus interferes with host-seeking behaviour of mosquito. *J. Exp. Biol.* **2017**, *220*, 3598–3603. [CrossRef] [PubMed]

16. Turell, M.J.; Gargan, T.P., 2nd; Bailey, C.L. Culex pipiens (Diptera: Culicidae) morbidity and mortality associated with Rift Valley fever virus infection. *J. Med. Entomol.* **1985**, *22*, 332–337. [CrossRef] [PubMed]

17. Bowers, D.F.; Abell, B.A.; Brown, D.T. Replication and tissue tropism of the alphavirus Sindbis in the mosquito Aedes albopictus. *Virology* **1995**, *212*, 1–12. [CrossRef] [PubMed]

18. Qualls, W.A.; Day, J.F.; Xue, R.D.; Bowers, D.F. Sindbis virus infection alters blood feeding responses and DEET repellency in Aedes aegypti (Diptera: Culicidae). *J. Med. Entomol.* **2012**, *49*, 418–423. [CrossRef] [PubMed]

19. Faran, M.E.; Turell, M.J.; Romoser, W.S.; Routier, R.G.; Gibbs, P.H.; Cannon, T.L.; Bailey, C.L. Reduced survival of adult Culex pipiens infected with Rift Valley fever virus. *Am. J. Trop. Med. Hyg.* **1987**, *37*, 403–409. [CrossRef]

20. Briant, L.; Despres, P.; Choumet, V.; Misse, D. Role of skin immune cells on the host susceptibility to mosquito-borne viruses. *Virology* **2014**, *464–465*, 26–32. [CrossRef]

21. Nuttall, P.A. Tick saliva and its role in pathogen transmission. *Wien. Klin. Wochenschr.* **2019**, 1–12. [CrossRef]

22. Pingen, M.; Bryden, S.R.; Pondeville, E.; Schnettler, E.; Kohl, A.; Merits, A.; Fazakerley, J.K.; Graham, G.J.; McKimmie, C.S. Host Inflammatory Response to Mosquito Bites Enhances the Severity of Arbovirus Infection. *Immunity* **2016**, *44*, 1455–1469. [CrossRef]

23. Styer, L.M.; Lim, P.Y.; Louie, K.L.; Albright, R.G.; Kramer, L.D.; Bernard, K.A. Mosquito saliva causes enhancement of West Nile virus infection in mice. *J. Virol.* **2011**, *85*, 1517–1527. [CrossRef] [PubMed]

24. Schmid, M.A.; Glasner, D.R.; Shah, S.; Michlmayr, D.; Kramer, L.D.; Harris, E. Mosquito Saliva Increases Endothelial Permeability in the Skin, Immune Cell Migration, and Dengue Pathogenesis during Antibody-Dependent Enhancement. *PLoS Pathog.* **2016**, *12*, e1005676. [CrossRef] [PubMed]

25. Jin, L.; Guo, X.; Shen, C.; Hao, X.; Sun, P.; Li, P.; Xu, T.; Hu, C.; Rose, O.; Zhou, H.; et al. Salivary factor LTRIN from Aedes aegypti facilitates the transmission of Zika virus by interfering with the lymphotoxin-beta receptor. *Nat. Immunol.* **2018**, *19*, 342–353. [CrossRef]

26. Moser, L.A.; Lim, P.Y.; Styer, L.M.; Kramer, L.D.; Bernard, K.A. Parameters of Mosquito-Enhanced West Nile Virus Infection. *J. Virol.* **2016**, *90*, 292–299. [CrossRef]

27. Faran, M.E.; Romoser, W.S.; Routier, R.G.; Bailey, C.L. The distribution of Rift Valley fever virus in the mosquito Culex pipiens as revealed by viral titration of dissected organs and tissues. *Am. J. Trop. Med. Hyg.* **1988**, *39*, 206–213. [CrossRef]

28. Hoch, A.L.; Gargan, T.P., 2nd; Bailey, C.L. Mechanical transmission of Rift Valley fever virus by hematophagous Diptera. *Am. J. Trop. Med. Hyg.* **1985**, *34*, 188–193. [CrossRef]

29. Le Coupanec, A.; Babin, D.; Fiette, L.; Jouvion, G.; Ave, P.; Misse, D.; Bouloy, M.; Choumet, V. Aedes mosquito saliva modulates Rift Valley fever virus pathogenicity. *PLoS Negl. Trop. Dis.* **2013**, *7*, e2237. [CrossRef]

30. Turell, M.J.; Gargan, T.P., 2nd; Bailey, C.L. Replication and dissemination of Rift Valley fever virus in Culex pipiens. *Am. J. Trop. Med. Hyg.* **1984**, *33*, 176–181. [CrossRef]

31. Hermance, M.E.; Thangamani, S. Tick Saliva Enhances Powassan Virus Transmission to the Host, Influencing Its Dissemination and the Course of Disease. *J. Virol.* **2015**, *89*, 7852–7860. [CrossRef]

32. Labuda, M.; Austyn, J.M.; Zuffova, E.; Kozuch, O.; Fuchsberger, N.; Lysy, J.; Nuttall, P.A. Importance of localized skin infection in tick-borne encephalitis virus transmission. *Virology* **1996**, *219*, 357–366. [CrossRef]

33. Jones, L.D.; Hodgson, E.; Nuttall, P.A. Enhancement of virus transmission by tick salivary glands. *J. Gen. Virol.* **1989**, 70, 1895–1898. [CrossRef]

34. Randolph, S.E.; Gern, L.; Nuttall, P.A. Co-feeding ticks: Epidemiological significance for tick-borne pathogen transmission. *Parasitol. Today* **1996**, 12, 472–479. [CrossRef]

35. McGee, C.E.; Schneider, B.S.; Girard, Y.A.; Vanlandingham, D.L.; Higgs, S. Nonviremic transmission of West Nile virus: Evaluation of the effects of space, time, and mosquito species. *Am. J. Trop. Med. Hyg.* **2007**, 76, 424–430. [CrossRef]

36. Reisen, W.K.; Fang, Y.; Martinez, V. Is nonviremic transmission of West Nile virus by Culex mosquitoes (Diptera: Culicidae) nonviremic? *J. Med. Entomol.* **2007**, 44, 299–302. [CrossRef]

37. Conway, M.J.; Watson, A.M.; Colpitts, T.M.; Dragovic, S.M.; Li, Z.; Wang, P.; Feitosa, F.; Shepherd, D.T.; Ryman, K.D.; Klimstra, W.B.; et al. Mosquito saliva serine protease enhances dissemination of dengue virus into the mammalian host. *J. Virol.* **2014**, 88, 164–175. [CrossRef]

38. Lieskovska, J.; Palenikova, J.; Sirmarova, J.; Elsterova, J.; Kotsyfakis, M.; Campos Chagas, A.; Calvo, E.; Ruzek, D.; Kopecky, J. Tick salivary cystatin sialostatin L2 suppresses IFN responses in mouse dendritic cells. *Parasite Immunol.* **2015**, 37, 70–78. [CrossRef]

39. Perner, J.; Sobotka, R.; Sima, R.; Konvickova, J.; Sojka, D.; Oliveira, P.L.; Hajdusek, O.; Kopacek, P. Acquisition of exogenous haem is essential for tick reproduction. *eLife* **2016**, 5. [CrossRef]

40. Shaw, W.R.; Catteruccia, F. Vector biology meets disease control: Using basic research to fight vector-borne diseases. *Nat. Microbiol.* **2019**, 4, 20–34. [CrossRef] [PubMed]

41. Suesdek, L. Microevolution of medically important mosquitoes—A review. *Acta Trop.* **2019**, 191, 162–171. [CrossRef]

42. Wahid, I.; Sunahara, T.; Mogi, M. Maxillae and mandibles of male mosquitoes and female autogenous mosquitoes (Diptera: Culicidae). *J. Med. Entomol.* **2003**, 40, 150–158. [CrossRef] [PubMed]

43. Choo, Y.M.; Buss, G.K.; Tan, K.; Leal, W.S. Multitasking roles of mosquito labrum in oviposition and blood feeding. *Front. Physiol.* **2015**, 6, 306. [CrossRef]

44. Lee, R. Structure and function of the fascicular stylets, and the la bral and cibarial sense organs of male and female Aedes aeg ypti (L.) (Diptera, Culicidae). *Quaest. Entomol.* **1974**, 10, 187–215.

45. Kong, X.Q.; Wu, C.W. Mosquito proboscis: An elegant biomicroelectromechanical system. *Phys. Rev. E Stat. Nonlin Soft Matter Phys.* **2010**, 82, 011910. [CrossRef] [PubMed]

46. Chadee, D.D.; Beier, J.C.; Mohammed, R.T. Fast and slow blood-feeding durations of Aedes aegypti mosquitoes in Trinidad. *J. Vector Ecol.* **2002**, 27, 172–177.

47. Wasinpiyamongkol, L.; Patramool, S.; Luplertlop, N.; Surasombatpattana, P.; Doucoure, S.; Mouchet, F.; Seveno, M.; Remoue, F.; Demettre, E.; Brizard, J.P.; et al. Blood-feeding and immunogenic Aedes aegypti saliva proteins. *Proteomics* **2010**, 10, 1906–1916. [CrossRef]

48. Eisen, L. Pathogen transmission in relation to duration of attachment by Ixodes scapularis ticks. *Ticks Tick Borne Dis.* **2018**, 9, 535–542. [CrossRef] [PubMed]

49. Ebel, G.D.; Kramer, L.D. Short report: Duration of tick attachment required for transmission of powassan virus by deer ticks. *Am. J. Trop. Med. Hyg.* **2004**, 71, 268–271. [CrossRef] [PubMed]

50. Kaufman, W.R. Tick-host interaction: A synthesis of current concepts. *Parasitol. Today* **1989**, 5, 47–56. [CrossRef]

51. Sonenshine, D.E.; Roe, R.M. *Biology of Ticks*, 2nd ed.; Oxford University Press: New York, NY, USA, 2014.

52. Pienaar, R.; de Klerk, D.G.; de Castro, M.H.; Featherston, J.; Mans, B.J. De novo assembled salivary gland transcriptome and expression pattern analyses for Rhipicephalus evertsi evertsi Neuman, 1897 male and female ticks. *Sci. Rep.* **2021**, 11, 1642. [CrossRef]

53. Strayer, E.C.; Lu, S.; Ribeiro, J.; Andersen, J.F. Salivary complement inhibitors from mosquitoes: Structure and mechanism of action. *J. Biol. Chem.* **2021**, 296, 100083. [CrossRef] [PubMed]

54. Tan, A.W.; Francischetti, I.M.; Slovak, M.; Kini, R.M.; Ribeiro, J.M. Sexual differences in the sialomes of the zebra tick, Rhipicephalus pulchellus. *J. Proteom.* **2015**, 117, 120–144. [CrossRef] [PubMed]

55. Gregson, J.D. Morphology and functioning of the mouthparts of Dermacentor andersoni Stiles. *Acta Trop.* **1960**, 17, 48–79. [PubMed]

56. Hermance, M.E.; Thangamani, S. Tick(-)Virus(-)Host Interactions at the Cutaneous Interface: The Nidus of Flavivirus Transmission. *Viruses* **2018**, 10, 362. [CrossRef]

57. Tatchell, R.J.; Moorhouse, D.E. The feeding processes of the cattle tick Boophilus microplus (Canestrini). II. The sequence of host-tissue changes. *Parasitology* **1968**, 58, 441–459. [CrossRef]

58. Anderson, J.F. The natural history of ticks. *Med. Clin. N. Am.* **2002**, 86, 205–218. [CrossRef]

59. Suppan, J.; Engel, B.; Marchetti-Deschmann, M.; Nurnberger, S. Tick attachment cement—Reviewing the mysteries of a biological skin plug system. *Biol. Rev.* **2018**, 93, 1056–1076. [CrossRef]

60. Meredith, J.; Kaufman, W.R. A proposed site of fluid secretion in the salivary gland of the ixodid tick Dermacentor andersoni. *Parasitology* **1973**, 67, 205–217. [CrossRef]

61. Simo, L.; Kazimirova, M.; Richardson, J.; Bonnet, S.I. The Essential Role of Tick Salivary Glands and Saliva in Tick Feeding and Pathogen Transmission. *Front. Cell. Infect. Microbiol.* **2017**, 7, 281. [CrossRef] [PubMed]

62. Ribeiro, J.M.; Alarcon-Chaidez, F.; Francischetti, I.M.; Mans, B.J.; Mather, T.N.; Valenzuela, J.G.; Wikel, S.K. An annotated catalog of salivary gland transcripts from Ixodes scapularis ticks. *Insect Biochem. Mol. Biol.* **2006**, 36, 111–129. [CrossRef]

63.	Raksakoon, C.; Potiwat, R. Current Arboviral Threats and Their Potential Vectors in Thailand. *Pathogens* **2021**, *10*, 80. [CrossRef]

64.	Downs, J.W.; Flood, D.T.; Orr, N.H.; Constantineau, J.A.; Caviness, J.W.B. Sandfly fever in Afghanistan-a sometimes overlooked disease of military importance: A case series and review of the literature. In *U.S. Army Medical Department Journal*; U.S. Army Medical Department Center & School: Fort Sam Houston, TX, USA, 2017; pp. 60–66.

65.	Lane, R.P. Sandflies (*Phlebotominae*). In *Medical Insects and Arachnids*; Lane, R.P., Crosskey, R.W., Eds.; Springer: Dordrecht, The Netherlands, 1993; pp. 78–119. [CrossRef]

66.	Jobling, B. *Anatomical Drawings of Biting Flies*; British Museum (Natural History): London, UK, 1987; p. 119.

67.	Lestinova, T.; Rohousova, I.; Sima, M.; de Oliveira, C.I.; Volf, P. Insights into the sand fly saliva: Blood-feeding and immune interactions between sand flies, hosts, and Leishmania. *PLoS Negl. Trop. Dis.* **2017**, *11*, e0005600. [CrossRef]

68.	Volf, P.; Tesarova, P.; Nohynkova, E.N. Salivary proteins and glycoproteins in phlebotomine sandflies of various species, sex and age. *Med. Vet. Entomol.* **2000**, *14*, 251–256. [CrossRef]

69.	Abdeladhim, M.; Kamhawi, S.; Valenzuela, J.G. What's behind a sand fly bite? The profound effect of sand fly saliva on host hemostasis, inflammation and immunity. *Infect. Genet. Evol.* **2014**, *28*, 691–703. [CrossRef] [PubMed]

70.	Jones, J.C. The feeding behavior of mosquitoes. *Sci. Am.* **1978**, *238*, 138–150. [CrossRef]

71.	Krenn, H.W.; Aspock, H. Form, function and evolution of the mouthparts of blood-feeding Arthropoda. *Arthropod Struct. Dev.* **2012**, *41*, 101–118. [CrossRef] [PubMed]

72.	Cerna, P.; Mikes, L.; Volf, P. Salivary gland hyaluronidase in various species of phlebotomine sand flies (Diptera: Psychodidae). *Insect Biochem. Mol. Biol.* **2002**, *32*, 1691–1697. [CrossRef]

73.	Martin-Martin, I.; Chagas, A.C.; Guimaraes-Costa, A.B.; Amo, L.; Oliveira, F.; Moore, I.N.; DeSouza-Vieira, T.S.; Sanchez, E.E.; Suntravat, M.; Valenzuela, J.G.; et al. Immunity to LuloHya and Lundep, the salivary spreading factors from Lutzomyia longipalpis, protects against Leishmania major infection. *PLoS Pathog.* **2018**, *14*, e1007006. [CrossRef]

74.	Rohousova, I.; Subrahmanyam, S.; Volfova, V.; Mu, J.; Volf, P.; Valenzuela, J.G.; Jochim, R.C. Salivary gland transcriptomes and proteomes of Phlebotomus tobbi and Phlebotomus sergenti, vectors of leishmaniasis. *PLoS Negl. Trop. Dis.* **2012**, *6*, e1660. [CrossRef] [PubMed]

75.	Rohousova, I.; Volfova, V.; Nova, S.; Volf, P. Individual variability of salivary gland proteins in three Phlebotomus species. *Acta Trop.* **2012**, *122*, 80–86. [CrossRef]

76.	Vlkova, M.; Sima, M.; Rohousova, I.; Kostalova, T.; Sumova, P.; Volfova, V.; Jaske, E.L.; Barbian, K.D.; Gebre-Michael, T.; Hailu, A.; et al. Comparative analysis of salivary gland transcriptomes of Phlebotomus orientalis sand flies from endemic and non-endemic foci of visceral leishmaniasis. *PLoS Negl. Trop. Dis.* **2014**, *8*, e2709. [CrossRef]

77.	Volfova, V.; Hostomska, J.; Cerny, M.; Votypka, J.; Volf, P. Hyaluronidase of bloodsucking insects and its enhancing effect on leishmania infection in mice. *PLoS Negl. Trop. Dis.* **2008**, *2*, e294. [CrossRef]

78.	Clayton, K.; Vallejo, A.F.; Davies, J.; Sirvent, S.; Polak, M.E. Langerhans Cells-Programmed by the Epidermis. *Front. Immunol.* **2017**, *8*, 1676. [CrossRef] [PubMed]

79.	Geherin, S.A.; Gomez, D.; Glabman, R.A.; Ruthel, G.; Hamann, A.; Debes, G.F. IL-10+ Innate-like B Cells Are Part of the Skin Immune System and Require alpha4beta1 Integrin To Migrate between the Peritoneum and Inflamed Skin. *J. Immunol.* **2016**, *196*, 2514–2525. [CrossRef] [PubMed]

80.	Ho, A.W.; Kupper, T.S. T cells and the skin: From protective immunity to inflammatory skin disorders. *Nat. Rev. Immunol.* **2019**, *19*, 490–502. [CrossRef] [PubMed]

81.	Cildir, G.; Akincilar, S.C.; Tergaonkar, V. Chronic adipose tissue inflammation: All immune cells on the stage. *Trends Mol. Med.* **2013**, *19*, 487–500. [CrossRef]

82.	Coates, M.; Blanchard, S.; MacLeod, A.S. Innate antimicrobial immunity in the skin: A protective barrier against bacteria, viruses, and fungi. *PLoS Pathog.* **2018**, *14*, e1007353. [CrossRef]

83.	Deckers, J.; Hammad, H.; Hoste, E. Langerhans Cells: Sensing the Environment in Health and Disease. *Front. Immunol.* **2018**, *9*, 93. [CrossRef]

84.	Nestle, F.O.; Di Meglio, P.; Qin, J.Z.; Nickoloff, B.J. Skin immune sentinels in health and disease. *Nat. Rev. Immunol.* **2009**, *9*, 679–691. [CrossRef]

85.	Tay, S.S.; Roediger, B.; Tong, P.L.; Tikoo, S.; Weninger, W. The Skin-Resident Immune Network. *Curr. Dermatol. Rep.* **2014**, *3*, 13–22. [CrossRef]

86.	Romani, N.; Koide, S.; Crowley, M.; Witmer-Pack, M.; Livingstone, A.M.; Fathman, C.G.; Inaba, K.; Steinman, R.M. Presentation of exogenous protein antigens by dendritic cells to T cell clones. Intact protein is presented best by immature, epidermal Langerhans cells. *J. Exp. Med.* **1989**, *169*, 1169–1178. [CrossRef]

87.	Limon-Flores, A.Y.; Perez-Tapia, M.; Estrada-Garcia, I.; Vaughan, G.; Escobar-Gutierrez, A.; Calderon-Amador, J.; Herrera-Rodriguez, S.E.; Brizuela-Garcia, A.; Heras-Chavarria, M.; Flores-Langarica, A.; et al. Dengue virus inoculation to human skin explants: An effective approach to assess in situ the early infection and the effects on cutaneous dendritic cells. *Int. J. Exp. Pathol.* **2005**, *86*, 323–334. [CrossRef]

88.	Surasombatpattana, P.; Hamel, R.; Patramool, S.; Luplertlop, N.; Thomas, F.; Despres, P.; Briant, L.; Yssel, H.; Misse, D. Dengue virus replication in infected human keratinocytes leads to activation of antiviral innate immune responses. *Infect. Genet. Evol.* **2011**, *11*, 1664–1673. [CrossRef]

89. Byrne, S.N.; Halliday, G.M.; Johnston, L.J.; King, N.J. Interleukin-1beta but not tumor necrosis factor is involved in West Nile virus-induced Langerhans cell migration from the skin in C57BL/6 mice. *J. Investig. Dermatol.* **2001**, *117*, 702–709. [CrossRef]

90. Wu, S.J.; Grouard-Vogel, G.; Sun, W.; Mascola, J.R.; Brachtel, E.; Putvatana, R.; Louder, M.K.; Filgueira, L.; Marovich, M.A.; Wong, H.K.; et al. Human skin Langerhans cells are targets of dengue virus infection. *Nat. Med.* **2000**, *6*, 816–820. [CrossRef]

91. Hermance, M.E.; Thangamani, S. Utilization of RNA in situ Hybridization to Understand the Cellular Localization of Powassan Virus RNA at the Tick-Virus-Host Interface. *Front. Cell. Infect. Microbiol.* **2020**, *10*, 172. [CrossRef]

92. Rodriguez, S.E.; McAuley, A.J.; Gargili, A.; Bente, D.A. Interactions of Human Dermal Dendritic Cells and Langerhans Cells Treated with Hyalomma Tick Saliva with Crimean-Congo Hemorrhagic Fever Virus. *Viruses* **2018**, *10*, 381. [CrossRef] [PubMed]

93. Dorrbecker, B.; Dobler, G.; Spiegel, M.; Hufert, F.T. Tick-borne encephalitis virus and the immune response of the mammalian host. *Travel Med. Infect. Dis.* **2010**, *8*, 213–222. [CrossRef]

94. Krylova, N.V.; Smolina, T.P.; Leonova, G.N. Molecular Mechanisms of Interaction Between Human Immune Cells and Far Eastern Tick-Borne Encephalitis Virus Strains. *Viral Immunol.* **2015**, *28*, 272–281. [CrossRef] [PubMed]

95. Wei, J.; Kang, X.; Li, Y.; Wu, X.; Zhang, Y.; Yang, Y. Pathogenicity of tick-borne encephalitis virus to monocytes. *Wei Sheng Wu Xue Bao* **2013**, *53*, 1221–1225.

96. Ayala-Nunez, N.V.; Follain, G.; Delalande, F.; Hirschler, A.; Partiot, E.; Hale, G.L.; Bollweg, B.C.; Roels, J.; Chazal, M.; Bakoa, F.; et al. Zika virus enhances monocyte adhesion and transmigration favoring viral dissemination to neural cells. *Nat. Commun.* **2019**, *10*, 4430. [CrossRef] [PubMed]

97. Foo, S.S.; Chen, W.; Chan, Y.; Bowman, J.W.; Chang, L.C.; Choi, Y.; Yoo, J.S.; Ge, J.; Cheng, G.; Bonnin, A.; et al. Asian Zika virus strains target CD14(+) blood monocytes and induce M2-skewed immunosuppression during pregnancy. *Nat. Microbiol.* **2017**, *2*, 1558–1570. [CrossRef]

98. Lum, F.M.; Lee, D.; Chua, T.K.; Tan, J.J.L.; Lee, C.Y.P.; Liu, X.; Fang, Y.; Lee, B.; Yee, W.X.; Rickett, N.Y.; et al. Zika Virus Infection Preferentially Counterbalances Human Peripheral Monocyte and/or NK Cell Activity. *mSphere* **2018**, *3*. [CrossRef]

99. Michlmayr, D.; Andrade, P.; Gonzalez, K.; Balmaseda, A.; Harris, E. CD14(+)CD16(+) monocytes are the main target of Zika virus infection in peripheral blood mononuclear cells in a paediatric study in Nicaragua. *Nat. Microbiol.* **2017**, *2*, 1462–1470. [CrossRef]

100. Castillo, J.A.; Naranjo, J.S.; Rojas, M.; Castano, D.; Velilla, P.A. Role of Monocytes in the Pathogenesis of Dengue. *Arch. Immunol. Ther. Exp.* **2019**, *67*, 27–40. [CrossRef]

101. Chapter 4—Arthropod Saliva and Its Role in Pathogen Transmission: Insect Saliva. In *Skin and Arthropod Vectors*; Boulanger, N. (Ed.) Academic Press: New York, NY, USA, 2018; pp. 83–119. [CrossRef]

102. Koyama, S.; Ishii, K.J.; Coban, C.; Akira, S. Innate immune response to viral infection. *Cytokine* **2008**, *43*, 336–341. [CrossRef]

103. Cavassani, K.A.; Aliberti, J.C.; Dias, A.R.; Silva, J.S.; Ferreira, B.R. Tick saliva inhibits differentiation, maturation and function of murine bone-marrow-derived dendritic cells. *Immunology* **2005**, *114*, 235–245. [CrossRef] [PubMed]

104. Skallova, A.; Iezzi, G.; Ampenberger, F.; Kopf, M.; Kopecky, J. Tick saliva inhibits dendritic cell migration, maturation, and function while promoting development of Th2 responses. *J. Immunol.* **2008**, *180*, 6186–6192. [CrossRef]

105. Sa-Nunes, A.; Bafica, A.; Lucas, D.A.; Conrads, T.P.; Veenstra, T.D.; Andersen, J.F.; Mather, T.N.; Ribeiro, J.M.; Francischetti, I.M. Prostaglandin E2 is a major inhibitor of dendritic cell maturation and function in Ixodes scapularis saliva. *J. Immunol.* **2007**, *179*, 1497–1505. [CrossRef] [PubMed]

106. Bizzarro, B.; Barros, M.S.; Maciel, C.; Gueroni, D.I.; Lino, C.N.; Campopiano, J.; Kotsyfakis, M.; Amarante-Mendes, G.P.; Calvo, E.; Capurro, M.L.; et al. Effects of Aedes aegypti salivary components on dendritic cell and lymphocyte biology. *Parasites Vectors* **2013**, *6*, 329. [CrossRef] [PubMed]

107. Hall, L.R.; Titus, R.G. Sand fly vector saliva selectively modulates macrophage functions that inhibit killing of Leishmania major and nitric oxide production. *J. Immunol.* **1995**, *155*, 3501–3506. [PubMed]

108. Schneider, B.S.; Soong, L.; Coffey, L.L.; Stevenson, H.L.; McGee, C.E.; Higgs, S. Aedes aegypti saliva alters leukocyte recruitment and cytokine signaling by antigen-presenting cells during West Nile virus infection. *PLoS ONE* **2010**, *5*, e11704. [CrossRef]

109. Slamova, M.; Skallova, A.; Palenikova, J.; Kopecky, J. Effect of tick saliva on immune interactions between Borrelia afzelii and murine dendritic cells. *Parasite Immunol.* **2011**, *33*, 654–660. [CrossRef]

110. Konradt, C.; Hunter, C.A. Pathogen interactions with endothelial cells and the induction of innate and adaptive immunity. *Eur. J. Immunol.* **2018**, *48*, 1607–1620. [CrossRef]

111. Mai, J.; Virtue, A.; Shen, J.; Wang, H.; Yang, X.F. An evolving new paradigm: Endothelial cells-conditional innate immune cells. *J. Hematol. Oncol.* **2013**, *6*, 61. [CrossRef] [PubMed]

112. Anisuzzaman, M.; Tsuji, N. Longistatin in tick-saliva targets RAGE. *Oncotarget* **2015**, *6*, 35133–35134. [CrossRef] [PubMed]

113. Morrison, T.E.; Heise, M.T. The host complement system and arbovirus pathogenesis. *Curr. Drug Targets* **2008**, *9*, 165–172. [CrossRef] [PubMed]

114. Hirsch, R.L.; Winkelstein, J.A.; Griffin, D.E. The role of complement in viral infections. III. Activation of the classical and alternative complement pathways by Sindbis virus. *J. Immunol.* **1980**, *124*, 2507–2510.

115. Mehlhop, E.; Diamond, M.S. Protective immune responses against West Nile virus are primed by distinct complement activation pathways. *J. Exp. Med.* **2006**, *203*, 1371–1381. [CrossRef]

116. Mehlhop, E.; Whitby, K.; Oliphant, T.; Marri, A.; Engle, M.; Diamond, M.S. Complement activation is required for induction of a protective antibody response against West Nile virus infection. *J. Virol.* **2005**, *79*, 7466–7477. [CrossRef]

117. Schlesinger, J.J.; Brandriss, M.W.; Walsh, E.E. Protection against 17D yellow fever encephalitis in mice by passive transfer of monoclonal antibodies to the nonstructural glycoprotein gp48 and by active immunization with gp48. *J. Immunol.* **1985**, *135*, 2805–2809.
118. Schlesinger, J.J.; Foltzer, M.; Chapman, S. The Fc portion of antibody to yellow fever virus NS1 is a determinant of protection against YF encephalitis in mice. *Virology* **1993**, *192*, 132–141. [CrossRef]
119. Nascimento, E.J.; Hottz, E.D.; Garcia-Bates, T.M.; Bozza, F.; Marques, E.T., Jr.; Barratt-Boyes, S.M. Emerging concepts in dengue pathogenesis: Interplay between plasmablasts, platelets, and complement in triggering vasculopathy. *Crit. Rev. Immunol.* **2014**, *34*, 227–240. [CrossRef]
120. Avirutnan, P.; Punyadee, N.; Noisakran, S.; Komoltri, C.; Thiemmeca, S.; Auethavornanan, K.; Jairungsri, A.; Kanlaya, R.; Tangthawornchaikul, N.; Puttikhunt, C.; et al. Vascular leakage in severe dengue virus infections: A potential role for the nonstructural viral protein NS1 and complement. *J. Infect. Dis.* **2006**, *193*, 1078–1088. [CrossRef]
121. Avirutnan, P.; Malasit, P.; Seliger, B.; Bhakdi, S.; Husmann, M. Dengue virus infection of human endothelial cells leads to chemokine production, complement activation, and apoptosis. *J. Immunol.* **1998**, *161*, 6338–6346.
122. Beatty, P.R.; Puerta-Guardo, H.; Killingbeck, S.S.; Glasner, D.R.; Hopkins, K.; Harris, E. Dengue virus NS1 triggers endothelial permeability and vascular leak that is prevented by NS1 vaccination. *Sci. Transl. Med.* **2015**, *7*, 304ra141. [CrossRef]
123. Mendes-Sousa, A.F.; do Vale, V.F.; Silva, N.C.S.; Guimaraes-Costa, A.B.; Pereira, M.H.; Sant'Anna, M.R.V.; Oliveira, F.; Kamhawi, S.; Ribeiro, J.M.C.; Andersen, J.F.; et al. The Sand Fly Salivary Protein Lufaxin Inhibits the Early Steps of the Alternative Pathway of Complement by Direct Binding to the Proconvertase C3b-B. *Front. Immunol.* **2017**, *8*, 1065. [CrossRef]
124. Ferreira, V.P.; Fazito Vale, V.; Pangburn, M.K.; Abdeladhim, M.; Mendes-Sousa, A.F.; Coutinho-Abreu, I.V.; Rasouli, M.; Brandt, E.A.; Meneses, C.; Lima, K.F.; et al. SALO, a novel classical pathway complement inhibitor from saliva of the sand fly Lutzomyia longipalpis. *Sci. Rep.* **2016**, *6*, 19300. [CrossRef]
125. Cavalcante, R.R.; Pereira, M.H.; Gontijo, N.F. Anti-complement activity in the saliva of phlebotomine sand flies and other haematophagous insects. *Parasitology* **2003**, *127*, 87–93. [CrossRef] [PubMed]
126. Mendes-Sousa, A.F.; Queiroz, D.C.; Vale, V.F.; Ribeiro, J.M.; Valenzuela, J.G.; Gontijo, N.F.; Andersen, J.F. An Inhibitor of the Alternative Pathway of Complement in Saliva of New World Anopheline Mosquitoes. *J. Immunol.* **2016**, *197*, 599–610. [CrossRef] [PubMed]
127. Aounallah, H.; Bensaoud, C.; M'Ghirbi, Y.; Faria, F.; Chmelar, J.I.; Kotsyfakis, M. Tick Salivary Compounds for Targeted Immunomodulatory Therapy. *Front. Immunol.* **2020**, *11*, 583845. [CrossRef] [PubMed]
128. Guo, X.; Booth, C.J.; Paley, M.A.; Wang, X.; DePonte, K.; Fikrig, E.; Narasimhan, S.; Montgomery, R.R. Inhibition of neutrophil function by two tick salivary proteins. *Infect. Immun.* **2009**, *77*, 2320–2329. [CrossRef]
129. Hidano, A.; Konnai, S.; Yamada, S.; Githaka, N.; Isezaki, M.; Higuchi, H.; Nagahata, H.; Ito, T.; Takano, A.; Ando, S.; et al. Suppressive effects of neutrophil by Salp16-like salivary gland proteins from Ixodes persulcatus Schulze tick. *Insect Mol. Biol.* **2014**, *23*, 466–474. [CrossRef] [PubMed]
130. Bowen, C.J.; Jaworski, D.C.; Wasala, N.B.; Coons, L.B. Macrophage migration inhibitory factor expression and protein localization in Amblyomma americanum (Ixodidae). *Exp. Appl. Acarol.* **2010**, *50*, 343–352. [CrossRef]
131. Jaworski, D.C.; Jasinskas, A.; Metz, C.N.; Bucala, R.; Barbour, A.G. Identification and characterization of a homologue of the pro-inflammatory cytokine Macrophage Migration Inhibitory Factor in the tick, Amblyomma americanum. *Insect Mol. Biol.* **2001**, *10*, 323–331. [CrossRef]
132. Umemiya, R.; Hatta, T.; Liao, M.; Tanaka, M.; Zhou, J.; Inoue, N.; Fujisaki, K. Haemaphysalis longicornis: Molecular characterization of a homologue of the macrophage migration inhibitory factor from the partially fed ticks. *Exp. Parasitol.* **2007**, *115*, 135–142. [CrossRef] [PubMed]
133. Wasala, N.B.; Bowen, C.J.; Jaworski, D.C. Expression and regulation of macrophage migration inhibitory factor (MIF) in feeding American dog ticks, Dermacentor variabilis. *Exp. Appl. Acarol.* **2012**, *57*, 179–187. [CrossRef]
134. Tian, Y.; Chen, W.; Mo, G.; Chen, R.; Fang, M.; Yedid, G.; Yan, X. An Immunosuppressant Peptide from the Hard Tick Amblyomma variegatum. *Toxins* **2016**, *8*, 133. [CrossRef]
135. Anjili, C.O.; Mbati, P.A.; Mwangi, R.W.; Githure, J.I.; Olobo, J.O.; Robert, L.L.; Koech, D.K. The chemotactic effect of Phlebotomus duboscqi (Diptera: Psychodidae) salivary gland lysates to murine monocytes. *Acta Trop.* **1995**, *60*, 97–100. [CrossRef]
136. Teixeira, C.R.; Teixeira, M.J.; Gomes, R.B.; Santos, C.S.; Andrade, B.B.; Raffaele-Netto, I.; Silva, J.S.; Guglielmotti, A.; Miranda, J.C.; Barral, A.; et al. Saliva from Lutzomyia longipalpis induces CC chemokine ligand 2/monocyte chemoattractant protein-1 expression and macrophage recruitment. *J. Immunol.* **2005**, *175*, 8346–8353. [CrossRef]
137. Mbow, M.L.; Bleyenberg, J.A.; Hall, L.R.; Titus, R.G. Phlebotomus papatasi sand fly salivary gland lysate down-regulates a Th1, but up-regulates a Th2, response in mice infected with Leishmania major. *J. Immunol.* **1998**, *161*, 5571–5577. [PubMed]
138. Teixeira, C.; Gomes, R.; Oliveira, F.; Meneses, C.; Gilmore, D.C.; Elnaiem, D.E.; Valenzuela, J.G.; Kamhawi, S. Characterization of the early inflammatory infiltrate at the feeding site of infected sand flies in mice protected from vector-transmitted Leishmania major by exposure to uninfected bites. *PLoS Negl. Trop. Dis.* **2014**, *8*, e2781. [CrossRef]
139. Wheat, W.H.; Arthun, E.N.; Spencer, J.S.; Regan, D.P.; Titus, R.G.; Dow, S.W. Immunization against full-length protein and peptides from the Lutzomyia longipalpis sand fly salivary component maxadilan protects against Leishmania major infection in a murine model. *Vaccine* **2017**, *35*, 6611–6619. [CrossRef]

140. Wanasen, N.; Nussenzveig, R.H.; Champagne, D.E.; Soong, L.; Higgs, S. Differential modulation of murine host immune response by salivary gland extracts from the mosquitoes Aedes aegypti and Culex quinquefasciatus. *Med. Vet. Entomol.* **2004**, *18*, 191–199. [CrossRef]
141. Mejri, N.; Franscini, N.; Rutti, B.; Brossard, M. Th2 polarization of the immune response of BALB/c mice to Ixodes ricinus instars, importance of several antigens in activation of specific Th2 subpopulations. *Parasite Immunol.* **2001**, *23*, 61–69. [CrossRef]
142. Mejri, N.; Brossard, M. Splenic dendritic cells pulsed with Ixodes ricinus tick saliva prime naive CD4+T to induce Th2 cell differentiation in vitro and in vivo. *Int. Immunol.* **2007**, *19*, 535–543. [CrossRef]
143. Zeidner, N.S.; Higgs, S.; Happ, C.M.; Beaty, B.J.; Miller, B.R. Mosquito feeding modulates Th1 and Th2 cytokines in flavivirus susceptible mice: An effect mimicked by injection of sialokinins, but not demonstrated in flavivirus resistant mice. *Parasite Immunol.* **1999**, *21*, 35–44. [CrossRef] [PubMed]
144. Kotal, J.; Stergiou, N.; Busa, M.; Chlastakova, A.; Berankova, Z.; Rezacova, P.; Langhansova, H.; Schwarz, A.; Calvo, E.; Kopecky, J.; et al. The structure and function of Iristatin, a novel immunosuppressive tick salivary cystatin. *Cell. Mol. Life Sci.* **2019**, *76*, 2003–2013. [CrossRef] [PubMed]
145. Kotsyfakis, M.; Sa-Nunes, A.; Francischetti, I.M.; Mather, T.N.; Andersen, J.F.; Ribeiro, J.M. Antiinflammatory and immunosuppressive activity of sialostatin L, a salivary cystatin from the tick Ixodes scapularis. *J. Biol. Chem.* **2006**, *281*, 26298–26307. [CrossRef] [PubMed]
146. Rogers, K.A.; Titus, R.G. Immunomodulatory effects of Maxadilan and Phlebotomus papatasi sand fly salivary gland lysates on human primary in vitro immune responses. *Parasite Immunol.* **2003**, *25*, 127–134. [CrossRef]

International Journal of
Molecular Sciences

Article

The First Molecular Detection of *Aedes albopictus* in Sudan Associates with Increased Outbreaks of Chikungunya and Dengue

Ayman Ahmed [1,2,3,4,*], Mustafa Abubakr [5], Hamza Sami [6], Isam Mahdi [6], Nouh S. Mohamed [4] and Jakob Zinsstag [2,3]

1   Institute of Endemic Diseases, University of Khartoum, Khartoum 11111, Sudan
2   Swiss Tropical and Public Health Institute (Swiss TPH), CH-4123 Allschwil, Switzerland
3   Faculty of Science, University of Basel, Petersplatz 1, CH-4001 Basel, Switzerland
4   Molecular Biology Unit, Sirius Training and Research Centre, Khartoum 11111, Sudan
5   Directorate of Environmental Health, Federal Ministry of Health, Khartoum 11111, Sudan
6   Directorate of the Integrated Vector Management (IVM), Federal Ministry of Health, Khartoum 11111, Sudan
*   Correspondence: ayman.ame.ahmed@gmail.com; Tel.: +249-123997091

Citation: Ahmed, A.; Abubakr, M.; Sami, H.; Mahdi, I.; Mohamed, N.S.; Zinsstag, J. The First Molecular Detection of *Aedes albopictus* in Sudan Associates with Increased Outbreaks of Chikungunya and Dengue. *Int. J. Mol. Sci.* **2022**, *23*, 11802. https://doi.org/10.3390/ijms231911802

Academic Editor: Michail Kotsyfakis

Received: 9 September 2022
Accepted: 29 September 2022
Published: 5 October 2022

Abstract: As part of our surveys of the invasive malaria vector *Anopheles stephensi* in four Sudanese states, including North and South Kordofan, Sennar, and White Nile, we collected 166 larvae. Our morphological identification confirmed that 30% of the collected mosquito samples were *Anopheles* species, namely *An. gambiae s.l.* and *An. stephensi*, while the 117 *Aedes* specimens were *Ae. luteocephalus* (39%), *Ae. aegypti* (32%), *Ae. vexans* (9%), *Ae. vittatus* (9%), *Ae. africanus* (6%), *Ae. metalicus* (3%), and *Ae. albopictus* (3%). Considering the serious threat of *Ae. albopictus* emergence for the public health in the area and our limited resources, we prioritized *Ae. albopictus* samples for further genomic analysis. We extracted the DNA from the three specimens and subsequently sequenced the cytochrome oxidase 1 (CO1) gene and confirmed their identity as *Aedes albopictus* and their potential origin by phylogenetic and haplotype analyses. *Aedes albopictus*, originating from Southeast Asia, is an invasive key vector of chikungunya and dengue. This is the first report and molecular characterization of *Ae. albopictus* from Sudan. Our sequences cluster with populations from the Central African Republic and La Réunion. Worryingly, this finding associates with a major increase in chikungunya and dengue outbreaks in rural areas of the study region and might be linked to the mosquito's spread across the region. The emergence of *Ae. albopictus* in Sudan is of serious public health concern and urges for the improvement of the vector surveillance and control system through the implementation of an integrated molecular xenosurveillance. The threat of major arboviral diseases in the region underlines the need for the institutionalization of the One Health strategy for the prevention and control of future pandemics.

Keywords: invasive diseases vectors; *Aedes aegypti*; *Aedes vexans*; *Aedes vittatus*; *Aedes africanus*; *Aedes metalicus*; *Aedes luteocephalus*; *Anopheles stephensi*; arboviruses; haplotype analysis; phylogenetic analysis; One Health; Sudan

1. Introduction

The global disease burden and distribution of arthropod-borne viruses (arboviruses) is rapidly growing. This growth is driven primarily by the spread of the two key invasive disease vectors, *Aedes aegypti* and *Ae. albopictus*, and by the spread of new and re-emerging viruses through international travel [1]. Arboviruses are infecting a wide range of hosts, including humans and other vertebrates, and arthropod disease vectors [2]. Some of the arboviruses are posing a serious global health threat, such as, chikungunya (CHIKV), Crimean–Congo hemorrhagic fever (CCHF), dengue (DENV), yellow fever (YFV), and Zika (ZIKV) virus infections [3,4]. Other arboviruses, such as Rift Valley fever (RVF), African swine fever, bluetongue, as well as Marburg (MBGV) and Schmallenberg viruses, are having devastating economic impacts because of their high morbidity and mortality among domestic animals, including cattle, sheep, and goats [5–7]. While some arboviruses, such as the Shuni virus, Wesselsbron virus, and West Nile virus (WNV), severely affect domestic and wildlife animals [8,9].

Aedes albopictus is a more recent species to Africa that only emerged in the area during the recent three decades [1,10]. *Aedes albopictus* is deemed to be the most invasive insect species and it is estimated that it will be present in 197 countries by 2080 [1]. Interestingly, the presence, spread, and abundance of these invasive disease vectors are mostly associated with socioeconomically disadvantaged communities [11,12]. In Europe, where *Ae. albopictus* is already widely spread, surveillance and control programmes were designed and implemented to slow down its spread and minimise the public health risk of diseases transmission [13–16]. Unfortunately, such a programme does not yet exist in low and middle-income countries due to resources limitations and the fact that health systems are heavily burdened by malaria, particularly in Africa [17]. Therefore, invasive vectors of diseases are spreading undetected until they are established locally and disease outbreaks occur [18–20].

During recent years, the distribution and burden of arboviruses in Sudan increased remarkably, with several annual outbreaks reported from across the country [21]. The emerging and re-emerging arboviral diseases in Sudan include chikungunya, CCHF, dengue, RVF, and WNV [4,5,17,22–26]. However, the surveillance system of arboviral disease vectors in the country is outdated, with only one comprehensive survey of the species vector composition and distribution of *Aedes* mosquitoes dating back to 1955 [27].

As part of our surveys of the invasive malaria vector *Anopheles stephensi*, we also identified, for the first time, the presence of *Ae. albopictus* in Sudan. Additionally, we confirmed the presence of other *Aedes* arbovirus vectors.

2. Results

We collected 166 larvae from human-made water containers in four Sudanese states, including North and South Kordofan, Sennar, and White Nile between 3 October and 29 December 2021. Thirty percent were *Anopheles* and 70% were *Aedes* mosquitoes. We morphologically identified the *Anopheles* species as *An. gambiae* s.l. and *An. stephensi*, while the 117 *Aedes* species were *Ae. luteocephalus* (39%), *Ae. aegypti* (32%), *Ae. vexans* (9%), *Ae. vittatus* (9%), *Ae. africanus* (6%), *Ae. metalicus* (3%), and—more importantly—3% were identified as *Ae. albopictus*. Figure 1 shows their geographical distribution per the states.

Considering its public health significance and our limited resources, following the morphological identification of emerged adult mosquitoes, we prioritised specimens that morphologically identified as *Ae. albopictus* for further sequence analysis. We sequenced an about 450 bp long PCR amplicon of the mitochondrial cytochrome oxidase 1 (CO1) gene and searched for similar sequences in the GenBank database at NCBI (https://blast.ncbi.nlm.nih.gov/nucleotide, accessed on 13 April 2022) using BLAST [28]. This confirmed the identity of the three morphologically identified *Ae. albopictus* samples, and we deposited the sequences in NCBI (accession numbers: ON248551-ON248553).

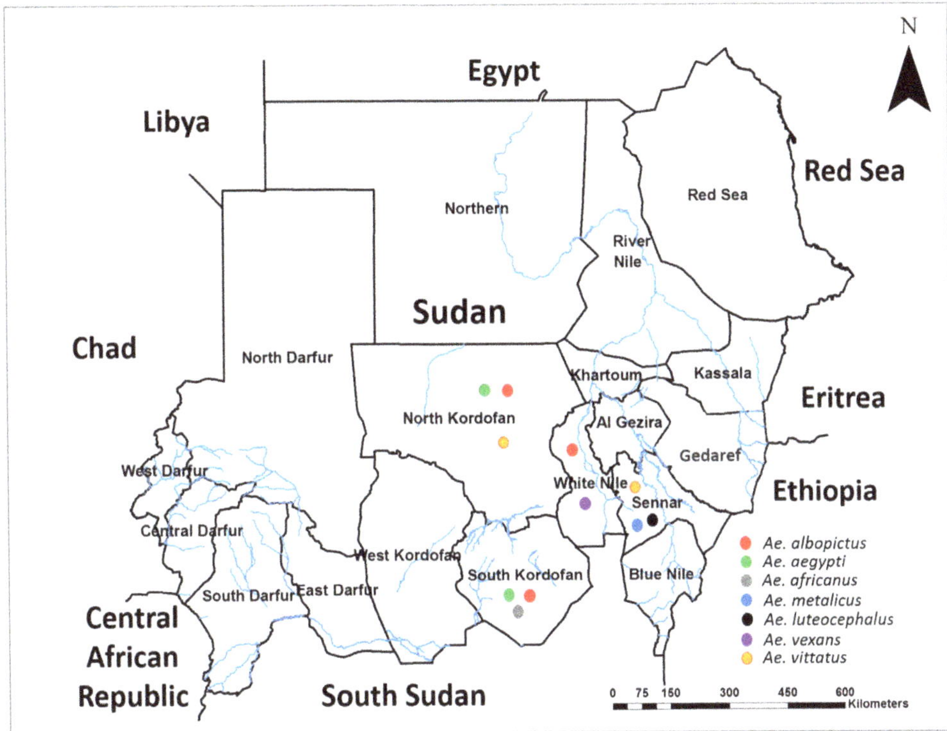

Figure 1. Map of Sudan showing the geographical distribution of identified *Aedes* mosquitoes in the study area.

2.1. Phylogenetic Analysis

We constructed a phylogenetic tree and found that our three sequences are mainly clustered with sequences originating from the Central African Republic and La Réunion (Figure 2).

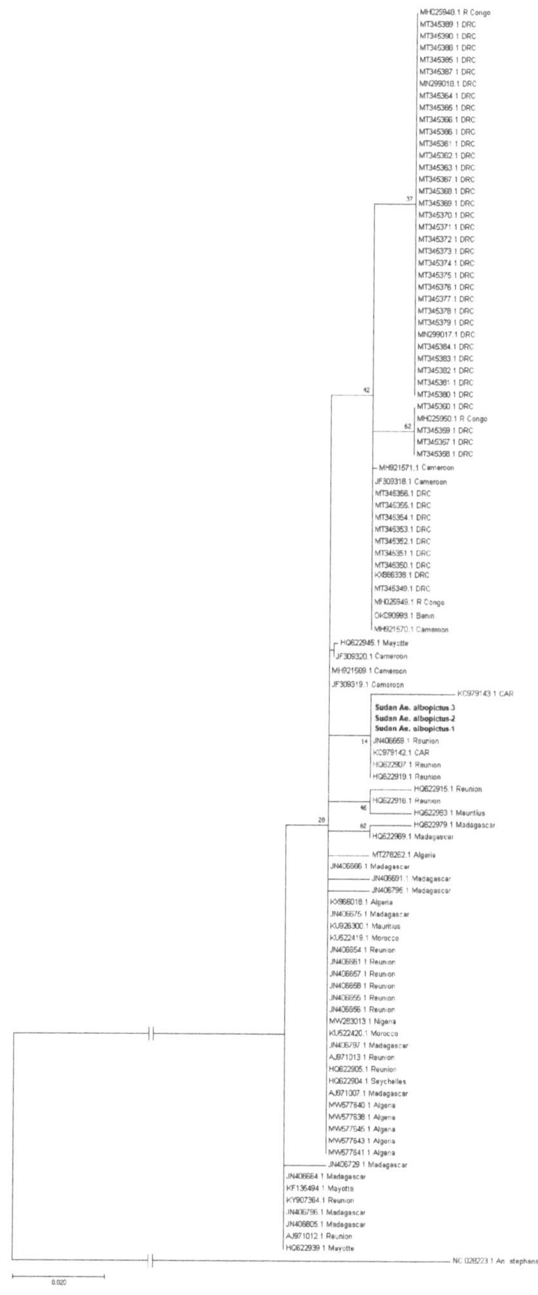

Figure 2. Maximum likelihood phylogenetic tree showing the relation between the Sudanese *Ae. albopictus* sequences and 103 African reference sequences. The Sudan *Ae. albopictus* sequences are highlighted in bold. The reference sequences along with their accession numbers and origin of the isolate are shown for each. DRC: Demoratic Republic of Congo, CAR: Central African Republic, R Congo: Republic of Congo. *An. stephensi* (Accession No. NC_028223.1) was used as an outgroup taxon. The branch to the outgroup was shortened by 0.020 substitutions per site. The bootstrap consensus tree inferred from 1000 replicates.

2.2. Haplotype Analysis and Global Network

The results of the haplotype numbers and haplotype diversity show the presence of 20 haplotypes among *Ae. albopictus* in Africa with a haplotype diversity of 0.83 ± 0.02 (mean $\pm$ sd), indicating a low diversity between the haplotypes. Tajima's D is -2.64, while Fu and Li's D statistic is -8.66 and Fu and Li's F statistic is -7.35. All tests were statistically significant, with the *p* value < 0.05. The distribution of haplotypes according to each country showed that H02 was reported from seven regions, followed by H03 in four regions (Figure 3).

	H01	H02	H03	H04	H05	H06	H07	H08	H09	H10	H11	H12	H13	H14	H15	H16	H17	H18	H19	H20
Democratic Republic of Congo	+	-	+	-	-	+	-	-	-	-	-	-	-	-	-	-	-	-	-	-
Republic of the Congo	+	-	+	-	-	+	-	-	-	-	-	-	-	-	-	-	-	-	-	-
Madagascar	-	+	-	-	+	-	+	+	+	+	-	+	-	-	-	-	-	-	-	-
Reunion	-	+	-	+	+	-	-	-	-	-	-	-	+	+	-	-	-	-	-	-
Algeria	-	+	-	-	-	-	-	-	-	-	-	-	-	-	-	-	+	-	-	-
Sudan	-	-	-	+	-	-	-	-	-	-	-	-	-	-	-	-	-	-	-	-
Central African Republic	-	-	-	+	-	-	-	-	-	-	-	-	-	-	-	+	-	-	-	-
Morocco	-	+	-	-	-	-	-	-	-	-	-	-	-	-	-	-	-	-	-	-
Cameroon	+	-	+	-	-	+	-	-	-	-	-	-	-	-	-	-	-	+	+	+
Nigeria	-	+	-	-	-	-	-	-	-	-	-	-	-	-	-	-	-	-	-	-
Benin	-	-	+	-	-	-	-	-	-	-	-	-	-	-	-	-	-	-	-	-
Mayotte	-	-	-	-	+	-	-	-	-	-	+	-	-	-	-	-	-	-	-	-
Seychelle	-	+	-	-	-	-	-	-	-	-	-	-	-	-	-	-	-	-	-	-
Mauritius	-	+	-	-	-	-	-	-	-	-	-	-	-	-	+	-	-	-	-	-

Figure 3. *Aedes albopictus* haplotype distribution among African countries. Countries or regions where a haplotype is present are indicated by (+), whereas, the absence of the haplotype is indicated by (−).

According to the haplotype diversity (Hd), Hd was high in Algeria and low in other counties, such as Madagascar, Mayotte, and La Réunion. Additionally, several counties presented very low haplotype diversity, such as Cameroon, the Central African Republic, Mauritius, and the Republic of the Congo. Although Sudan is presented with three sequences, all three sequences are one haplotype. Evolutionary analysis using Tajima D and Fu's statistics suggests different scenarios of a population bottleneck event and population expansion in the Democratic Republic of the Congo. The evolutionary analysis was not applicable for Sudan mainly because of the limited sample size; the three sequences constituted one haplotype (Table 1). However, analyzing the haplotypes of *Ae. albopictus* in other African countries indicates that *Ae. albopictus* populations are having a wide diversity and evolutionary dynamic.

Table 1. Diversity and neutrality indices for *Ae. albopictus* populations in Africa calculated from the nucleotide datasets of the sequences available in the GenBank database at NCBI.

Populations *	N	S	H	Hd ± VarHd	Pi	Tajima's D	Fu Li's D	Fu Li's F
MadagasCAR	12	6	7	0.864 ± 0.00618	0.00321	-1.0217	-1.1084	-1.2312
Algeria	7	1	2	0.286 ± 0.03856	0.00063	-1.0062	-1.0488	-1.1015
CAR	2	2	2	1.0 ± 0.25	0.00442	n.d.	n.d.	n.d.
DRC	45	2	3	0.457 ± 0.00562	0.00129	0.5144	0.7583	0.7967
Mauritius	2	2	2	1.0 ± 0.25	0.00442	n.d.	n.d.	n.d.
Mayotte	3	2	2	0.667 ± 0.09877	0.00294	n.d.	n.d.	n.d.
Cameroon	9	5	6	0.917 ± 0.00526	0.00343	-0.6542	-0.5973	-0.6796
Morocco	2	0	1	0.0 ± 0.00	0.0	n.d.	n.d.	n.d.
R Congo	3	2	3	1.0 ± 0.07407	0.00294	n.d.	n.d.	n.d.
La Réunion	14	3	4	0.648 ± 0.1353	0.0017	-0.5651	0.0168	-0.1524
Sudan	3	0	1	0.0 ± 0.00	0.0	n.d.	n.d.	n.d.
Nigeria	1	0	n.a.	0.0 ± 0.00	0.0	n.d.	n.d.	n.d.
Seychelles	1	0	n.a.	0.0 ± 0.00	0.0	n.d.	n.d.	n.d.
Benin	1	0	n.a.	0.0 ± 0.00	0.0	n.d.	n.d.	n.d.

* CAR: Central African Republic, DRC: Democratic Republic of the Congo, R Congo: Republic of the Congo. N: number of sequences, S: number of segregating sites, H: number of haplotypes, Hd ± VarHd: haplotype diversity ± variance of haplotype diversity, Pi: nucleotide diversity per site. n.a.: not applicable, and n.d.: not determined.

2.3. Population Diversity and Evolutionary Characteristics

Further investigation of the population diversity showed that the calculated Fst values confirm the wide genetic variation among the African *Ae. albopictus* populations (Table 2).

Table 2. Pairwise fixation index (Fst test values) between the African populations of *Ae. albopictus* calculated from the nucleotide datasets available in the GenBank.

	DRC	Madagascar	Mauritius	Mayotte	Morocco	R Congo	Sudan	Cameroon	Reunion	CAR
DRC	-	-	-	-	-	-	-	-	-	-
Madagascar	0.613	-	-	-	-	-	-	-	-	-
Mauritius	0.538	0.058	-	-	-	-	-	-	-	-
Mayotte	0.657	**0.004**	0.167	-	-	-	-	-	-	-
Morocco	0.837	0.127	**0.000**	0.333	-	-	-	-	-	-
R Congo	**0.028**	0.442	0.375	0.500	0.600	-	-	-	-	-
Sudan	0.787	0.603	0.500	0.667	1.000	0.667	-	-	-	-
Cameroon	0.051	0.062	0.021	0.178	0.842	0.054	0.662	-	-	-
Reunion	0.648	**0.047**	**0.020**	0.151	0.103	0.462	0.615	0.0721	-	-
CAR	0.456	0.390	0.333	0.444	0.500	0.444	**0.000**	0.493	0.308	-
Algeria	0.776	0.098	**0.000**	0.292	**0.000**	0.553	0.875	0.943	0.077	0.467

CAR: Central African Republic, DRC: Democratic Republic of Congo, and R Congo: Republic of the Congo. Populations where small genetic differentiation is reported were written in bold. Population consisted of only one sequence were excluded: Nigeria, Benin, and Seychelles.

The constructed haplotypes network reveals the expansion of H02 in a star-like network, suggesting that H02 is the main haplotype from which the other haplotypes evolutionary expanded (Figure 4).

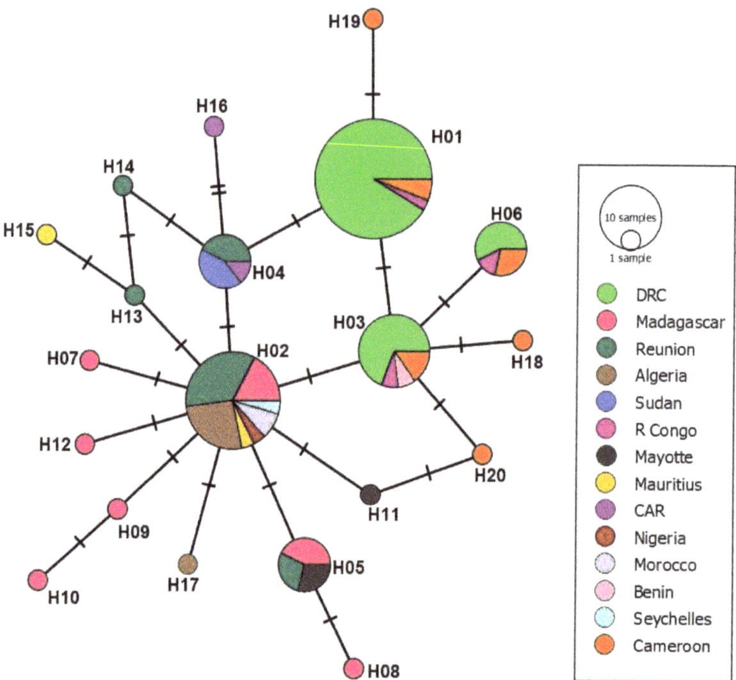

Figure 4. Parsimony haplotype network of the 103 African *Ae. albopictus* reference sequences and the 3 Sudanese *Ae. albopictus* sequences. Haplotypes of each region are presented using color coding. The number of each haplotype is written next to its representing node. Hatch marks in lines linking between the haplotypes are indicative of the numbers of nucleotide diversity. DRC: Democratic Republic of the Congo, R Congo: Republic of the Congo, and CAR: Central African Republic.

2.4. Recent Increase in the Burden and Spread of CHIKV in Sudan and the Neighboring Countries

Reviewing the national outbreak reports, the literature and online records indicate that before 2020, no case of CHIKV was reported from any states in West Sudan (i.e., Central, East, North, and West Darfur and West Kordofan) [21]. However, between 2020 and 2022, outbreaks of CHIKV were reported in these states (Figure 5). This underscores the rapid expansion of the geographical distribution of this virus in the country. Additionally, we identified several outbreaks of CHIKV that occurred during the recent two years in the region. These outbreaks include 45 cases in Kenya in 2018; 48,734 cases in Sudan in 2018–2019; 40,340 and 11,230 cases in Ethiopia and the Republic of Congo in 2019, respectively, and 30,220 cases in Chad in 2020.

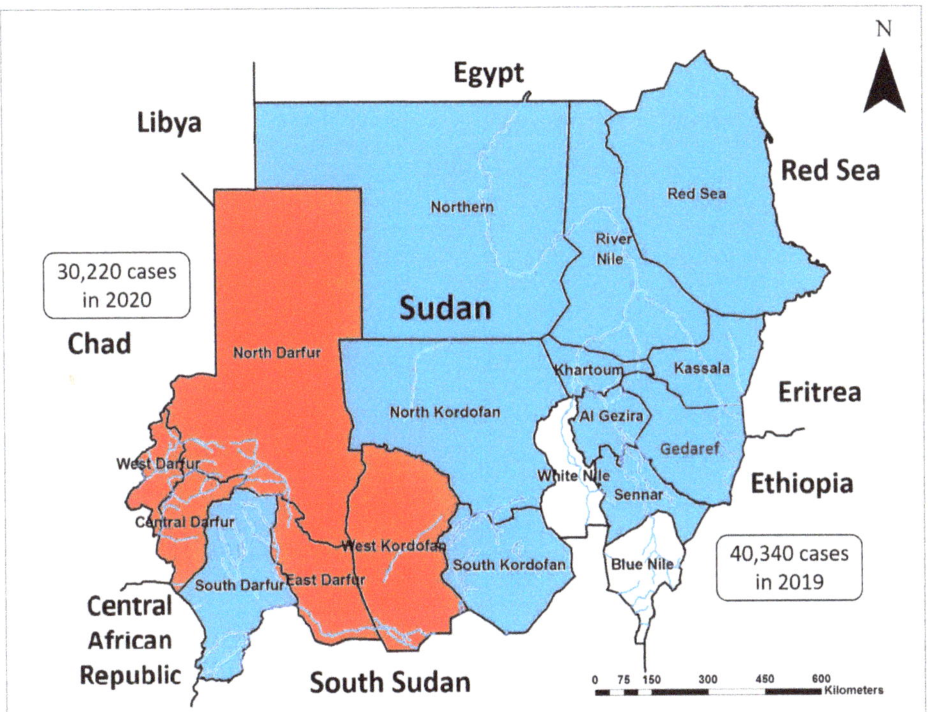

Figure 5. Map of Sudan showing the geographical distribution of chikungunya virus infections before (blue) and after (red) 2020 throughout the country. Up to now, no transmission is reported in the two states without color: White Nile and Blue Nile states. The number of cases reported during two outbreaks in the neighboring countries, Chad and Ethiopia, are included in the textboxes.

3. Discussion

This study offers the first molecular confirmation about the presence of the invasive arbovirus vector *Ae. albopictus* in Sudan. However, there were epidemiological indicators about the spread of this invasive vector in the area. Some of these indicators are regional, such as the recent substantial growth of chikungunya outbreaks in East Africa, including Chad, Ethiopia, Kenya, Republic of Congo, and Sudan [17,29–32]. Additionally, local indicators include the geographical expansion of arboviral disease distributions in the country indicated by the emergence and re-emergence of several *Aedes*-borne arboviral diseases and development of outbreaks in rural areas throughout the country [5,17,21–25]. This is further underscored by the emergence of chikungunya and dengue fever outbreaks in several states in Western Sudan, including the study area for the first time during the last five years [5,17,21–25,33].

The role of the other species of *Aedes* mosquitoes, the presence of which was confirmed in this study, in the outbreaks of arboviral diseases that recently developed throughout the country should not be ignored because they are known vectors of several arboviruses. For instance, *Ae. aegypti* is an endemic vector in Africa that transmits major human viruses, such as CHIKV, DENV, YFV, and ZIKV [21]. However, it is a globally known invasive vector for important viruses including CHIKV, DENV, MBGV, RVF, and WNV, African horse sickness virus (AHSV), Epizootic hemorrhagic disease virus (EHDV), and Mayaro virus (MAYV), as well as Eastern Equine Encephalitis virus (EEEV), Japanese Encephalitis virus (JBEV), Murray Valley Encephalitis virus (MVEV), and Venezuelan Equine Encephalitis virus (VEEV) [34]. Interestingly, *Ae. africanus* is an African native vector of arboviruses that is also associated with the transmission of CHIKV, DENV, RVF, YFV, and ZIKV [35]. Furthermore, *Ae. luteocephalus* is involved in the transmission of chikungunya, dengue, yellow fever, and Zika viruses [36]. Moreover, *Ae. vittatus* is another African vector of arboviruses that is mainly involved in the transmission of CHIKV, DENV, JBEV, WNV, YFV, and ZIKV [37]. In addition to transmitting the EEEV, JBEV, St. Louis Encephalitis virus (SLEV), VEEV, WEEV, WNV, and ZIKV [38], while *Ae. vexans* is known as the initial vector for RVF outbreaks because RVF virus can be maintained vital in the mosquitoes eggs that can withstand desiccation in dry land for years [39]. This is the reason behind the inter-epizootic and inter-epidemic periods between RVF outbreaks, because *Ae. vexans* act as an alternative host [39–41]. However, a main limitation of this study is that we did not have enough resources to collect the adult mosquitoes of these confirmed species of *Aedes* vectors to incriminate them and investigate their role in the local transmission of arboviruses. Unfortunately, the current surveillance and control systems for human and animal diseases, as well as their vectors in the country, are separate and inadequate, limiting their ability to detect and report diseases in a timely manner in order to take effective control measures [21,42,43]. Furthermore, in addition to the heavy burden of malaria that overwhelmingly occupies the healthcare system, the country suffers from armed conflict and climate change, as evidenced by the recent frequency of extreme weather events across the country [33,44,45]. These factors together favor the spread and increased abundance of disease vectors, particularly the invasive vectors, such as *Ae. albopictus* and *An. stephensi* [46–50]. Nonetheless, the delay in detecting the emergence of these vectors of arboviral diseases that we report here could be attributed to the partial freeze in environmental and public health services due to the global COVID-19 pandemic [51]. Therefore, the need for a One Health-integrated surveillance and response system for diseases and their vectors is extreme and urgent [50,52–54].

The phylogenetic analyses show that the Sudanese specimens of *Ae. albopictus* closely cluster with sequences from La Réunion and the Central African Republic. Considering the open borders and high dynamics of human and animal populations between the Central African Republic and Sudan, and that all the current three samples were collected relatively close to these borders, it is likely that it was introduced into Sudan from the Central African Republic. In addition, the western borders of Sudan had suffered from an armed-conflict that persisted since 2003, and several diseases and disease vectors emerged in the area. The emergence and spread of several zoonotic diseases, such as CCHF, dengue, hepatitis E virus, RVF, YF, and WNV in Sudan, were associated with the environmental and socioeconomic changes due to the war-induced humanitarian crisis [4,5,21–25,33,55–57]. Climate change is also playing a major role in the spread of infectious diseases and their vectors in the country and globally [17,33,58–60]. Our haplotype analysis identified a single haplotype in the country, mainly because of the small sample size; therefore, we cannot exclude the presence of other haplotypes in the country, particularly since at least around 20 haplotypes exist among the African populations of *Ae. albopictus* (Figure 4) [61–66]. Our haplotype network analysis suggested that H02 is the central haplotype from which other haplotypes were driven. Interestingly, this is further supported by the fact that H02 is exclusively reported from islands and coastal countries where the initial introduction of this invasive vector is highly likely (Figure 4).

Unlike *Ae. aegypti*, the other identified species of *Aedes* mosquitoes are zoophilic, which means they prefer to feed on a non-human host, yet readily feed on humans according to host availability, making them of high risk for the transmission of zoonoses. Particularly, nearly a third of the 545 suspected arboviruses are zoonotic in nature, readily infecting humans, domestic, and wildlife animals [3]. Some of these arboviruses have a wide range of hosts and vectors, making them challenging to prevent and control [2]. However, due to their evolution, high rate of mutations, and capacity to spillover and spill back, the host range and competency of different arthropods to transmit the different arboviruses is persistently expanding [67,68]. Considering the climate change, rapid unplanned urbanization, globalization, conflicts, and increased size of population lives in humanitarian crisis with limited services, now more than ever, there is a severe need for a countrywide One Health-integrated surveillance and response system to prevent future pandemics [52,54,69]. Particularly, the geographical location of Sudan, its size spanning over different ecological zones, including coastal, desert, poor and rich Savanah, and forest, its wide open international borders with many neighboring countries, and the high variety and size of domestic and wildlife animal populations in the country make it very prone for the emergence of pandemics of viral hemorrhagic fevers [70]. Additionally, molecular xenosurveillance is a robust tool for the surveillance of emerging infectious diseases and their vectors that leverages the blood-fed vectors as collectors of blood samples from both humans and animals [71–73]. Therefore, public health authorities, particularly in poor resource settings, should utilize this approach to reduce the cost of sampling human and animal populations. Particularly, samples of the same species of vectors with similar meta-data (place and date of collection) could be pooled together to further reduce the molecular testing cost. Furthermore, through the implementation of the integrated One Health, molecular xenosurveillance, and response system, there is less delay between identifying diseases circulating in the area and the implementation of control measures to prevent potential epidemics, outbreaks, and pandemics [52,54,72,73]. The added value of this approach, which includes but is not limited to, less morbidity and mortality among humans, less social disturbance, as well as less economic and environmental loss, makes it worth the investment [52,74].

4. Materials and Methods

4.1. Exploratory Surveys

In response to the emergence and further spread of the invasive malaria vector, *Anopheles stephensi* in Sudan [20,60], we initiated exploratory surveys targeting to delineate the distribution of *An. stephensi* that co-breed with *Aedes* mosquitoes in human-made water containers [50,75]. Household container surveys confirmed the presence of *An. stephensi* in the eastern, western, and northern international borders; therefore, we focused our second survey on the southern borders of the country, namely North and South Kordofan, Sennar, and White Nile states (Figure 6) [59,75].

4.2. Mosquito Samples Collection and Morphological Identification

During the household surveys, we used disposable dippers for the collection of aquatic stages (larvae and pupae) of mosquitoes from the human-made water containers. Then, we transferred the collected larvae to the laboratory at Sirius Training and Research Centre in Khartoum for morphological identification and molecular investigations. We reared the larvae to adults and morphologically identified the adults *Aedes* mosquitoes using standard morphological keys and consulted the online available database of the Walter Reed Biosystematics Unit (WRBU) for further confirmation (https://www.wrbu.si.edu/index.php/) [27,76]. *Anopheles* mosquitoes were identified using the recently published standard key for the morphological identification of Afrotropical *Anopheles* [76].

Figure 6. Map of Sudan shows the study area shaded in grey.

4.3. DNA Extraction from Mosquito and Polymerase Chain Reaction

The genomic DNA was extracted from the mosquito samples following the manufacturer instructions using QiaAmp tissue extraction kits (QIAGEN Diagnostics GmbH, Hilden, Germany). Extracted DNA quality was checked using a nanodrop spectrophotometer (Implen, München, Germany), then preserved at -20 °C until molecular examination.

To amplify the cytochrome oxidase 1 (CO1) region of the mitochondrial DNA of the mosquito genome, the Folmer primers (LCO1490 and HCO2198) were used [77]. PCR reaction mixture containing 2 µL of the extracted DNA was added to a 4 µL PCR master mix (Solis Biodyne, Tartu, Estonia) containing 1 U DNA polymerase, 12.5 mM MgCl2, and 4 mM dNTPs. PCR thermal conditions were as follows: initial denaturation at 95 °C for 5 min, followed by 35 cycles of denaturation at 95 °C for 30 s, annealing at 58 °C for 30 s, and extension at 72 °C for 30 s, and a final extension step at 72 °C for 10 min. Thermal conditions were performed using the 2721 Thermocycler (Applied Biosystems, ThermoFisher Scientific, Budapest, Hungary). Following PCR, amplicons were visualized using 2% gel electrophoresis (Major Science Co., Ltd, Taoyuan City, Taiwan).

4.4. PCR Amplicons Sequencing and Sequences' Identity Confirmation

The amplified PCR amplicons were sequenced in duplicates based on both directions' primers (LCO1490 and HCO2198) by the Sanger Deoxyribonucleic acid sequencing method using the 3730XL DNA analyzer (Life Technologies Corporation, California, United States) through Macrogen (Macrogen Inc., Amsterdam, The Netherlands). Prior to checking sequences' identities, sequences of each mosquito sample were aligned on GENtel software to check for any base calling errors during the sequencing process. Following peaks correction, a consensus sequence was constructed from each duplicate. Then, we used the online BLAST nucleotide algorithm to compare our sequences with the previously published *Aedes* species consensus sequences available in the NCBI GenBank database (https://blast.ncbi.nlm.nih.gov/Blast.cgi, accessed on 13 April 2022) [28]. The obtained DNA sequences during this study were deposited into the GenBank database at NCBI.

4.5. Bioinformatics and Phylogenetic Analysis

Phylogenetic analysis was conducted using the maximum likelihood method to determine the close relatedness of sequences to the previously published sequences available at the NCBI GenBank database [28]. To construct the phylogenetic tree, all sequences available for the *Ae. albopictus* species were downloaded and each of the data that were related to each sequence were sorted based on country of isolation (Supplementary Materials Table S1). A total of 103 sequences representing African countries were selected for constructing a maximum likelihood phylogenetic tree. The selected sequences and the study sequences were aligned and trimmed to produce similar sequences in length prior to constructing the phylogenetic tree using MEGA7 software. The nucleotide substitution model with the lowest Bayesian Information Criterion (BIC) scores was considered as the best-fit model. The non-uniformity of evolutionary rates among sites was modelled using a discrete Gamma distribution with 1000 bootstraps during the phylogenetic tree construction [78].

Bioinformatics analysis was conducted for all sequences included in the study, including the previously published sequences to obtain sequence diversity parameters, including numbers of haplotypes (H) and haplotypes diversity (Hd) according to each country using the DnaSP v5.10 software [79]. Additionally, a haplotype network was constructed using the median-joining network using popART software (v4.8) (http://popart.otago.ac.nz, accessed on 29 April 2022). Further, to investigate the estimated genetic differentiation between the study sequences and the African countries' sequences, the pairwise fixation index (Fst) value was calculated using DnaSP [79]. Fst values were interpreted, according to [79], as follows: 0.00–0.05, indicating a small genetic differentiation between the populations; 0.05–0.15, indicating a moderate genetic differentiation among the populations; 0.15–0.25, indicating a large genetic differentiation; and Fst more than 0.25 was indicative of a great genetic differentiation between the populations [79].

4.6. Exploring the Potential Health Impacts of the Emergence of This Invasive Vector in the Area

To investigate the potential role of this invasive vector, *Aedes albopictus* and other vectors on the mergence and geographical spread of arboviral disease outbreaks, particularly CHIK in the region, we reviewed the publicly available records about CHIKV outbreaks, particularly rural areas. We compared our findings about outbreaks of relevant *Aedes*-borne arboviruses in the country with the previously mapped distribution of arboviral diseases in Sudan up to 2020 [21].

5. Conclusions

Here we provide the first molecular evidence about the emergence of the rural invasive arboviral diseases vector, *Aedes albopictus* in Sudan. This finding was made accidentally; however, this indicates that this vector might be very prevalent in the country. Records about recent outbreaks of chikungunya in the region of East Africa are very alarming due to their massive magnitude and rapid development, and this could be attributed to the contribution of this undetected invasive vector, which is very competent in transmitting several arboviruses, and its adaptive nature to rural areas. An integrated One Health approach is urgently needed in the country to prevent a dire situation in the future, particularly that the country health system is overwhelmingly challenged by the burden of other infectious diseases such as malaria. It would be strategic, less resource-demanding, and less invasive to humans and animals to incorporate molecular xenosurveillance as a proxy for monitoring the dynamics of zoonotic diseases and their vectors in the area.

Supplementary Materials: The supporting information can be downloaded at: https://www.mdpi.com/article/10.3390/ijms231911802/s1.

Author Contributions: Conceptualization, A.A.; methodology, A.A., M.A. and N.S.M.; formal analysis, A.A. and N.S.M.; investigation, A.A., M.A., N.S.M., H.S. and I.M.; resources, A.A., M.A. and J.Z.; data curation, A.A.; writing—original draft preparation, A.A.; writing—review and editing, A.A., N.S.M. and J.Z.; visualization, A.A. and N.S.M.; supervision, J.Z.; project administration, A.A.; funding acquisition, A.A. and J.Z. All authors have read and agreed to the published version of the manuscript.

Funding: This research was funded by the small grant from R. Geigy-Stiftung and the Swiss Government Excellence Scholarships for Foreign Scholars and Artists that awarded to A.A. by the Federal Commission for Scholarships for Foreign Students (FCS) (Personal ESKAS-Nr: 2021.0671).

Institutional Review Board Statement: Not applicable.

Informed Consent Statement: Not applicable.

Data Availability Statement: Sequences of *Ae. albopictus* mosquitoes that were generated during this study were deposited into the GenBank database at NCBI (https://blast.ncbi.nlm.nih.gov/Blast.cgi, accessed 18 April 2022) under the accession numbers ON248551-ON248553. The accession numbers of all *Ae. albopictus* sequences sorted based on country of isolation that were downloaded from the NCBI GenBank database and used in the phylogenetic and haplotype analyses are available in Supplementary Materials Table S1.

Acknowledgments: We are thankful to our colleagues who helped in the samples collection and Pie Müller for his insightful comments.

Conflicts of Interest: The authors declare no conflict of interest.

References

1. Kraemer, M.U.G.; Reiner, R.C., Jr.; Brady, O.J.; Messina, J.P.; Gilbert, M.; Pigott, D.M.; Yi, D.; Johnson, K.; Earl, L.; Marczak, L.B.; et al. Past and future spread of the arbovirus vectors Aedes aegypti and Aedes albopictus. *Nat. Microbiol.* **2019**, *4*, 854–863. [CrossRef]
2. Weaver, S.C.; Barrett, A.D.T. Transmission cycles, host range, evolution and emergence of arboviral disease. *Nat. Rev. Microbiol.* **2004**, *2*, 789–801. [CrossRef] [PubMed]
3. Cleton, N.; Koopmans, M.; Reimerink, J.; Godeke, G.-J.; Reusken, C. Come fly with me: Review of clinically important arboviruses for global travelers. *J. Clin. Virol.* **2012**, *55*, 191–203. [CrossRef] [PubMed]
4. Ahmed, A.; Ali, Y.; Salim, B.; Dietrich, I.; Zinsstag, J. Epidemics of Crimean-Congo Hemorrhagic Fever (CCHF) in Sudan between 2010 and 2020. *Microorganisms* **2022**, *10*, 928. [CrossRef] [PubMed]
5. Ahmed, A.; Ali, Y.; Elduma, A.; Eldigail, M.H.; Mhmoud, R.A.; Mohamed, N.S.; Ksiazek, T.G.; Dietrich, I.; Weaver, S.C. Unique Outbreak of Rift Valley Fever in Sudan, 2019. *Emerg. Infect. Dis.* **2020**, *26*, 3030–3033. [CrossRef]
6. Zientara, S.; Ponsart, C. Viral emergence and consequences for reproductive performance in ruminants: Two recent examples (bluetongue and Schmallenberg viruses). *Reprod. Fertil. Dev.* **2015**, *27*, 63–71. [CrossRef]
7. Pereira De Oliveira, R.; Hutet, E.; Lancelot, R.; Paboeuf, F.; Duhayon, M.; Boinas, F.; Pérez de León, A.A.; Filatov, S.; Le Potier, M.F.; Vial, L. Differential vector competence of Ornithodoros soft ticks for African swine fever virus: What if it involves more than just crossing organic barriers in ticks? *Parasites Vectors* **2020**, *13*, 618. [CrossRef]
8. Blahove, M.R.; Carter, J.R. Flavivirus persistence in wildlife populations. *Viruses* **2021**, *13*, 2099. [CrossRef]
9. Braack, L.; Gouveia de Almeida, A.P.; Cornel, A.J.; Swanepoel, R.; de Jager, C. Mosquito-borne arboviruses of African origin: Review of key viruses and vectors. *Parasites Vectors* **2018**, *11*, 29. [CrossRef]
10. Weetman, D.; Kamgang, B.; Badolo, A.; Moyes, C.L.; Shearer, F.M.; Coulibaly, M.; Pinto, J.; Lambrechts, L.; McCall, P.J. Aedes Mosquitoes and Aedes-Borne Arboviruses in Africa: Current and Future Threats. *Int. J. Environ. Res. Public Health* **2018**, *15*, 220. [CrossRef]
11. Whiteman, A.; Loaiza, J.R.; Yee, D.A.; Poh, K.C.; Watkins, A.S.; Lucas, K.J.; Rapp, T.J.; Kline, L.; Ahmed, A.; Chen, S.; et al. Do socioeconomic factors drive Aedes mosquito vectors and their arboviral diseases? A systematic review of dengue, chikungunya, yellow fever, and Zika Virus. *One Health* **2020**, *11*, 100188. [CrossRef]
12. Elaagip, A.; Alsedig, K.; Altahir, O.; Ageep, T.; Ahmed, A.; Siam, H.A.; Samy, A.M.; Mohamed, W.; Khalid, F.; Gumaa, S.; et al. Seroprevalence and associated risk factors of Dengue fever in Kassala state, eastern Sudan. *PLoS Negl. Trop. Dis.* **2020**, *14*, e0008918. [CrossRef]
13. Müller, P.; Engeler, L.; Vavassori, L.; Suter, T.; Guidi, V.; Gschwind, M.; Tonolla, M.; Flacio, E. Surveillance of invasive Aedes mosquitoes along Swiss traffic axes reveals different dispersal modes for Aedes albopictus and Ae. japonicus. *PLoS Negl. Trop. Dis.* **2020**, *14*, e0008705. [CrossRef]
14. Flacio, E.; Engeler, L.; Tonolla, M.; Lüthy, P.; Patocchi, N. Strategies of a thirteen year surveillance programme on Aedes albopictus (*Stegomyia albopicta*) in southern Switzerland. *Parasites Vectors* **2015**, *8*, 208. [CrossRef]

15. Carrieri, M.; Albieri, A.; Angelini, P.; Baldacchini, F.; Venturelli, C.; Zeo, S.M.; Bellini, R. Surveillance of the chikungunya vector Aedes albopictus (Skuse) in Emilia-Romagna (northern Italy): Organizational and technical aspects of a large scale monitoring system. *J. Vector Ecol.* **2011**, *36*, 108–116. [CrossRef]

16. Canali, M.; Rivas-Morales, S.; Beutels, P.; Venturelli, C. The Cost of Arbovirus Disease Prevention in Europe: Area-Wide Integrated Control of Tiger Mosquito, Aedes albopictus, in Emilia-Romagna, Northern Italy. *Int. J. Environ. Res. Public Health* **2017**, *14*, 444. [CrossRef]

17. Ahmed, A.; Ali, Y.; Mohamed, N.S. Arboviral diseases: The emergence of a major yet ignored public health threat in Africa. *Lancet Planet Health* **2020**, *4*, e555. [CrossRef]

18. de Santi, V.P.; Khaireh, B.A.; Chiniard, T.; Pradines, B.; Taudon, N.; Larréché, S.; Mohamed, A.B.; de Laval, F.; Berger, F.; Gala, F.; et al. Role of Anopheles stephensi Mosquitoes in Malaria Outbreak, Djibouti, 2019. *Emerg. Infect. Dis.* **2021**, *27*, 1697–1700. [CrossRef]

19. Faulde, M.K.; Rueda, L.M.; Khaireh, B.A. First record of the Asian malaria vector Anopheles stephensi and its possible role in the resurgence of malaria in Djibouti, Horn of Africa. *Acta Trop.* **2014**, *139*, 39–43. [CrossRef]

20. Ahmed, A.; Pignatelli, P.; Elaagip, A.; Hamid, M.M.A.; Alrahman, O.F.; Weetman, D. Invasive Malaria Vector Anopheles stephensi Mosquitoes in Sudan, 2016–2018. *Emerg. Infect. Dis. J.-CDC* **2021**, *27*, 2952–2954. [CrossRef]

21. Ahmed, A.; Dietrich, I.; LaBeaud, A.D.; Lindsay, S.W.; Musa, A.; Weaver, S.C. Risks and Challenges of Arboviral Diseases in Sudan: The Urgent Need for Actions. *Viruses* **2020**, *12*, 81. [CrossRef]

22. Ahmed, A.; Ali, Y.; Elmagboul, B.; Mohamed, O.; Elduma, A.; Bashab, H.; Mahamoud, A.; Khogali, H.; Elaagip, A.; Higazi, T. Dengue Fever in the Darfur Area, Western Sudan. *Emerg. Infect. Dis.* **2019**, *25*, 2126. [CrossRef]

23. Ahmed, A.; Eldigail, M.; Elduma, A.; Breima, T.; Dietrich, I.; Ali, Y.; Weaver, S.C. First report of epidemic dengue fever and malaria co-infections among internally displaced persons in humanitarian camps of North Darfur, Sudan. *Int. J. Infect. Dis.* **2021**, *108*, 513–516. [CrossRef]

24. Ahmed, A.; Elduma, A.; Magboul, B.; Higazi, T.; Ali, Y. The First Outbreak of Dengue Fever in Greater Darfur, Western Sudan. *Trop. Med. Infect. Dis.* **2019**, *4*, 43. [CrossRef]

25. Ahmed, A.; Mahmoud, I.; Eldigail, M.; Elhassan, R.M.; Weaver, S.C. The Emergence of Rift Valley Fever in Gedaref State Urges the Need for a Cross-Border One Health Strategy and Enforcement of the International Health Regulations. *Pathogens* **2021**, *10*, 885. [CrossRef]

26. Hamid, Z.; Hamid, T.; Alsedig, K.; Abdallah, T.; Elaagip, A.; Ahmed, A.; Khalid, F.; Abdel Hamid, M. Molecular Investigation of Dengue virus serotype 2 Circulation in Kassala State, Sudan. *Jpn. J. Infect. Dis.* **2019**, *72*, 58–61. [CrossRef]

27. Lewis, D.J. The Aēdes mosquitoes of the Sudan. *Ann. Trop. Med. Parasitol.* **1955**, *49*, 164–173. [CrossRef]

28. Altschul, S.F.; Gish, W.; Miller, W.; Myers, E.W.; Lipman, D.J. Basic local alignment search tool. *J. Mol. Biol.* **1990**, *215*, 403–410. [CrossRef]

29. World Health Organization. Regional Office for Africa. Weekly Bulletin on Outbreak and Other Emergencies: Week 38: 16–22 September 2019. 2019. Available online: https://apps.who.int/iris/handle/10665/327836 (accessed on 27 August 2022).

30. World Health Organization. Regional Office for Africa, Health Emergencies Programme. Weekly Bulletin on Outbreaks and other Emergencies: Week 02: 06–12 January 2018. Health Emergency Information and Risk Assessment. 2018. Available online: https://apps.who.int/iris/handle/10665/259850 (accessed on 24 August 2022).

31. World Health Organization. Regional Office for Africa. Weekly Bulletin on Outbreak and other Emergencies: Week 39: 21–27 September 2020. 2020. Available online: https://apps.who.int/iris/handle/10665/335723 (accessed on 21 August 2022).

32. World Health Organization. Regional Office for Africa. Weekly Bulletin on Outbreak and other Emergencies: Week 32: 05–11 August 2019. 2019. Available online: https://apps.who.int/iris/handle/10665/326304 (accessed on 24 August 2022).

33. Ahmed, A.; Mohamed, N.S.; Siddig, E.E.; Algaily, T.; Sulaiman, S.; Ali, Y. The impacts of climate change on displaced populations: A call for actions. *J. Clim. Change Health* **2021**, *3*, 100057. [CrossRef]

34. Walter Reed Biosystematics Unit Website. Aedes Aegypti Species Page. Available online: http://wrbu.si.edu/vectorspecies/mosquitoes/aegypti (accessed on 27 August 2022).

35. Walter Reed Biosystematics Unit Website. Aedes Africanus Species Page. Available online: http://wrbu.si.edu/vectorspecies/mosquitoes/africanus (accessed on 7 August 2022).

36. Walter Reed Biosystematics Unit Website. Aedes Luteocephalus Species Page. Available online: http://wrbu.si.edu/vectorspecies/mosquitoes/luteocephalus (accessed on 15 August 2022).

37. Walter Reed Biosystematics Unit Website. Aedes Vittatus Species Page. Available online: http://wrbu.si.edu/vectorspecies/mosquitoes/ae_vittatus (accessed on 8 August 2022).

38. Walter Reed Biosystematics Unit Website. Aedes Vexans Species Page. Available online: http://wrbu.si.edu/vectorspecies/mosquitoes/vexans (accessed on 8 August 2022).

39. Himeidan, Y.E.; Kweka, E.J.; Mahgoub, M.M.; El Rayah, E.A.; Ouma, J.O. Recent Outbreaks of Rift Valley Fever in East Africa and the Middle East. *Front. Public Health* **2014**, *2*, 169. [CrossRef]

40. Davies, F.G.; Linthicum, K.J.; James, A.D. Rainfall and epizootic Rift Valley fever. *Bull. World Health Organ.* **1985**, *63*, 941–943. [PubMed]

41. Gerdes, G.H. Rift Valley fever. *Rev. Sci. Tech.* **2004**, *23*, 613–623. [CrossRef] [PubMed]

42. Ahmed, A. Urgent call for a global enforcement of the public sharing of health emergencies data: Lesson learned from serious arboviral disease epidemics in Sudan. *Int. Health* **2020**, *12*, 238–240. [CrossRef] [PubMed]
43. Ahmed, A. Current status of Mosquito-borne Arboviruses in Sudan, and challenges of surveillance and responses. In Proceedings of the Mosquito-Borne Arboviruses: The Rising Global Threat, Malaria Consortium Webinar, Online, 10 February 2021. [CrossRef]
44. Elagali, A.; Ahmed, A.; Makki, N.; Ismail, H.; Ajak, M.; Alene, K.A.; Weiss, D.J.; Mohammed, A.A.; Abubakr, M.; Cameron, E.; et al. Spatiotemporal mapping of malaria incidence in Sudan using routine surveillance data. *Sci. Rep.* **2022**, *12*, 14114. [CrossRef]
45. Mohamed, N.S.; Ali, Y.; Muneer, M.S.; Siddig, E.E.; Sibley, C.H.; Ahmed, A. Malaria epidemic in humanitarian crisis settings the case of South Kordofan state, Sudan. *J. Infect. Dev. Ctries.* **2021**, *15*, 168–171. [CrossRef]
46. Franklinos, L.H.V.; Jones, K.E.; Redding, D.W.; Abubakar, I. The effect of global change on mosquito-borne disease. *Lancet Infect. Dis.* **2019**, *19*, e302–e312. [CrossRef]
47. Caminade, C.; McIntyre, K.M.; Jones, A.E. Impact of recent and future climate change on vector-borne diseases. *Ann. N. Y. Acad. Sci.* **2019**, *1436*, 157–173. [CrossRef]
48. Crowl, T.A.; Crist, T.O.; Parmenter, R.R.; Belovsky, G.; Lugo, A.E. The spread of invasive species and infectious disease as drivers of ecosystem change. *Front. Ecol. Environ.* **2008**, *6*, 238–246. [CrossRef]
49. Gubler, D.J.; Reiter, P.; Ebi, K.L.; Yap, W.; Nasci, R.; Patz, J.A. Climate variability and change in the United States: Potential impacts on vector- and rodent-borne diseases. *Environ. Health Perspect.* **2001**, *109*, 223–233. [CrossRef]
50. Ahmed, A.; Abubakr, M.; Ali, Y.; Siddig, E.E.; Mohamed, N.S. Vector control strategy for Anopheles stephensi in Africa. *Lancet Microbe* **2022**, *3*, e403. [CrossRef]
51. Ahmed, A.; Mohamed, N.S.; EL-Sadig, S.M.; Fahal, L.A.; Abelrahim, Z.B.; Ahmed, E.S.; Siddig, E.E. COVID-19 in Sudan. *J. Infect. Dev. Ctries.* **2021**, *15*, 204–208. [CrossRef]
52. Zinsstag, J.; Utzinger, J.; Probst-Hensch, N.; Shan, L.; Zhou, X.-N. Towards integrated surveillance-response systems for the prevention of future pandemics. *Infect. Dis. Poverty* **2020**, *9*, 87–92. [CrossRef]
53. Ali, Y.; Ahmed, A.; Siddig, E.E.; Mohamed, N.S. The role of integrated programs in the prevention of COVID-19 in a humanitarian setting. *Trans. R. Soc. Trop. Med. Hyg.* **2021**, *116*, 193–196. [CrossRef]
54. Zinsstag, J.; Hediger, K.; Osman, Y.M.; Abukhattab, S.; Crump, L.; Kaiser-Grolimund, A.; Mauti, S.; Ahmed, A.; Hattendorf, J.; Bonfoh, B.; et al. The Promotion and Development of One Health at Swiss TPH and Its Greater Potential. *Diseases* **2022**, *10*, 65. [CrossRef]
55. Elduma, A.H.; LaBeaud, A.D.; Plante, J.A.; Plante, K.S.; Ahmed, A. High Seroprevalence of Dengue Virus Infection in Sudan: Systematic Review and Meta-Analysis. *Trop. Med. Infect. Dis.* **2020**, *5*, 120. [CrossRef]
56. Markoff, L. Yellow Fever Outbreak in Sudan. *N. Engl. J. Med.* **2013**, *368*, 689–691. [CrossRef]
57. Ahmed, A.; Ali, Y.; Siddig, E.E.; Hamed, J.; Mohamed, N.S.; Khairy, A.; Zinsstag, J. Hepatitis E Virus Outbreak among Tigray War Refugees from Ethiopia, Sudan. *Emerg. Infect. Dis.* **2022**, *28*, 1722. [CrossRef]
58. Mordecai, E.A.; Ryan, S.J.; Caldwell, J.M.; Shah, M.M.; LaBeaud, A.D. Climate change could shift disease burden from malaria to arboviruses in Africa. *Lancet Planet. Health* **2020**, *4*, e416–e423. [CrossRef]
59. Abubakr, M.; Sami, H.; Mahdi, I.; Altahir, O.; Abdelbagi, H.; Mohamed, N.S.; Ahmed, A. The Phylodynamic and Spread of the Invasive Asian Malaria Vectors, Anopheles stephensi, in Sudan. *Biology* **2022**, *11*, 409. [CrossRef]
60. Ahmed, A.; Khogali, R.; Elnour, M.-A.B.; Nakao, R.; Salim, B. Emergence of the invasive malaria vector Anopheles stephensi in Khartoum State, Central Sudan. *Parasites Vectors* **2021**, *14*, 511. [CrossRef]
61. Gubler, D.J. Aedes albopictus in Africa. *Lancet Infect. Dis.* **2003**, *3*, 751–752. [CrossRef]
62. Ngoagouni, C.; Kamgang, B.; Nakouné, E.; Paupy, C.; Kazanji, M. Invasion of Aedes albopictus (Diptera: Culicidae) into central Africa: What consequences for emerging diseases? *Parasites Vectors* **2015**, *8*, 191. [CrossRef]
63. Kamgang, B.; Wilson-Bahun, T.A.; Irving, H.; Kusimo, M.O.; Lenga, A.; Wondji, C.S. Geographical distribution of Aedes aegypti and Aedes albopictus (Diptera: Culicidae) and genetic diversity of invading population of Ae. albopictus in the Republic of the Congo. *Wellcome Open Res.* **2018**, *3*, 79. [CrossRef]
64. Kamgang, B.; Brengues, C.; Fontenille, D.; Njiokou, F.; Simard, F.; Paupy, C. Genetic structure of the tiger mosquito, Aedes albopictus, in Cameroon (Central Africa). *PLoS ONE* **2011**, *6*, e20257. [CrossRef]
65. Tedjou, A.N.; Kamgang, B.; Yougang, A.P.; Njiokou, F.; Wondji, C.S. Update on the geographical distribution and prevalence of Aedes aegypti and Aedes albopictus (Diptera: Culicidae), two major arbovirus vectors in Cameroon. *PLoS Negl. Trop. Dis.* **2019**, *13*, e0007137. [CrossRef]
66. Ruiling, Z.; Tongkai, L.; Dezhen, M.; Zhong, Z. Genetic characters of the globally spread tiger mosquito, Aedes albopictus (Diptera, Culicidae): Implications from mitochondrial gene COI. *J. Vector Ecol.* **2018**, *43*, 89–97. [CrossRef]
67. Weaver, S.C. Prediction and prevention of urban arbovirus epidemics: A challenge for the global virology community. *Antivir. Res.* **2018**, *156*, 80–84. [CrossRef]
68. Weaver, S.C.; Reisen, W.K. Present and future arboviral threats. *Antivir. Res.* **2010**, *85*, 328–345. [CrossRef]
69. Zinsstag, J.; Schelling, E.; Crump, L.; Whittaker, M.; Tanner, M.; Stephen, C. *One Health: The Theory and Practice of Integrated Health Approaches*, 2nd ed.; CABI: Wallingford, UK, 2020; p. 459.
70. Pigott, D.M.; Deshpande, A.; Letourneau, I.; Morozoff, C.; Reiner, R.C.; Kraemer, M.U.G.; Brent, S.E.; Bogoch, I.I.; Khan, K.; Biehl, M.H.; et al. Local, national, and regional viral haemorrhagic fever pandemic potential in Africa: A multistage analysis. *Lancet* **2017**, *390*, 2662–2672. [CrossRef]

71. Fauver, J.R.; Gendernalik, A.; Weger-Lucarelli, J.; Grubaugh, N.D.; Brackney, D.E.; Foy, B.D.; Ebel, G.D. The Use of Xenosurveillance to Detect Human Bacteria, Parasites, and Viruses in Mosquito Bloodmeals. *Am. J. Trop. Med. Hyg.* **2017**, *97*, 324–329. [CrossRef]
72. Brinkmann, A.; Nitsche, A.; Kohl, C. Viral Metagenomics on Blood-Feeding Arthropods as a Tool for Human Disease Surveillance. *Int. J. Mol. Sci.* **2016**, *17*, 1743. [CrossRef] [PubMed]
73. Grubaugh, N.D.; Sharma, S.; Krajacich, B.J.; Iii, L.S.F.; Bolay, F.K.; Ii, J.W.D.; Johnson, W.E.; Ebel, G.D.; Foy, B.D.; Brackney, D.E. Xenosurveillance: A Novel Mosquito-Based Approach for Examining the Human-Pathogen Landscape. *PLOS Negl. Trop. Dis.* **2015**, *9*, e0003628. [CrossRef] [PubMed]
74. Zinsstag, J.; Crump, L.; Schelling, E.; Hattendorf, J.; Maidane, Y.O.; Ali, K.O.; Muhummed, A.; Umer, A.A.; Aliyi, F.; Nooh, F. Climate change and one health. *FEMS Microbiol. Lett.* **2018**, *365*, fny085. [CrossRef] [PubMed]
75. Ahmed, A.; Irish, S.R.; Zohdy, S.; Yoshimizu, M.; Tadesse, F.G. Strategies for conducting Anopheles stephensi surveys in non-endemic areas. *Acta Trop.* **2022**, *236*, 106671. [CrossRef]
76. Coetzee, M. Key to the females of Afrotropical Anopheles mosquitoes (Diptera: Culicidae). *Malar. J.* **2020**, *19*, 70. [CrossRef]
77. Folmer, O.; Black, M.; Hoeh, W.; Lutz, R.; Vrijenhoek, R. DNA primers for amplification of mitochondrial cytochrome c oxidase subunit I from diverse metazoan invertebrates. *Mol. Mar. Biol. Biotechnol.* **1994**, *3*, 294–299.
78. Tamura, K. Estimation of the number of nucleotide substitutions when there are strong transition-transversion and G+C-content biases. *Mol. Biol. Evol.* **1992**, *9*, 678–687. [CrossRef]
79. Rozas, J.; Sánchez-DelBarrio, J.C.; Messeguer, X.; Rozas, R. DnaSP, DNA polymorphism analyses by the coalescent and other methods. *Bioinformatics* **2003**, *19*, 2496–2497. [CrossRef]

International Journal of
Molecular Sciences

Article

Trypsin-like Inhibitor Domain (TIL)-Harboring Protein Is Essential for *Aedes aegypti* Reproduction

Chinmay Vijay Tikhe [1,2], Victor Cardoso-Jaime [1,2], Shengzhang Dong [1,2], Natalie Rutkowski [1,2] and George Dimopoulos [1,2,*]

[1] W. Harry Feinstone Department of Molecular Microbiology and Immunology, Johns Hopkins Bloomberg School of Public Health, Baltimore, MD 21205, USA; ctikhe1@jhu.edu (C.V.T.); vcardos1@jhu.edu (V.C.-J.); sdong13@jhu.edu (S.D.); nrutkow1@jhmi.edu (N.R.)
[2] Johns Hopkins Malaria Research Institute, Johns Hopkins Bloomberg School of Public Health, Baltimore, MD 21205, USA
* Correspondence: gdimopo1@jhu.edu

Abstract: Cysteine-rich trypsin inhibitor-like domain (TIL)-harboring proteins are broadly distributed in nature but remain understudied in vector mosquitoes. Here we have explored the biology of a TIL domain-containing protein of the arbovirus vector *Aedes aegypti*, cysteine-rich venom protein 379 (CRVP379). CRVP379 was previously shown to be essential for dengue virus infection in *Ae. aegypti* mosquitoes. Gene expression analysis showed CRVP379 to be highly expressed in pupal stages, male testes, and female ovaries. CRVP379 expression is also increased in the ovaries at 48 h post-blood feeding. We used CRISPR-Cas9 genome editing to generate two mutant lines of CRVP379 with mutations inside or outside the TIL domain. Female mosquitoes from both mutant lines showed severe defects in their reproductive capability; mutant females also showed differences in their follicular cell morphology. However, the CRVP379 line with a mutation outside the TIL domain did not affect male reproductive performance, suggesting that some CRVP379 residues may have sexually dimorphic functions. In contrast to previous reports, we did not observe a noticeable difference in dengue virus infection between the wild-type and any of the mutant lines. The importance of CRVP379 in *Ae. aegypti* reproductive biology makes it an interesting candidate for the development of *Ae. aegypti* population control methods.

Keywords: *Aedes aegypti*; reproduction; trypsin inhibitor; mosquitoes; CRISPR-Cas9; dengue

Citation: Tikhe, C.V.; Cardoso-Jaime, V.; Dong, S.; Rutkowski, N.; Dimopoulos, G. Trypsin-like Inhibitor Domain (TIL)-Harboring Protein Is Essential for *Aedes aegypti* Reproduction. *Int. J. Mol. Sci.* **2022**, 23, 7736. https://doi.org/10.3390/ijms23147736

Academic Editor: Michail Kotsyfakis

Received: 7 June 2022
Accepted: 5 July 2022
Published: 13 July 2022

Publisher's Note: MDPI stays neutral with regard to jurisdictional claims in published maps and institutional affiliations.

1. Introduction

The anthropophilic yellow fever mosquito *Aedes aegypti* is considered as the principal vector of multiple arboviruses, being responsible for the transmission of yellow fever virus (YFV), dengue virus (DENV), Zika virus (ZIKV), Chikungunya virus (CHIKV), and a newly emerging pathogen, Mayaro virus (MAYV). Dengue fever is the most widespread mosquito-borne viral disease, reported in at least 130 countries. In the last 20 years, the prevalence of dengue fever has increased eightfold. In 2019 alone, approximately 5.2 million cases of dengue were reported globally [1]. Cases are mainly reported from tropical and subtropical countries in Asia and South America. The staggering number of dengue fever cases adds a significant burden to the medical and economic infrastructure of many developing countries [2,3]. The lack of effective medical interventions such as vaccines and drugs make the control of dengue fever challenging [4,5].

Mosquito control with chemical insecticides remains the primary method for controlling *Ae. aegypti* mosquitoes. However, in recent years, multiple mechanisms of insecticide resistance against a majority of insecticides have been reported in *Ae. aegypti* mosquitoes around the globe [6–9]. Given the limited arsenal of insecticides and increased resistance, there is a dire need to develop novel control strategies against *Ae. aegypti*.

Aedes aegypti are primarily anautogenous, requiring a blood meal for every gonotrophic cycle. This dependence on blood meals makes them an ideal vector for multiple viral pathogens, including the dengue virus (DENV). The dengue transmission cycle begins when a female *Ae. aegypti* mosquito takes a blood meal from an infected host. The virus first infects and replicates in the mosquito midgut, eventually being released in the hemocoel. From this stage, the virus causes systemic infection in multiple mosquito tissues including the salivary gland. At this stage, the mosquito becomes infectious and can transmit DENV to a new human host when the mosquito takes another blood meal. The virus encounters multiple tissue barriers at each stage of the infection [10]. During the infection cycle in the vector, DENV interacts with multiple proteins that either positively or negatively affect the outcome of the infection [11,12]. One such reported protein is the *Ae. aegypti* cysteine-rich venom protein 379 (CRVP379, AAEL000379), which has been shown to interact with DENV and to be important for the establishment of infection [13].

The name cysteine-rich venom protein 379 was derived from similar cysteine-rich secreted proteins that were initially characterized from reptilian venoms. These proteins, broadly known as CRISPs, act as ion channel inhibitors and anticoagulants in reptile venoms [14]. CRISPs also play important roles in reproduction and immune systems in various organisms of the animal kingdom [15–17]. Despite being rich in cysteine residues, *Ae. aegypti* CRVP379 belongs to a broad group of proteins known as the cysteine-rich trypsin inhibitor-like proteins (TIL; PF01826.) since they contain a conserved cysteine-rich trypsin-like inhibitor domain (TIL). Apart from the conserved TIL domain, the proteins have diverged significantly. Even though these proteins possess a conserved TIL domain, very few have been experimentally shown to possess protease inhibitor activity. These proteins are widely distributed in the animal and plant kingdoms and play indispensable roles in multiple biological processes [18–20].

TIL domain-containing proteins remain understudied in vector mosquitoes. CRVP379 is a TIL domain-containing protein that has been studied in *Ae. aegypti* in the context of DENV infection. This protein was shown to be upregulated upon DENV infection, and RNA interference (RNAi)-mediated silencing of CRVP379 reduces DENV titers in *Ae. aegypti* [13].

In recent years, CRISPR-Cas9-based advances in genome editing have helped expand our understanding of the biological roles of many mosquito proteins. For example, CRISPR-Cas9-mediated knockout of both *Ae. aegypti* and *Anopheles gambiae* genes has more convincingly revealed their functions in mosquito biology than did RNAi-mediated gene silencing, which often resulted in a partial hylomorphic phenotype because of insufficient protein depletion [21–23].

In the present study we have analyzed the gene expression and effects of the CRVP379 protein in various tissues and life stages of *Ae. aegypti*. We also generated two lines of *Ae. aegypti* with different mutations in CRVP379, utilizing CRISPR-Cas9-mediated gene knockout to decipher the biological role of CRVP379 in *Ae. aegypti*.

2. Results

As a first step towards understanding the role of CRVP379, we analyzed its DNA and protein sequences and compared them to other similar sequences available in the NCBI and VectorBase datasets. CRVP379 is located on chromosome 2 in the *Ae. aegypti* genome and has one intron and two exons. The gene encodes a 128-aa protein (13.7 kDa). BLAST results for CRVP379 showed multiple similar proteins present in *Ae. aegypti*, *Ae. albopictus*, and *Anopheles* and *Culex* genomes. CRVP379 showed sequence similarity to a wide variety of proteins, ranging from a small 82-aa protein from *Anopheles epiroticus* (AEPI007239) to a larger protein of 669 aa from *Culex quinquefasciatus* (CQUJHB008149.R12566). A phylogenetic analysis revealed that CRVP379 is closely related to the *Aedes albopictus* hypothetical protein (KXJ80742.1). Interestingly, in addition to mosquitoes, closely related proteins were also found in multiple spider species (Supplementary Figure S1A).

CRVP379 contains a predicted signal peptide (1–19 aa) and a predicted TIL domain (23–79 aa), which consists of ten cysteine residues that form five disulfide bonds (1–7, 2–6, 3–5, 4–10 and 8–9, Supplementary Figure S1B). This arrangement is conserved in most of the proteins belonging to the TIL protein superfamily. All of the proteins with sequence similarity to CRVP379 (from the BLAST results) showed the presence of this conserved TIL domain.

2.1. CRVP379 mRNA Abundance

To gain insight into the possible functions of CRVP379 in *Ae. aegypti* biology, we first analyzed its mRNA abundance at various life stages and in various tissues under a range of conditions.

CRVP379 was expressed at low levels during the first three larval instars. Larval instar L3 had the lowest comparative CRVP379 mRNA abundance among the assayed life stages, having a relative gene expression (RGE) of 0.7166 ± 0.1046 (an RGE of 1.0 being the average RGE across all the life stages) (Figure 1A). CRVP379 mRNA abundance increased in the fourth instar, with a statistically significant increase in the RGE to between L3 and L4 (Figure 1A). When compared to the larval stages, the pupal stage in both males and females showed an elevated abundance of CRVP379 mRNA, with mean RGE values of 1.363 ± 0.1994 and 1.696 ± 0.3665, respectively (Figure 1A). There was a statistically significant increase in the CRVP379 RGE between the pupal stage and the first three larval instars in females. In adult mosquitoes, both adult males and females showed reduced gene CRVP379 expression as compared to their respective pupal stages. Male adults showed a lower CRVP379 RGE (0.9195 ± 0.07066) than did female adults (1.244 ± 0.04201, Figure 1A).

Next, we analyzed the CRVP379 mRNA abundance in various tissues of male and female adults. CRVP379 was expressed at very low levels in the female midgut, fat body, and Malpighian tubules, with RGEs of 0.7267 ± 0.09050, 0.7007 ± 0.03540, and 0.7166 ± 0.1046, respectively. Male testes showed the second highest RGE, 1.316 ± 0.1933. Female ovaries showed a significantly higher abundance of CRVP379 mRNA than did all the other assayed tissues (RGE 2.572 ± 0.6676, Figure 1B).

CRVP379 was previously reported to play an important role in the midgut during DENV infection [13]. Therefore, we investigated the abundance of CRVP379 mRNA in the midgut at 24 and 48 h post-DENV infection. There was no significant difference in the CRVP379 mRNA abundance before blood feeding and 24 h after blood feeding in the midgut, regardless of infection status. CRVP379 mRNA abundance was significantly higher at 48 h after blood feeding than either before blood feeding or 24 h afterwards. However, this increase was seen in both the control and DENV-infected mosquitoes, with a significant difference between them (Figure 1C). These results indicate that CRVP379 transcription is regulated by blood feeding and not DENV infection.

Since our expression analyses showed that CRVP379 was highly expressed in *Ae. aegypti* female ovaries, we sought to investigate the influence of blood feeding on the transcriptional regulation of ovarian CRVP379 by blood feeding. We saw no difference in the CRVP379 mRNA abundance between unfed and blood-fed (6 h post-blood feeding) females, with RGEs of 0.9399 ± 0.02709 and 0.9459 ± 0.04592, respectively. CRVP379 mRNA abundance decreased slightly at 24 h after blood feeding in the ovaries, with an RGE of 0.7922 ± 0.1814. However, this decrease was not statistically significantly different from the values for unfed females or at 6 h after blood feeding. Interestingly, at 48 h after blood feeding, the CRVP379 mRNA abundance levels increased significantly (RGE 2.003 ± 0.7814) in the ovaries when compared to the pre-feeding and the 6- and 24-h post-feeding time points (Figure 1D).

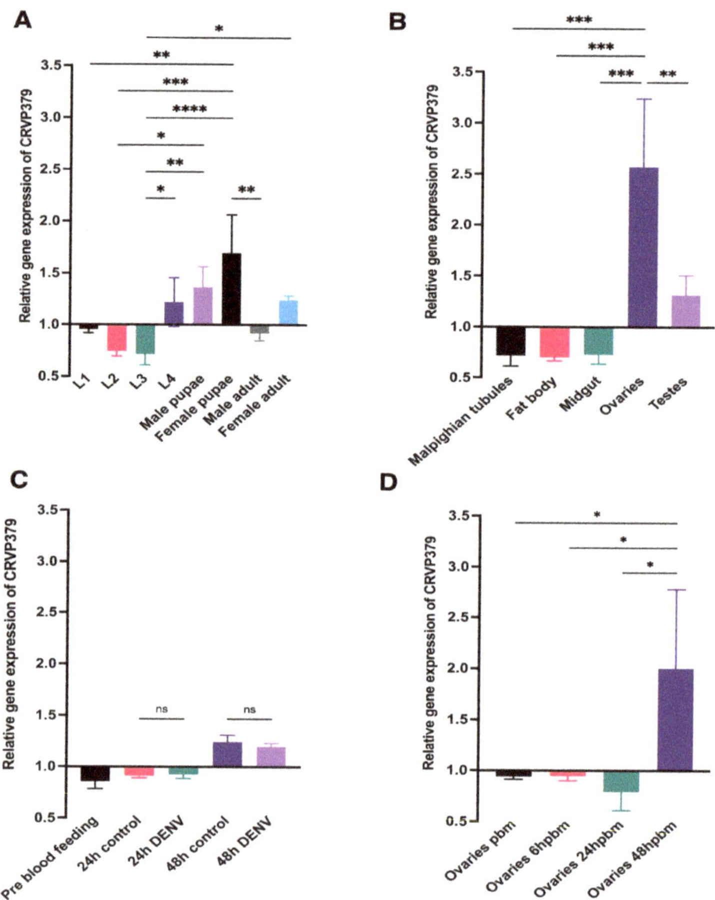

Figure 1. Relative gene expression of CRVP379 in different life stages and tissues of *Aedes aegypti*. (**A**) CRVP379 relative gene expression in the whole bodies of larval (L1–L4) and both male and female pupal and adult stages. (**B**) CRVP379 relative gene expression in various tissues of adult mosquitoes. (**C**) CRVP379 relative gene expression in the midgut before blood feeding and at 24 and 48 h after DENV infection. (**D**) CRVP379 relative gene expression in the ovaries before blood feeding and at 24 and 48 h after blood feeding. Means were compared with Tukey's multiple comparisons test; * = $p < 0.05$, ** = $p < 0.01$, *** = $p < 0.005$, **** = $p < 0.0001$. Relative gene expression (RGE) was calculated by dividing the delta ct for an individual life stage or condition by the average delta ct for that experiment. ns—non-significant.

2.2. CRVP379 Is Present in Ae. aegypti Ovaries but Not Secreted in the Hemolymph

We further analyzed CRVP379 protein abundance in *Ae. aegypti* mosquitoes with western blotting, utilizing a polyclonal antibody against CRVP379 protein. CRVP379 is a 128-aa, 13.7-kDa protein, and whole-body western blot analysis showed that CRVP379 is more abundant in adult female mosquitoes than in males (Figure 2A). The CRVP379 polyclonal antibody was not highly specific and also recognized other putative proteins, as indicated by the observance of multiple bands. Our mRNA abundance analysis showed that CRVP379 is highly expressed in the ovaries. Western blotting with proteins from female ovaries showed a high enrichment of CRVP379. The protein also has a predicted signal peptide, and we therefore investigated whether CRVP379 is secreted into the hemolymph. We did not detect CRVP379 in either the hemolymph or female bodies lacking ovaries, suggesting that CRVP379 is not secreted from the ovaries (Figure 2B). We also did not detect the presence of CRVP379 in female midguts (Supplementary Figure S2). Upon analyzing protein extracts from male testes, we observed non-specific binding to multiple proteins and one intense band at about 10 kDa, which was lower than the molecular size of CRVP379 (Supplementary Figure S2). This band may represent CRVP379 without its native signal peptide; however, the results of these western blots of the testes must be considered inconclusive because of the lack of specificity of the antibody available (Supplementary Figure S2).

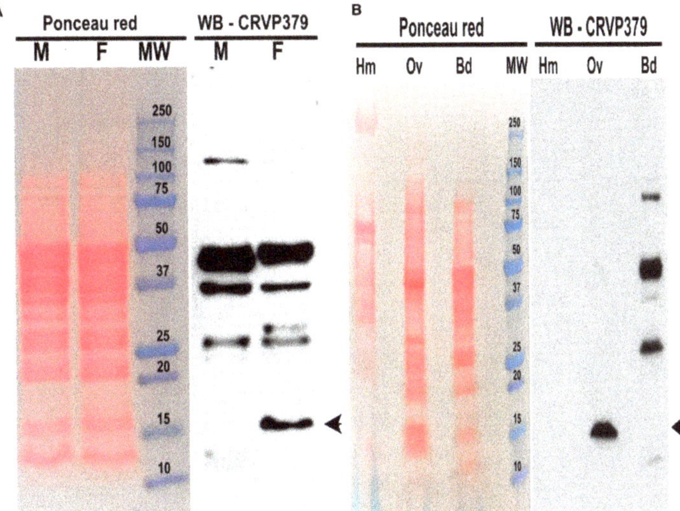

Figure 2. Western blot of CRVP379: the left panel shows protein staining with Ponceau red; the right panel shows western blotting with the CRVP379 antibody. In (**A**) M: male whole body; F: female whole body, MW: molecular weight marker. (**B**) Hm: female hemolymph; Ov: female ovaries; Bd: remaining female body without ovaries or hemolymph. Arrowhead—CRVP379.

2.3. CRVP379 Gene Silencing Does Not Influence DENV Infection in the Midgut

A previous study reported that silencing CRVP379 decreased DENV titers in the *Ae. aegypti* midgut [13]. Since we observed low CRVP379 mRNA and protein abundance in the midgut, we decided to retest the effect of CRVP379 silencing on DENV infection in the midgut. Three days after CRVP379 dsRNA injection, CRVP739 mRNA abundance was reduced by 50.71% to 70.53% when compared to the GFP dsRNA-injected controls. Our results showed that at seven days post-infection, the mean log DENV titer in the midgut of control mosquitoes injected with GFP dsRNA was 3.107 pfu/midgut (n = 72), and for CRVP379-silenced mosquitoes it was 2.966 pfu/midgut (n = 78). Even though this difference represented a slight reduction in DENV titers in the midgut upon silencing of the CRVP379 gene, this reduction was not statistically significant (Mann-Whitney test, p = 0.3484, Figure 3). There was also a slight reduction in DENV prevalence in CRVP379-silenced mosquitoes (59.40%) as compared to the GFP controls (65.49 %), although it also was not statistically significant (Fisher's exact test, p = 0.3572).

In the previous study the *Ae. aegypti* Rockefeller strain was used in gene silencing experiments, whereas here we used the Liverpool strain. Therefore, we also performed gene silencing experiments in the *Ae. aegypti* Rockefeller strain after DENV infection. As we observed for the Liverpool strain, we did not see any significant difference in the DENV titers in the midgut at four- or seven-days post-infection when the CRVP379 gene was silenced (Mann-Whitney test, p = 0.2645 for day four, p = 0.1239 for day seven; Supplementary Figure S3).

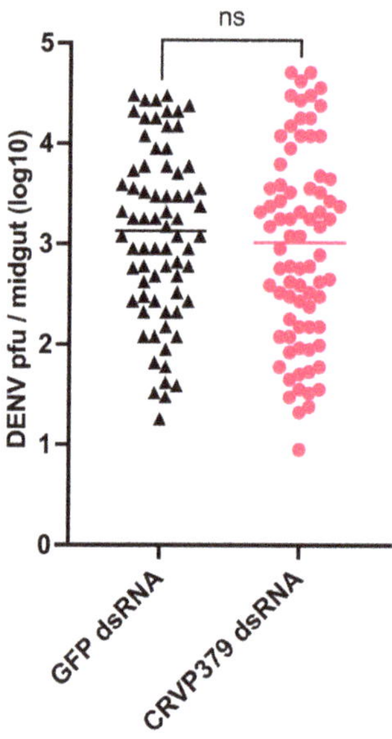

Figure 3. Effect of CRVP379 gene silencing on DENV titers in *Ae. aegypti* Liverpool midguts seven days after infection. Control mosquitoes injected with GFP dsRNA (*n* = 72), CRVP379 dsRNA injected misquotes (*n* = 78). ns—non-significant.

2.4. Generation of CRVP379 Knockout Lines Using the CRISPR-Cas9 System

To gain further insight into the role of CRVP379 in the biology of *Ae. aegypti*, we generated knockout lines utilizing our previously established CRISPR-Cas9 gene-editing method [21]. We used the *Ae. aegypti* Exu-Cas9 line, which expresses Cas9 driven by an embryo-specific promoter [24]. We performed two independent rounds of embryo injections with two separate mixes of guide RNAs (gRNAs, Figure 4A). The first embryo injection experiment involved a gRNA mix of gRNA 8, 4, and 1, and the injection of a total of 427 eggs, yielding a hatching rate of 59.25%. The second experiment involved injection of a gRNA mix comprising gRNA 8 and gRNA 4. For the second experiment we injected 245 eggs and obtained a hatching rate of 33.87%. Mutations in the surviving G0 were assessed by PCR and Sanger sequencing. From embryo injection experiment 1 we recovered 24 males and 24 females from G0 that were analyzed for CRVP379 mutations by PCR. All of the mosquitoes tested showed two distinct bands on agarose gels, which proved to represent a deleted region of the gene (Figure 4B). From embryo injection experiment 2, we recovered 12 males and 12 females that were analyzed for CRVP379 mutations. A total of 14 mosquitoes (six males, eight females) showed two distinct PCR bands on agarose gels, which were shown to indicate insertions in the gene (Figure 4B). Putative mutants from G0 were outcrossed with virgin *Ae. aegypti* Liverpool mosquitoes, and after confirming the mutations in G1, we maintained and outcrossed two independent lines with *Ae. aegypti* Liverpool mosquitoes for five generations. After five generations, a loss of the Exu-cas9 cassette was confirmed by screening the larvae under red fluorescent light. Sequencing of CRVP379 fragments from these mosquitoes confirmed a 105-bp deletion between gRNA 1 and gRNA 4 in one line. This deletion was in-frame with the coding sequence toward the end of the TIL domain, so this line was therefore designated as a CRVP379-partial knockout (CRVP379-pKO). CRVP379-pKO was predicted to produce a mutant protein with a missing 35-aa region towards the C-terminus. This CRVP-pKO line had most of its predicted TIL domain intact, with only one cysteine and 5 aa missing from it (Figure 4B,C).

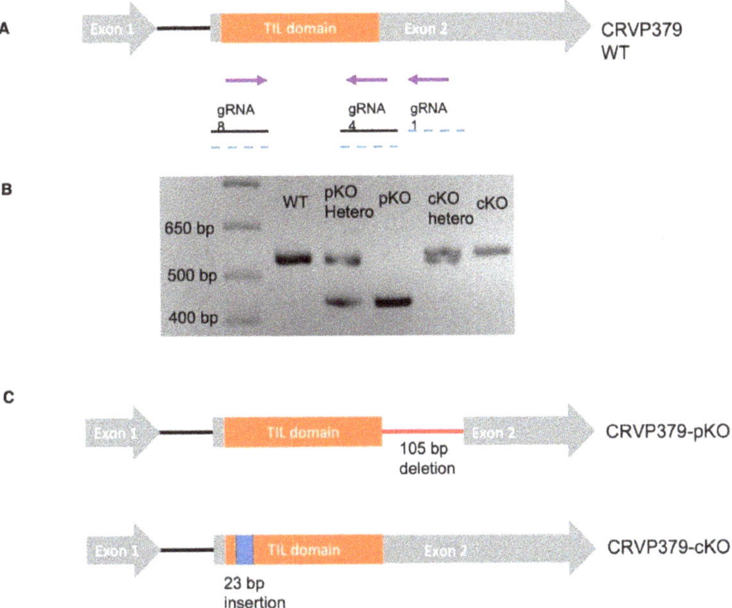

Figure 4. Schematic representation of CRISPR-Cas9-mediated knockout of CRVP379. (**A**) Guide RNAs used for embryo injection; guide RNAs highlighted with a dashed blue line were used together for injection 1; guide RNAs highlighted with a solid black line were used together for injection 2. (**B**) Agarose gel electrophoresis of PCR products of the CRVP379 gene with the CRVP Seq F/R primer set, from left to right: molecular weight ladder, wild type Liverpool, CRVP-pKO heterozygous, CRVP-pKO homozygous (105-bp deletion), CRVP-cKO heterozygous, CRVP-cKO homozygous (23-bp insertion). (**C**) Schematic representation of CRVP-pKO and CRVP-cKo mutations.

Sequencing of CRVP379 from the second KO line confirmed a 23-bp insertion near gRNA 8. This insertion was a result of a duplication of a 23-bp sequence adjacent to gRNA8 (Figure 4B,C). This insertion was in the predicted TIL domain, and it disrupted the open reading frame of the gene; hence, this line was designated CRVP379 complete knockout (CRVP379-cKO). This line was predicted to produce a 90-aa mutant protein with minimal similarity to wild-type CRVP379 and no intact TIL domain. The CRVP-cKO line did not yield hatchable eggs in its homozygous state, so this line was maintained in a heterozygous state. The predicted mutant protein sequences from the CRVP-pKO and cKO lines are provided in Supplementary file S1.

2.5. CRVP379 Is Important for Follicular Morphology in Ae. aegypti Ovaries

Next, we performed Western blots with protein extracts from adult females of both the mutant pKO and cKo lines to detect CRVP379. We observed a significant reduction in detectable CRVP379 protein in both the pKO and cKO lines (Supplementary Figure S4). To investigate whether CRVP379 was present in the ovaries, we performed immunofluorescent staining on dissected ovaries from wild-type and mutant lines. In the wild-type mosquitoes, CRVP379 was clearly present in the follicular cells (Figure 5A), and specifically in the periphery of these cells (Figure 5B). CRVP379 was not detected in either the pKO or cKO lines in the ovarian follicular cells (Figure 5C,D). Both western blot analysis and confocal microscopy confirmed the loss of CRVP379 from the mutant lines. We also investigated the presence of CRVP379 in male testes by immunofluorescent staining. However, CRVP379 could not be clearly localized in the testes because of the presence of non-specific background fluorescence in both the wild-type and mutant lines (Supplementary Figure S5). The confocal microscopy for testes is in accordance with our western blot results.

Our microscopy studies also revealed striking differences in the general morphology of the follicles between wild-type and mutant mosquitoes. In the wild-type mosquitoes, the follicles had a well-defined follicular epithelial cell lining, whereas in both mutant lines the follicular cell lining was less defined and showed an abnormal morphology (Supplementary Figure S6). The size of the follicles also appeared to be smaller in the mutant lines than in the wild-type mosquitoes. Using the ImageJ Fiji software (area in pixels), we confirmed that the CRVP379-cKO line had significantly smaller follicles than did either the wild-type or the CRVP379-pKO line (Supplementary Figure S6).

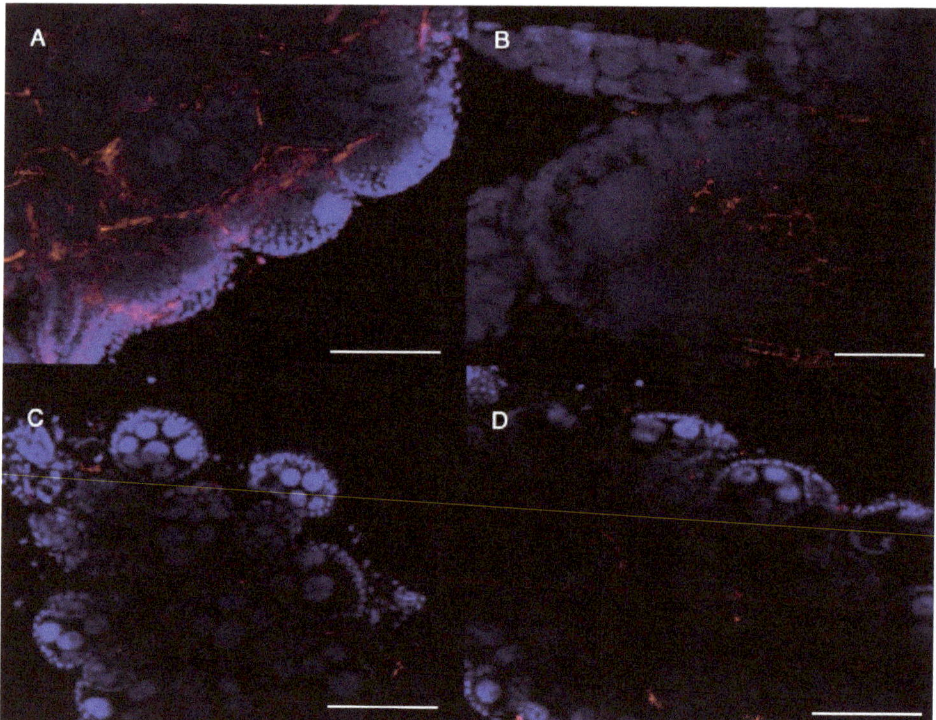

Figure 5. Immunofluorescent detection of CRVP379 in the ovaries of *Ae. aegypti*. Anti-CRVP379 antibody in red, DAPI (nuclei) in blue. (**A**) *Ae. aegypti* Liverpool ovaries, 20× magnification. (**B**) *Ae. aegypti* Liverpool ovaries, 63× magnification, in both (**A**,**B**). CRVP379 can be seen in the follicular epithelial cells. (**C**) CRVP-pKO ovaries, 20× magnification. (**D**) CRVP-cKO ovaries, 20× magnification. In both CRVP-pKO and CRVP-cKO ovaries, a clear localization of CRVP379 was not observed. Images are representative images of ovaries from each line. Images were obtained by confocal microscopy; Z-slides were stacked to get a single projection. Bar scale (**A**,**C**,**D**) = 100 μm, (**B**) = 20 μm.

2.6. CRVP379 Plays an Important Role in the Male and Female Reproductive Systems

As described above, CRVP379 was highly expressed in the ovaries, and the CRVP379-pKO and cKO lines showed abnormal follicular morphology. To investigate the role of CRVP379 in *Ae. aegypti* reproductive biology, we compared the fecundity of the CRVP379 mutant lines to wild-type mosquitoes. For this purpose, we performed crosses with both female and male wild-type Liverpool and CRVP-pKO males and females, respectively. For the pKO line, there was no significant difference in the number of eggs laid by female mosquitoes irrespective of the genotype of the male (Figure 6A). When we compared the hatching rate, we saw no difference between wild-type Liverpool male-female crosses (mean hatching rate, 79.38%) and wild-type female-CRVP379-pKO male crosses (mean hatching rate, 80.30%). These data indicate that CRVP-pKO males are reproductively comparable to Liverpool wild-type males (Figure 6B). However, in the case of females, CRVP379-pKO females showed a reduced hatching rate irrespective of the male genotype with which they were crossed. CRVP379 pKO females crossed with wild-type Liverpool males had a mean egg hatching rate of 47.01%, which was significantly lower than that for the Liverpool line or for crosses between Liverpool females and CRVP-pKO males (Dunn's multiple comparison test, $p < 0.006$). Crossing of CRVP379-pKO females with CRVP-pKO males produced an even lower mean hatching rate of 38.97% (Dunn's multiple comparison test) when compared to Liverpool crosses and Liverpool female-CRVP-pKO male crosses ($p < 0.0001$, Figure 6B). A simplified graphical version of various CRVP-pKO crosses is presented in the supplementary files (Supplementary Figure S7A). This result suggests that even a partial deletion of CRVP379 affects female reproductive capabilities, and the WT male genotype is unable to rescue this phenotype.

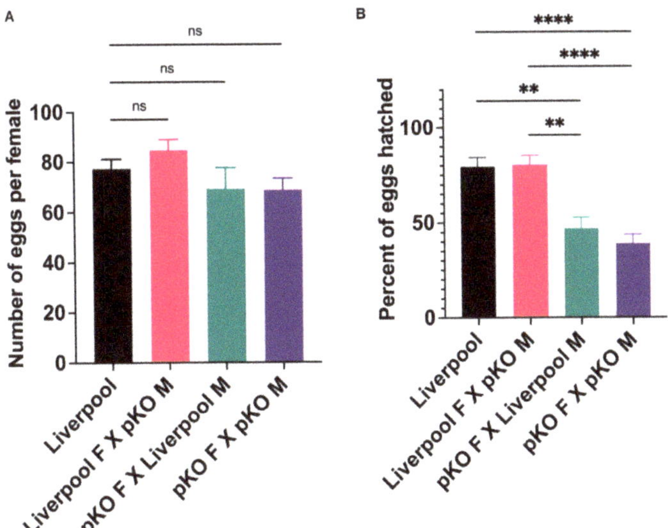

Figure 6. (**A**) Number of eggs laid per female with various crosses of the CRVP-pKO line; M = male, F = female. (**B**) Percentage of eggs hatched from various crosses of CRVP-pKO. Statistically significant p values are depicted by an asterisk (*); ** = $p < 0.006$, **** = $p < 0.0001$, error bars represent standard error of mean (Dunn's multiple comparisons test, ns—non-significant).

For the CRVP-cKO line we were unable to maintain a homozygous genotype. Multiple crosses with CRVP-cKO homozygous mosquitoes yielded eggs that did not hatch (confirmed by leg PCR and sequencing). For this reason, every single mosquito had to be genotyped by leg PCR and sequencing before any experiment. Given its hampered reproduction, the CRVP-cKO line was maintained as a heterozygous genotype. Crosses between heterozygotes yielded wild-type, homozygous, and heterozygous mosquitoes. To assess the effect of complete CRVP379 knockout on male and female fecundity, we set up multiple crosses with wild-type, heterozygous, and homozygous mosquitoes. All but one of the CRVP-cKO crosses showed a significant difference from the wild-type in the number of eggs laid by an individual female. Only crosses between CRVP370-cKO heterozygous females and CRVP-cKO homozygous males produced significantly lower numbers of eggs when compared to the Liverpool strain (Dunn's multiple comparisons test, $p = 0.0066$, Figure 7A). However, we did see significant differences in hatching rates with the various CRVP-cKO crosses. When wild-type Liverpool females were crossed with CRVP-cKO homozygous males, there was a significant reduction in the hatching rate (33.94%, $p = 0.0072$) when compared to the Liverpool WT mosquitoes (76.84% hatching rate, Figure 7B). When CRVP-cKO heterozygous females were crossed with wild-type Liverpool males, there was no difference in the hatching rate. However, when CRVP-cKO heterozygous females were crossed with either heterozygous or homozygous males, we again observed a significant reduction in hatching rate when compared to the wild-type Liverpool crosses (Dunn's multiple comparisons test, $p = 0.0317$ and $p < 0.0001$ respectively; Figure 7B). When CRVP-cKO homozygous females were crossed with wild-type Liverpool males, the hatching rate was comparable to that for the wild-type Liverpool crosses. Like the previous results, when homozygous females were crossed with heterozygous males, a significant reduction in the hatching rate was observed (Dunn's multiple comparisons test, $p < 0.0001$; Figure 7B). For homozygous CRVP-cKO crosses, we were unable to obtain viable eggs from the crosses. None of eggs laid by the CRVP-cKO homozygous cross females hatched (Figure 7B). A simplified graphical version of various CRVP-cKO crosses is presented in the supplementary files (Supplementary Figure S7B). Preliminary observations indicate that the eggs from this cross had a collapsed chorion, typical of nonviable eggs previously reported from *Ae. aegypti* [25]. These experiments clearly show that CRVP379 plays an important role in both the male and female reproductive systems, since crosses involving both males and females from the CRVP-cKO line had a significantly lowered egg hatching rate.

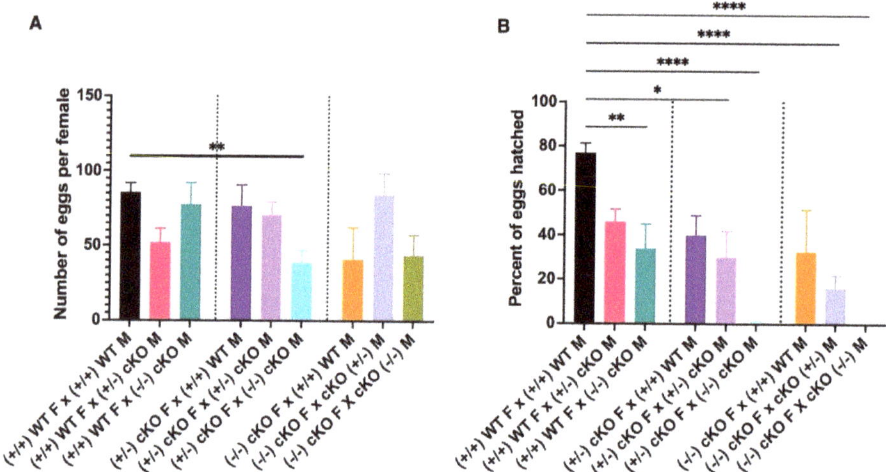

Figure 7. (**A**) Number of eggs laid per female for various crosses of the CRVP-cKO line. M = male, F = female. (**B**) Percentage of eggs hatched from various crosses of CRVP-cKO. Statistically significant *p* values are depicted with an asterisk (*); * = $p < 0.05$, ** = $p < 0.007$, **** = $p < 0.0001$, (Dunn's multiple comparisons test), error bars represent standard error of mean. WT Liverpool +/+, CRVP379-cKO heterozygous genotype +/−, CRVP379-cKO homozygous genotype −/−.

2.7. CRVP379 Does Not Influence DENV Infection in the Midgut

Next, we investigated whether the CRVP-pKO and cKO lines showed altered permissiveness to DENV infection in the midgut at seven days after infection, but the CRVP-pKO and wild-type Liverpool mosquitoes did not differ in their DENV infection prevalence (Liverpool 80.21%, CRVP-pKO 82.54%, Fisher's exact test, $p = 0.8366$). There was also no significant difference in DENV infection intensity (titer) in the midgut at seven days after infection (Mann-Whitney test, $p = 0.6501$; Figure 8A). Liverpool mosquitoes had a mean log DENV titer of 2.754, while CRVP-pKO mosquitoes had a mean log DENV titer of 2.832. For the CRVP-cKO line, there was no significant difference in prevalence among the wild-type Liverpool, CRVP-cKO heterozygous, and CRVP-cKO homozygous lines (Fisher's exact test, $p > 0.999$). There was also no significant difference in DENV infection intensity among these lines. The mean log DENV titer in the midgut at seven days after infection for the Liverpool, CRVP-cKO heterozygous, and CRVP-cKO homozygous lines was 2.505, 2.700, and 2.717, respectively (Figure 8B). It should be noted that the sample size for the infection experiments with CRVP-cKO were relatively smaller because of the difficulty in maintaining this line. Thus, these data corroborate our previous gene silencing experiments showing that CRVP379 does not serve as a DENV host or restriction factor.

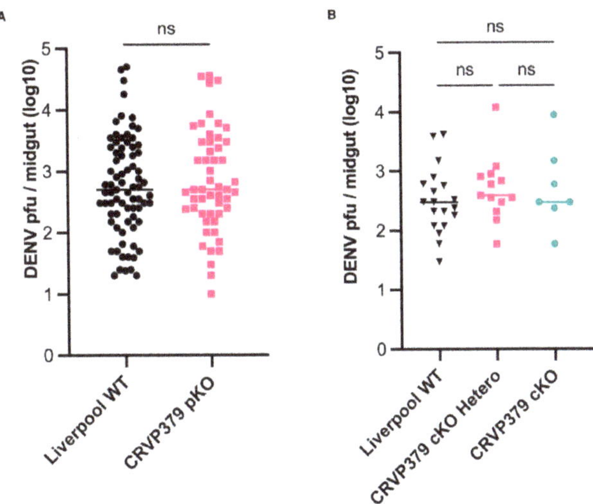

Figure 8. DENV titers in *Ae. aegypti* midguts at 7 days after infection, as determined by plaque assay. (**A**) CRVP-pKO compared with wild-type Liverpool mosquitoes. (**B**) CRVP-cKO heterozygous and homozygous mosquitoes compared with wild-type Liverpool mosquitoes. (Dunn's multiple comparisons test, ns = not statistically significant).

2.8. CRVP-pKO Females Have a Significantly Reduced Lifespan

Next, we asked whether the CRVP379 mutations influenced the lifespan of male and female mosquitoes. Because of the limited number of mosquitoes available from the CRVP-cKO line, we were unable to perform statistically meaningful lifespan experiments with this line. All of the lifespan experiments were carried out with the CRVP-pKO line. The male CRVP-pKO mosquitoes were not significantly different in lifespan than the wild-type mosquitoes. The median lifespan for the Liverpool males was 28 days, and that of the CRVP-pKO males was 29 days (log-rank Mantel-Cox test, $p = 0.4545$, Supplementary Figure S8A). The CRVP-pKO females displayed a significantly reduced lifespan when compared to the Liverpool wild-type mosquitoes. The median life span of the Liverpool females was 44 days, whereas the CRVP-pKO mosquitoes had a median life span of only 21 days (log-rank Mantel-Cox test, $p < 0.0001$, Supplementary Figure S8B). This result demonstrates that even a partial deletion in CRVP379 outside of its TIL domain has a significant impact on female lifespan but no effect on male lifespan.

3. Discussion

Here we sought to understand the role of cysteine-rich venom protein 379 (CRVP379) in *Ae. aegypti* biology by utilizing a CRISPR-Cas9-mediated gene knockout approach. Our study showed that CRVP379 is highly expressed in male testes and female ovaries. We also observed that CRVP379 expression is upregulated in the ovaries 48 h after blood feeding.

Immunohistological assays localized CRVP379 to the periphery of follicular epithelial cells in *Ae. aegypti* ovaries. We further showed that RNAi-mediated gene silencing and gene knockout of CRVP379 did not have an impact on DENV infection in *Ae. aegypti* mosquitoes. Experiments utilizing our CRISPR-Cas9-generated mutant lines suggested that CRVP379 can be at least partially functional when the deletion occurs outside the TIL domain. A CRVP379-pKO line with a partial in-frame deletion that left most of the TIL domain intact showed no difference with regards to male lifespan or male reproductive capabilities. However, female mosquitoes from this line had a significantly lower lifespan and egg hatching rate. When CRVP379 was completely inactivated in our CRVP379-cKO line, both male and female reproduction were severely hampered. We observed a smaller follicular size in both CRVP379 mutant lines. Overall, our study highlights the importance of CRVP379, a TIL domain-containing protein, in *Ae. aegypti* reproduction and possibly in other processes that regulate the female life span.

Even though TIL domain-containing proteins are distributed throughout the animal kingdom, very few proteins from insects have been reported or studied. The endoparasitoid wasp *Cotesia vestalis* has been shown to secrete a TIL protein from its teratocytes that blocks its host's melanization activity [26]. A cysteine rich TIL domain-containing protein from *Bombus ignitus* bumblebee venom has been shown to possess protease inhibitor and antifibrinolytic activity [27]. Similar cysteine-rich TIL domain-containing proteins from *Bombyx mori* have been shown to possess antifungal activity and effectively inhibit multiple microbial proteases [28,29]. These studies indicate that proteins with a conserved TIL domain are likely to have diverse tissue- and life stage-specific roles in different insects.

Phylogenetic analysis of *Ae. aegypti* CRVP379 has revealed multiple TIL domain-containing proteins in both *Anopheline* and *Culicine* mosquitoes. These proteins showed a high degree of conservation in the TIL domain but were otherwise diverse in terms of sequence and length. In humans, a TIL domain-containing protein, Von Willebrand factor, is translated as a massive 2818-aa protein and plays a key role in blood clotting [30,31]. At the same time, in the frog *Lepidobatrachus laevis*, a small 55-aa peptide containing the same TIL domain has protease-inhibiting activity [32]. These examples highlight the fact that despite the presence of a conserved TIL domain, there is functional and spatial diversity in TIL-containing proteins.

Our gene expression analysis indicated an increased expression of CRVP379 during the pupal stage. These results are similar to those from a previous study in the cotton bollworm *Helicoverpa armigera*, in which a TIL domain-containing protein, HaTIL2, was highly expressed in the pupal stage [33]. Multiple *Ae. aegypti* and *Ae. albopictus* trypsins are known to play an important role during the pupal stage [34–36]. It is likely that CRVP379-like trypsin inhibitors may similarly be playing an important role in trypsin regulation during the pupal stage. We also observed CRVP379 expression in male testes and high expression in female ovaries. These data are in accord with the *Ae. aegypti* atlas and *Ae. aegypti* developmental transcriptome by Akbari et al., in which CRVP379 is shown to be enriched in the ovaries and testes [37,38]. This finding was further corroborated by our western blotting and immunostaining experiments. We found CRVP379 to be localized to the periphery of the follicular epithelial cells. However, despite the presence of a signal peptide, we did not observe CRVP379 to be secreted into the hemolymph. This observation raises the possibility that the signal peptide plays a role in localizing the protein around the follicular epithelial cells rather than in extracellular secretion. The expression of CRVP379 was significantly increased in the ovaries at 48 h after blood feeding. All of these findings suggest that CRVP379 plays a key role in *Ae. aegypti* reproduction. An increase in the expression during pupal stages and in the ovaries after blood feeding strongly suggests a hormonal regulation of CRVP379. During these two stages, various hormones, such as 20E and juvenile hormone (JH) regulate multiple genes in mosquitoes [39]. Juvenile hormone also plays a key role in previtellogenic ovarian development [40,41]. A chymotrypsin-like serine protease in *Ae. aegypti* has been shown to be regulated by JH [42]. It would be interesting to study the possibility of hormonal regulation of CRVP379 in more detail in the future.

Interestingly, in contrast to the previously reported findings by Londono-Renteria et al., we did not detect a significant CRVP379 presence in the midgut [13]. We also did not observe any changes in CRVP379 gene expression after DENV infection. Similarly, gene silencing experiments also had no effect on DENV titers in two different *Ae. aegypti* lab strains. Because we did not observe CRVP379 in the midgut, it is unlikely to play a role in the DENV infection of that tissue. However, our results are not perfectly comparable to the study by Londono-Renteria et al. because of inherent differences in methodology and a lack of sufficient information concerning experimental procedures, such as their primer sequences used for gene expression and silencing [13].

Other than the signal peptide, the only notable feature of CRVP379 is the TIL domain, although its importance had never been experimentally validated. The importance of the C-terminus of this protein, which does not have any notable features, is also unknown. We were able to create two distinct CRVP379 mutant lines, one with a mutation inside and one outside the TIL domain. The CRVP379 partial knockout line (CRVP379-pKO) had a mutation outside the TIL domain and could still produce a 93-aa protein with most of its TIL domain intact. The partial knockout line (CRVP-pKO) did not show any altered permissiveness to DENV infection in the female midgut or effects on male lifespan or male reproduction. However, females from this line had a significantly shorter lifespan and reduced fecundity. These results raise two possibilities: (a) the CRVP379 residues between aa76 and aa110 play an important role in female *Ae. aegypti* but not in males; and (b) CRVP-pKO produces a completely inactive product, and CRVP379 does not play a role in either male reproduction or male lifespan. If the latter hypothesis were true, the other mutant line with a complete inactivation CRVP379-cKO should have shown a similar phenotype. However, the CRVP-cKO line had severe defects in both male and female reproduction. In fact, we were unable to maintain a homozygous knockout line because none of the eggs from this cross hatched. We also observed that CRVP-cKO homozygous males and females could produce viable eggs if the other partner had a wild-type CRVP379 genotype. This result suggests that CRVP379 may be important for fertilization and/or embryonic development. Localization of CRVP379 to the male reproductive system was not possible with our polyclonal antibody, since it bound to multiple proteins (i.e., was nonspecific). A more specific monoclonal antibody is likely to shed light on the localization of CRVP379 in the male reproductive tissues. In the case of both mutant lines, we observed smaller follicles in the females when compared to wild-type mosquitoes, with the cKO line showing the smallest follicles. Localization of CRVP379 to the follicular epithelial cells further supports a role for CRVP379 in the *Ae. aegypti* female reproductive system. These results are similar to another study carried out in the brown plant hopper *Nilaparvata lugens*, in which an inter-alpha-trypsin inhibitor has been shown to play an important role in reproduction and ovarian development [43].

It should be noted that it was challenging to maintain the CRVP mutant lines. The CRVP-pKO line demonstrated a very short lifespan and reduced fecundity, and the CRVP-cKO line could not be maintained as a homozygous genotype. To study the CRVP379-cKO line, heterozygous mosquitoes were crossed to obtain different genotypes of CRVP-cKO. Each mosquito among the progeny was then genotyped by leg PCR and the sequencing of the PCR product. Some of the progeny had to be utilized for maintenance of the line, and this necessity effectively limited the experiments we were able to perform with this line. Alternative strategies such as the use of conditional knockouts might permit a more comprehensive study of CRVP379.

Finally, we showed that CRVP379 does not play a role in DENV infection in the midgut, given that our data showed that both the pKO and cKO lines had the same DENV titer in the midgut at seven days after infection. Our gene expression analysis, gene silencing experiments, and DENV infection assays in our mutant lines comprehensively show that CRVP379 does not play a role in DENV infection.

Our study highlights the role of TIL domain-containing proteins in *Ae. aegypti* re productive biology. We also show that certain residues in CRVP379 may play different roles in male and female mosquitoes. Given its importance in the biology of *Ae. aegypti* mosquitoes, which are key vectors of many viral diseases, CRVP379 deserves to be evaluated more fully in terms of its potential usefulness in developing novel mosquito population control strategies.

4. Materials and Methods

4.1. Ethics Statement

This study was carried out in strict accordance with the recommendations in the Guide for the Care and Use of Laboratory Animals of the National Institutes of Health and the Animal Care and Use Committee of the Johns Hopkins University (Permit Number: M006H300). Mice were only used for mosquito rearing as a blood source, according to approved protocol. Commercial anonymous human blood was used for DENV infection assays in mosquitoes, and informed consent was therefore not applicable. The Institutional Animal Care and Use Committee (IACUC) approved the protocol.

4.2. Mosquito Rearing

Ae. aegypti Liverpool, Rockefeller, and Exu-Cas9 strains were used in this study. *Ae. aegypti* Exu-Cas9 strain was a gift from Omar Akbari (University of California San Diego). All of the mosquitoes were maintained on a 10% sucrose solution at 27 °C and 80% relative humidity with a 14:10 h light-dark cycle in the Johns Hopkins Malaria Research Institute's insectary.

4.3. Cell Culture

Ae. albopictus cells C6/36 (ATCC CRL-1660) were cultured in MEM (Gibco) supplemented with 10% FBS, 1% L-glutamine, 1% non-essential amino acids, and 1% penicillin/streptomycin. Cells were maintained at 32 °C with 5% CO_2. Baby hamster kidney cells (BHK-21, ATCC CCL-10) were cultured in DMEM (Gibco) supplemented with 10% FBS, 1% L-glutamine, 1% penicillin/streptomycin, and 5 µg/mL Plasmocin (Invivogen, San Diego, CA, USA). Cells were maintained at 37 °C with 5% CO_2.

4.4. Dengue Virus

DENV2 strain New Guinea C (NGC) was use in the study. The virus was propagated in C6/36 cells and stored at −80 °C. The virus titer was determined by plaque assay on BHK21 cells.

4.5. dsRNA Synthesis

Total RNA was extracted from a pool of five one-week-old *Ae. aegypti* Liverpool and Rockefeller females using the Trizol RNA extraction method. cDNA was synthesized from the total RNA using an oligo dT primer with the MMLV reverse transcriptase kit. Total cDNA was used to amplify *Ae. aegypti* CRVP379 (AAEL000379) using the primer set CRVP-RNAi-F and CRVP-RNAi-R (Supplementary Table S1). GFP gene was amplified using primer set GFP-RNAi-F and GFP-RNAi-R with GFP plasmid as a template (Supplementary Table S1). Amplified PCR products were purified using a Zymo DNA concentrator and cleanup kit. A total of 1 µg of PCR product was used to synthesize dsRNA using a NEB T7 HiScribe kit according to the manufacturer's instructions. dsRNA was purified using isopropanol-sodium acetate precipitation. dsRNA was dissolved in sterile distilled water, and the concentration was measured with a Nanodrop spectrophotometer. The final concentration of the dsRNA was adjusted to 3 µg/µL.

4.6. Gene Silencing

One-week old female *Ae. aegypti* mosquitoes (Liverpool and Rockefeller) were anesthetized on a cold block and intrathoracically injected with 69 nL of 3 µg/µL dsRNA (either

CRVP379 or GFP as a control). *Ae. aegypti* Liverpool and Rockefeller females were injected with dsRNA generated from the same strain. A pool of five females was collected at 72 h post-injection prior to infection with DENV2. RNA was extracted from this pool to quantify gene silencing efficiency.

4.7. Gene Expression

Ae. aegypti (Liverpool) mosquitoes were used to analyze the gene expression of CRVP379 at various life stages and in various tissues. Ten L1-L4 larvae, five male and five female pupae, and five male and five female adults were collected in Trizol and used for RNA extraction. The midgut, fat bodies, ovaries, and Malpighian tubules were dissected out and collected from a total of 10 one-week-old females. Testes were collected from a total of 20 one-week-old males. To study the CRVP379 gene expression in the ovaries after blood feeding, ovaries from 10 females were collected before blood feeding and at 6, 24, and 48 h after blood feeding. To quantify CRVP379 gene expression in the midguts after DENV2 infection, 20 female midguts were dissected and collected prior to DENV infection, and at 24 and 48 h after DENV infection. All of the tissues were collected in Trizol, and RNA was isolated using the Trizol extraction method. All of the experiments were carried out independently in triplicate. Gene expression was measured using TaqMan™ Universal Master Mix with CRVP qPCR primers. The *Ae. aegypti* S17 gene was used as a housekeeping gene control. Relative gene expression (RGE) was calculated by dividing the delta ct for an individual life stage or condition by the average delta ct for that experiment.

4.8. Generation of CRVP379 Knockout Lines

To generate CRVP379 knockout lines, the CRISPR-Cas9-mediated knockout method using the *Ae. aegypti* Exu-Cas9 line was utilized. *Ae. aegypti* Exu-Cas9 expresses Cas9 driven by an embryo-specific promoter. Guide RNAs targeting *Ae. aegypti* CRVP379 were designed using the CHOPCHOP web tool [44]. Guide RNAs were synthesized in vitro according to the previously described protocol using a gene-specific forward primer with a T7 promoter and a common reverse primer (Supplementary Table S1).

4.9. Embryo Injections

Early embryos (within 30 min of egg laying) of *Ae. aegypti*-Exu-Cas9 were injected with 100 ng/μL of each guide RNA. Embryos were injected independently with two or three different guide RNA mixes (Supplementary Table S1), as described previously. The surviving G0 adults were sexed, and mutations were confirmed by leg PCR followed by sequencing with CRVP primers (Supplementary Table S1).

4.10. Leg PCR

A quarter of a hind leg of a four-day-old adult mosquito was dissected and suspended in 19 μL of squash buffer and 1 μL of proteinase K (Promega). The legs were incubated at 56 °C for 30 min, followed by incubation for 10 min at 95 °C. Five microliters of the above solution was then used for PCR with NEB OneTaq 2X master mix in a 25-μL reaction with the primer set. PCR products were sequenced, and gene sequences and peaks were compared to the *Ae. aegypti* Liverpool CRVP379 gene sequence. Mosquitoes with mutations were individually outcrossed with the *Ae. aegypti* Liverpool strain. After five generations of outcrossing with the Liverpool strain, mutant lines were in-crossed, and homozygous mutant lines were obtained. Two *Ae. aegypti* lines with a different mutation in the CRVP379 gene were generated. One line was designated as CRVP partial knockout (CRVP-pKO, 105-bp deletion), and the other was designated as CRVP379 complete knockout (CRVP-cKO, 23-bp insertion).

4.11. Line Maintenance, Fecundity, and Life Span

The CRVP379 complete knockout line (CRVP379-cKO) was maintained as a heterozygous genotype. For each generation, mosquitoes were genotyped by leg PCR as described above. Multiple crosses were set up to include both males and females of wildtype CRVP379, heterozygous CRVP379-cKO mutant, and homozygous CRVP379-cKO genotypes. Female mosquitoes were fed on anesthetized mice, and fully engorged females were selected for fecundity experiments. Three days after blood feeding, individual females were added to 50-mL conical tubes lined with moist filter paper. Individual females were allowed to lay eggs in one tube for 48 h. After 48 h, the filter papers were removed, and the eggs were allowed to dry for five days. Five days later, the eggs were hatched in sterile hatching broth, and the larvae were counted three days later. The life spans of the CRVP379-pKO and Liverpool parental lines were measured as described previously. In brief, male and female mosquitoes were sexed at the pupal stage. A total of 30 virgin males and females were maintained in a paper cup and given 10% sucrose solution on a piece of cotton. Mosquito mortality was measured every day.

4.12. Immunoblotting

Mosquitoes (either complete mosquitoes or individual tissues) were lysed in 40 µL of RIPA Buffer (Sigma, St. Louis, MO, USA, R0278) containing protease inhibitor cocktail (Complete, EDTA-free; Roche, Basel, Switzerland). Protein concentrations were measured using a Micro BCA Protein Assay Kit (Thermo Scientific, Waltham, MA, USA, # 23235), and the concentration of the samples was adjusted to 5 µg/µL. Bolt LDS Sample buffer (Invitrogen, Waltham, MA, USA, NP0007) and Bolt/NuPAGE reducing agent load (Invitrogen, NP0009) were added, and samples were heated at 100 °C for 5 min. Samples (30 µg of protein) were run in Tris-glycine gel, 4–20% (Invitrogen, XP04205BOX) at a constant 25 mA for 75 min. Proteins were electro-transferred to a nitrocellulose membrane (0.2-µm pore size; BIO-RAD, 1704271) using a Trans-Blot Turbo Transfer System (BIO-RAD, Hercules, CA, USA, 1704150) at 2.5 mA for 7 min. The membrane was stained with Ponceau S staining solution (Cell Signaling Technology, Danvers, MA, USA, 54803S) to corroborate the quality of transfer, followed by three washes with PBS for 10 min each. The membrane was blocked with Blocker Solution (blotting-grade blocker; BIO-RAD, 170604) at 5% on PBS for 60 min at room temperature (RT). The membrane was incubated overnight with 0.5× of blocker solution containing 1:2000 of rabbit anti-*Aedes aegypti* CRVP379 antibody (Boster Biological Technology, Pleasanton CA, USA, Catalog # DZ41086) at 4 °C. CRVP379 antibody at a dilution of 1:2000 was observed to be optimum for western blots. The membrane was washed three times with PBST (PBS with 0.2% Tween 20), then incubated with 0.5× of blocker solution containing 1:10,000 Anti-Rabbit IgG, HRP-linked Antibody (Cell Signaling Technology, 7074) for 60 min at RT, washed again, and developed using ECL Prime Western Blotting Detection Reagent (GE Healthcare, Chicago, IL, USA, RPN2232) and High Performance Chemiluminescence Film (GE Healthcare).

4.13. Immunohistochemical Staining and Microscopy

Ae. aegypti Liverpool, CRVP379-pKO, and CRVP379-cKO males and females were cold-anesthetized on ice. Ovaries and testes from four-day old mosquitoes were dissected using sharp forceps and entomological needles. The ovaries were stained following a method previously described [45]. CRVP379 antibody was used at a dilution of 1:200, and AlexaFluor568-goat anti-rabbit antibody (Life Technologies, A11011) was used at a dilution of 1:200. Samples were mounted using Fluoromount-G with DAPI (Invitrogen, 00-4959-52) and examined with a Zeiss LSM 1710 confocal microscope. Images were analyzed with ImageJ Fiji software [46]. The Z-stack (all slides) data sets were analyzed and presented as a single 2D projection. Brightfield and DAPI stained ovaries were also observed under a LEICA DM2500 microscope, and images were captured with a DFC310 FX camera. Three ovaries from each line were used to measure the follicular cell size using ImageJ Fiji software, License GPLv3+, Release 2.5.0.

Supplementary Materials: The following supporting information can be downloaded at: https://www.mdpi.com/article/10.3390/ijms23147736/s1.

Author Contributions: C.V.T., G.D., V.C.-J. and S.D. conceived and designed the experiments. C.V.T., V.C.-J. and S.D. performed the experiments. C.V.T., V.C.-J., S.D. and N.R. analyzed the data. C.V.T. and G.D. wrote the manuscript. All authors have read and agreed to the published version of the manuscript.

Funding: This work has been supported by the National Institutes of Health grant R01AI141532, the Bloomberg Philanthropies and Johns Hopkins Malaria Institute postdoctoral fellowship awarded to C.V.T.

Institutional Review Board Statement: Not applicable.

Informed Consent Statement: Not applicable.

Data Availability Statement: Not applicable.

Acknowledgments: We thank the Johns Hopkins Malaria Research Institute Insectary for providing the mosquito-rearing facility and the Parasitology Core facilities for providing naive human blood. We would like to thank Deborah McClellan for editorial assistance.

Conflicts of Interest: The authors declare that they have no conflict of interest.

References

1. WHO Dengue Fact Sheet. Available online: https://www.who.int/news-room/fact-sheets/detail/dengue-and-severe-dengue (accessed on 4 July 2022).
2. Gubler, D.J. The Economic Burden of Dengue. *Am. J. Trop. Med. Hyg.* **2012**, *86*, 743. [CrossRef]
3. Shepard, D.S.; Coudeville, L.; Halasa, Y.A.; Zambrano, B.; Dayan, G.H. Economic Impact of Dengue Illness in the Americas. *Am. J. Trop. Med. Hyg.* **2011**, *84*, 200. [CrossRef] [PubMed]
4. Lim, S.P. Dengue Drug Discovery: Progress, Challenges and Outlook. *Antiviral Res.* **2019**, *163*, 156–178. [CrossRef] [PubMed]
5. Wilder-Smith, A. Dengue Vaccine Development by the Year 2020: Challenges and Prospects. *Curr. Opin. Virol.* **2020**, *43*, 71–78. [CrossRef]
6. Badolo, A.; Sombié, A.; Pignatelli, P.M.; Sanon, A.; Yaméogo, F.; Wangrawa, D.W.; Sanon, A.; Kanuka, H.; McCall, P.J.; Weetman, D. Insecticide Resistance Levels and Mechanisms in Aedes Aegypti Populations in and around Ouagadougou, Burkina Faso. *PLoS Negl. Trop. Dis.* **2019**, *13*, e0007439. [CrossRef] [PubMed]
7. Marcombe, S.; Fustec, B.; Cattel, J.; Chonephetsarath, S.; Thammavong, P.; Phommavanh, N.; David, J.P.; Corbel, V.; Sutherland, I.W.; Hertz, J.C.; et al. Distribution of Insecticide Resistance and Mechanisms Involved in the Arbovirus Vector Aedes Aegypti in Laos and Implication for Vector Control. *PLoS Negl. Trop. Dis.* **2019**, *13*, e0007852. [CrossRef]
8. Sene, N.M.; Mavridis, K.; Ndiaye, E.H.; Diagne, C.T.; Gaye, A.; Ngom, E.H.M.; Ba, Y.; Diallo, D.; Vontas, J.; Dia, I.; et al. Insecticide Resistance Status and Mechanisms in Aedes Aegypti Populations from Senegal. *PLoS Negl. Trop. Dis.* **2021**, *15*, e0009393. [CrossRef]
9. Vontas, J.; Kioulos, E.; Pavlidi, N.; Morou, E.; della Torre, A.; Ranson, H. Insecticide Resistance in the Major Dengue Vectors Aedes Albopictus and Aedes Aegypti. *Pestic. Biochem. Physiol.* **2012**, *104*, 126–131. [CrossRef]
10. Mukherjee, D.; Das, S.; Begum, F.; Mal, S.; Ray, U. The Mosquito Immune System and the Life of Dengue Virus: What We Know and Do Not Know. *Pathogens* **2019**, *8*, 77. [CrossRef]
11. Simões, M.L.; Caragata, E.P.; Dimopoulos, G. Diverse Host and Restriction Factors Regulate Mosquito-Pathogen Interactions. *Trends Parasitol.* **2018**, *34*, 603–616. [CrossRef]
12. Tikhe, C.V.; Dimopoulos, G. Mosquito Antiviral Immune Pathways. *Dev. Comp. Immunol.* **2021**, *116*, 103964. [CrossRef]
13. Londono-Renteria, B.; Troupin, A.; Conway, M.J.; Vesely, D.; Ledizet, M.; Roundy, C.M.; Cloherty, E.; Jameson, S.; Vanlandingham, D.; Higgs, S.; et al. Dengue Virus Infection of Aedes Aegypti Requires a Putative Cysteine Rich Venom Protein. *PLoS Pathog.* **2015**, *11*, e1005202. [CrossRef] [PubMed]
14. Yamazaki, Y.; Morita, T. Structure and Function of Snake Venom Cysteine-Rich Secretory Proteins. *Toxicon* **2004**, *44*, 227–231. [CrossRef] [PubMed]
15. Gibbs, G.M.; Roelants, K.; O'Bryan, M.K. The CAP Superfamily: Cysteine-Rich Secretory Proteins, Antigen 5, and Pathogenesis-Related 1 Proteins—Roles in Reproduction, Cancer, and Immune Defense. *Endocr. Rev.* **2008**, *29*, 865–897. [CrossRef] [PubMed]
16. Koppers, A.J.; Reddy, T.; O'Bryan, M.K. The Role of Cysteine-Rich Secretory Proteins in Male Fertility. *Asian J. Androl.* **2011**, *13*, 111. [CrossRef]
17. Maróti, G.; Downie, J.A.; Kondorosi, É. Plant Cysteine-Rich Peptides That Inhibit Pathogen Growth and Control Rhizobial Differentiation in Legume Nodules. *Curr. Opin. Plant Biol.* **2015**, *26*, 57–63. [CrossRef]
18. Lavergne, V.; JTaft, R.J.; Alewood, P.F. Cysteine-Rich Mini-Proteins in Human Biology. *Curr. Top. Med. Chem.* **2012**, *12*, 1514–1533. [CrossRef]

19. Srivastava, S.; Dashora, K.; Ameta, K.L.; Singh, N.P.; El-Enshasy, H.A.; Pagano, M.C.; Hesham, A.E.-L.; Sharma, G.D.; Sharma, M. Bhargava, A. Cysteine-Rich Antimicrobial Peptides from Plants: The Future of Antimicrobial Therapy. *Phyther. Res.* **2021**, *35* 256–277. [CrossRef]
20. Zeng, X.C.; Liu, Y.; Shi, W.; Zhang, L.; Luo, X.; Nie, Y.; Yang, Y. Genome-Wide Search and Comparative Genomic Analysis of the Trypsin Inhibitor-like Cysteine-Rich Domain-Containing Peptides. *Peptides* **2014**, *53*, 106–114. [CrossRef]
21. Dong, S.; Ye, Z.; Tikhe, C.V.; Tu, Z.J.; Zwiebel, L.J.; Dimopoulos, G. Pleiotropic Odorant-Binding Proteins Promote Aedes Aegypti Reproduction and Flavivirus Transmission. *MBio* **2021**, *12*, e0253121. [CrossRef]
22. Dong, Y.; Simões, M.L.; Marois, E.; Dimopoulos, G. CRISPR/Cas9 -Mediated Gene Knockout of Anopheles Gambiae FREP1 Suppresses Malaria Parasite Infection. *PLoS Pathog.* **2018**, *14*, e1006898. [CrossRef] [PubMed]
23. Simões, M.L.; Dong, Y.; Mlambo, G.; Dimopoulos, G. C-Type Lectin 4 Regulates Broad-Spectrum Melanization-Based Refractoriness to Malaria Parasites. *PLoS Biol.* **2022**, *20*, e3001515. [CrossRef] [PubMed]
24. Li, M.; Bui, M.; Yang, T.; Bowman, C.S.; White, B.J.; Akbari, O.S. Germline Cas9 Expression Yields Highly Efficient Genome Engineering in a Major Worldwide Disease Vector, Aedes Aegypti. *Proc. Natl. Acad. Sci. USA* **2017**, *114*, E10540–E10549. [CrossRef] [PubMed]
25. Isoe, J.; Koch, L.E.; Isoe, Y.E.; Rascón, A.A.; Brown, H.E.; Massani, B.B.; Miesfeld, R.L. Identification and Characterization of a Mosquito-Specific Eggshell Organizing Factor in Aedes Aegypti Mosquitoes. *PLoS Biol.* **2019**, *17*, e3000068. [CrossRef] [PubMed]
26. Gu, Q.-J.; Zhou, S.-M.; Zhou, Y.-N.; Huang, J.-H.; Shi, M.; Chen, X.-X. A Trypsin Inhibitor-like Protein Secreted by Cotesia Vestalis Teratocytes Inhibits Hemolymph Prophenoloxidase Activation of Plutella Xylostella. *J. Insect Physiol.* **2019**, *116*, 41–48. [CrossRef] [PubMed]
27. Kim, B.Y.; Lee, K.S.; Lee, K.Y.; Yoon, H.J.; Jin, B.R. Anti-Fibrinolytic Activity of a Metalloprotease Inhibitor from Bumblebee (Bombus Ignitus) Venom. *Comp. Biochem. Physiol. Part C Toxicol. Pharmacol.* **2021**, *245*, 109042. [CrossRef]
28. Li, Y.; Zhao, P.; Liu, S.; Dong, Z.; Chen, J.; Xiang, Z.; Xia, Q. A Novel Protease Inhibitor in Bombyx Mori Is Involved in Defense against Beauveria Bassiana. *Insect Biochem. Mol. Biol.* **2012**, *42*, 766–775. [CrossRef]
29. Li, Y.; Liu, H.; Zhu, R.; Xia, Q.; Zhao, P. Loss of Second and Sixth Conserved Cysteine Residues from Trypsin Inhibitor-like Cysteine-Rich Domain-Type Protease Inhibitors in Bombyx Mori May Induce Activity against Microbial Proteases. *Peptides* **2016**, *86*, 13–23. [CrossRef]
30. Mojzisch, A.; Brehm, M.A. The Manifold Cellular Functions of von Willebrand Factor. *Cells* **2021**, *10*, 2351. [CrossRef]
31. Zhou, Y.F.; Eng, E.T.; Zhu, J.; Lu, C.; Walz, T.; Springer, T.A. Sequence and Structure Relationships within von Willebrand Factor. *Blood* **2012**, *120*, 449–458. [CrossRef]
32. Wang, Y.W.; Tan, J.M.; Du, C.W.; Luan, N.; Yan, X.W.; Lai, R.; Lu, Q.M. A Novel Trypsin Inhibitor-Like Cysteine-Rich Peptide from the Frog Lepidobatrachus Laevis Containing Proteinase-Inhibiting Activity. *Nat. Products Bioprospect.* **2015**, *5*, 209–214. [CrossRef] [PubMed]
33. Shakeel, M.; Zafar, J. Molecular Identification, Characterization, and Expression Analysis of a Novel Trypsin Inhibitor-like Cysteine-Rich Peptide from the Cotton Bollworm, Helicoverpa Armigera (Hübner) (Lepidoptera: Noctuidae). *Egypt. J. Biol. Pest Control* **2020**, *30*, 1–7. [CrossRef]
34. Mesquita-Rodrigues, C.; Saboia-Vahia, L.; Cuervo, P.; Masini d'Avila Levy, C.; Alves Honorio, N.; Domont, G.B.; Batista de Jesus, J. Expression of Trypsin-like Serine Peptidases in Pre-Imaginal Stages of Aedes Aegypti (Diptera: Culicidae). *Arch. Insect Biochem. Physiol.* **2011**, *76*, 223–235. [CrossRef]
35. Saboia-Vahia, L.; Borges-Veloso, A.; Mesquita-Rodrigues, C.; Cuervo, P.; Dias-Lopes, G.; Britto, C.; De Barros Silva, A.P.; De Jesus, J.B. Trypsin-like Serine Peptidase Profiles in the Egg, Larval, and Pupal Stages of Aedes Albopictus. *Parasites Vectors* **2013**, *6*, 1–11. [CrossRef] [PubMed]
36. Yang, Y.J.; Davies, D.M. Trypsin and Chymotrypsin during Metamorphosis in Aedes Aegypti and Properties of the Chymotrypsin. *J. Insect Physiol.* **1971**, *17*, 117–131. [CrossRef]
37. Akbari, O.S.; Antoshechkin, I.; Amrhein, H.; Williams, B.; Diloreto, R.; Sandler, J.; Hay, B.A. The Developmental Transcriptome of the Mosquito Aedes Aegypti, an Invasive Species and Major Arbovirus Vector. *G3 (Bethesda)* **2013**, *3*, 1493–1509. [CrossRef]
38. Hixson, B.; Bing, X.-L.; Yang, X.; Bonfini, A.; Nagy, P.; Buchon, N. A Transcriptomic Atlas of Aedes Aegypti Reveals Detailed Functional Organization of Major Body Parts and Gut Regional Specializations in Sugar-Fed and Blood-Fed Adult Females. *Elife* **2022**, *11*, e76132. [CrossRef]
39. Ling, L.; Raikhel, A.S. Cross-Talk of Insulin-like Peptides, Juvenile Hormone, and 20-Hydroxyecdysone in Regulation of Metabolism in the Mosquito Aedes Aegypti. *Proc. Natl. Acad. Sci. USA* **2021**, *118*, e2023470118. [CrossRef]
40. Clifton, M.E.; Noriega, F.G. The Fate of Follicles after a Blood Meal Is Dependent on Previtellogenic Nutrition and Juvenile Hormone in Aedes Aegypti. *J. Insect Physiol.* **2012**, *58*, 1007–1019. [CrossRef]
41. Hernández-Martínez, S.; Cardoso-Jaime, V.; Nouzova, M.; Michalkova, V.; Ramirez, C.E.; Fernandez-Lima, F.; Noriega, F.G. Juvenile Hormone Controls Ovarian Development in Female Anopheles Albimanus Mosquitoes. *Sci. Rep.* **2019**, *9*, 1–10. [CrossRef]
42. Bian, G.; Raikhel, A.S.; Zhu, J. Characterization of a Juvenile Hormone-Regulated Chymotrypsin-like Serine Protease Gene in Aedes Aegypti Mosquito. *Insect Biochem. Mol. Biol.* **2008**, *38*, 190–200. [CrossRef] [PubMed]
43. Ji, J.-L.; Han, S.-J.; Zhang, R.-J.; Yu, J.-B.; Li, Y.-B.; Yu, X.-P.; Liu, G.-F.; Xu, Y.-P. Inter-Alpha-Trypsin Inhibitor Heavy Chain 4 Plays an Important Role in the Development and Reproduction of Nilaparvata Lugens. *Insects* **2022**, *13*, 303. [CrossRef] [PubMed]

44. Labun, K.; Montague, T.G.; Krause, M.; Torres Cleuren, Y.N.; Tjeldnes, H.; Valen, E. CHOPCHOP v3: Expanding the CRISPR Web Toolbox beyond Genome Editing. *Nucleic Acids Res.* **2019**, *47*, W171–W174. [CrossRef]
45. Maya-Maldonado, K.; Cardoso-Jaime, V.; Hernández-Martínez, S.; Vázquez-Calzada, C.; Hernández-Hernández, F.C.; Lanz-Mendoza, H. DNA Synthesis Increases during the First Hours Post-Emergence in Anopheles Albimanus Mosquito Midgut. *Dev. Comp. Immunol.* **2020**, *112*, 103753. [CrossRef] [PubMed]
46. Schindelin, J.; Arganda-Carreras, I.; Frise, E.; Kaynig, V.; Longair, M.; Pietzsch, T.; Preibisch, S.; Rueden, C.; Saalfeld, S.; Schmid, B.; et al. Fiji: An Open-Source Platform for Biological-Image Analysis. *Nat. Methods* **2012**, *9*, 676–682. [CrossRef]

International Journal of
Molecular Sciences

Aedes aegypti Piwi4 Structural Features Are Necessary for RNA Binding and Nuclear Localization

Adeline E. Williams [1,2], Gaurav Shrivastava [1], Apostolos G. Gittis [1], Sundar Ganesan [1], Ines Martin-Martin [1], Paola Carolina Valenzuela Leon [1], Ken E. Olson [2,*] and Eric Calvo [1,*]

[1] Laboratory of Malaria and Vector Research, National Institute of Allergy and Infectious Diseases, National Institutes of Health, Rockville, MD 20852, USA; adeline.williams@colostate.edu (A.E.W.); gaurav.shrivastava@nih.gov (G.S.); gittisa@niaid.nih.gov (A.G.G.); sundar.ganesan@nih.gov (S.G.); ines.martin-martin@nih.gov (I.M.-M.); paolacarolina.valenzuelaleon@nih.gov (P.C.V.L.)

[2] Center for Vector-Borne Infectious Diseases, Department of Microbiology, Immunology, and Pathology, Colorado State University, Fort Collins, CO 80523, USA

* Correspondence: kenneth.olson@colostate.edu (K.E.O.); ecalvo@od.nih.gov (E.C.)

Citation: Williams, A.E.; Shrivastava, G.; Gittis, A.G.; Ganesan, S.; Martin-Martin, I.; Valenzuela Leon, P.C.; Olson, K.E.; Calvo, E. *Aedes aegypti* Piwi4 Structural Features Are Necessary for RNA Binding and Nuclear Localization. *Int. J. Mol. Sci.* **2021**, *22*, 12733. https://doi.org/10.3390/ijms222312733

Academic Editor: Michail Kotsyfakis

Received: 31 October 2021
Accepted: 23 November 2021
Published: 25 November 2021

Publisher's Note: MDPI stays neutral with regard to jurisdictional claims in published maps and institutional affiliations.

Abstract: The PIWI-interacting RNA (piRNA) pathway provides an RNA interference (RNAi) mechanism known from *Drosophila* studies to maintain the integrity of the germline genome by silencing transposable elements (TE). *Aedes aegypti* mosquitoes, which are the key vectors of several arthropod-borne viruses, exhibit an expanded repertoire of Piwi proteins involved in the piRNA pathway, suggesting functional divergence. Here, we investigate RNA-binding dynamics and subcellular localization of *A. aegypti* Piwi4 (AePiwi4), a Piwi protein involved in antiviral immunity and embryonic development, to better understand its function. We found that AePiwi4 PAZ (Piwi/Argonaute/Zwille), the domain that binds the 3′ ends of piRNAs, bound to mature (3′ 2′ O-methylated) and unmethylated RNAs with similar micromolar affinities ($K_D = 1.7 \pm 0.8$ μM and K_D of 5.0 ± 2.2 μM, respectively; $p = 0.05$) in a sequence independent manner. Through site-directed mutagenesis studies, we identified highly conserved residues involved in RNA binding and found that subtle changes in the amino acids flanking the binding pocket across PAZ proteins have significant impacts on binding behaviors, likely by impacting the protein secondary structure. We also analyzed AePiwi4 subcellular localization in mosquito tissues. We found that the protein is both cytoplasmic and nuclear, and we identified an AePiwi4 nuclear localization signal (NLS) in the N-terminal region of the protein. Taken together, these studies provide insights on the dynamic role of AePiwi4 in RNAi and pave the way for future studies aimed at understanding Piwi interactions with diverse RNA populations.

Keywords: Piwi4; *Aedes aegypti*; piRNA; RNA interference; RNAi; Piwi; NLS; arbovirus; mosquito

1. Introduction

The P-element-induced wimpy testis (PIWI)-interacting RNA (piRNA) pathway is an RNA interference (RNAi) mechanism that is traditionally known to silence transposable elements (TEs) that can integrate into the germline genome and threaten its integrity [1–4]. piRNAs, 23–30 nucleotide (nt) small RNAs (sRNAs), bind Piwi proteins, a subfamily of the Argonautes. piRNA-bound Piwis assemble into piRNA-induced silencing complexes (piRISCs), where effector Piwis are targeted to complementary RNA substrates [5–7]. *Drosophila* express three Piwis (Piwi, Aubergine (Aub), and Argonaute-3 (Ago3)), where Aub and Ago3 are expressed exclusively in the germline, while Piwi is also expressed in neighboring somatic cells [8–10]. Aub and Ago3 silence their targets post-transcriptionally in the cytoplasm, while Piwi translocates into the nucleus, forms a nuclear effector complex, and silences its targets co-transcriptionally [11–13]. Depletion of the piRNA pathway in *Drosophila* leads to TE insertion accumulation, DNA damage, defects in embryonic development, and female sterility [14–19].

The functions of the piRNA pathway are more diverse than initially thought [3,20–23] For example, Piwis display differential expression patterns in the germline or soma as well as in the cytoplasm or the nucleus across organisms [21], suggesting that their roles and functions are broad. A good example is in arthropods, where both somatic and germline piRNAs are common [24]. Culicine mosquitoes, specifically, have undergone an expansion of their Piwi protein repertoire, suggesting functional divergence [25]. For example, *Aedes* express seven Piwis, Piwi2-7, and Ago3, where Piwi4-6 and Ago3 are abundantly expressed in both the soma and germline [25,26]. Many studies have shown that the piRNA pathway in *Aedes* is multi-functional and important for, in addition to transposon silencing, antiviral immunity [25,27–38], embryonic development [39,40], and gene regulation [41,42].

A particularly interesting *A. aegypti* Piwi of unknown function is Piwi4 (henceforth termed "AePiwi4"). Although unnecessary for small RNA production, AePiwi4 associated with Semliki Forest virus-specific small-interfering RNAs (vsiRNAs) and virus-specific piRNAs (vpiRNAs) in infected cells, as well as with several protein players involved in both the siRNA (Ago2 and Dcr2) and piRNA (Ago3, Piwi5, Piwi6, Yb, vreteno, Tejas, and minotaur) pathways [26,43]. Furthermore, silencing *AePiwi4* depleted 3′ 2′ O-methylated (mature) Sindbis virus (SINV)-specific vsiRNAs and vpiRNAs and increased SINV, dengue (DENV2), and chikungunya (CHIKV) virus replication in infected cells [32]. This phenotype was recapitulated in DENV2-infected *A. aegypti* mosquitoes, where silencing *AePiwi4* increased infectious virus titers 5–10 days post-infection (dpi) [32]. AePiwi4 also associated with highly conserved satellite repeat-derived piRNAs (tapiR1 and tapiR2) that were 3′ 2′ O-methylated [39]. Knocking down *AePiwi4* reduced tapiR1 and tapiR2 transcripts, and depleting tapiR1 in embryos arrested their development and prevented the degradation of maternally deposited transcripts [39]. Taken together, the role(s) of AePiwi4 appear to be diverse and span across several different RNAi pathways.

AePiwi4 has been consistently associated with long (28–30 nt), mature 3′ 2′ O-methylated (henceforth termed "3′m") piRNAs, and it was found in both the cytoplasmic and nuclear fractions in an embryonic *Aedes aegypti* cell line (Aag2) [26,32,39]. We therefore set out to characterize AePiwi4 structural motifs involved in piRNA binding and nuclear localization to gain further insights on AePiwi4 function. In human Piwis (Hiwi1, Hiwi2, and Hili), the PAZ (Piwi/Argonaute/Zwille) domain preferentially binds 3′m piRNAs because of a hydrophobic-binding pocket that is flexible enough to accommodate the methyl group [44]. This contrasts with the human Argonaute-1 (Ago1) PAZ domain where its more restrictive RNA-binding pocket exhibits preferential binding to 3′ 2′-OH (henceforth termed "3′nm") groups present on microRNAs [44]. We hypothesized that a flexible AePiwi4 PAZ domain would also determine AePiwi4 preferential binding to mature, long piRNA populations.

We first compared PAZ sequences across previously crystalized Piwis to determine AePiwi4 PAZ structural features and binding pockets involved in 3′ end piRNA recognition. We then characterized recombinant AePiwi4 PAZ-binding dynamics with the 3′ ends of mature and un-methylated piRNAs by surface plasmon resonance (SPR). We found that mutating putative RNA-binding residues depleted or significantly impacted binding to both 3′m and 3′nm sRNAs, while a T41R change, present in *A. aegypti* Ago3, significantly improved binding. Finally, we characterized a functional nuclear localization signal (NLS) in the N-terminal region of the AePiwi4 protein. We found that subtle structural differences across Piwi proteins may have important impacts on preferential RNA-binding behaviors and subcellular localization.

2. Results

2.1. Biophysical Properties of AePiwi4 by Structural Modeling and Alignment

Using I-TASSER and Chimera, we first modeled AePiwi4 against the recently crystalized *Drosophila melanogaster* Piwi protein [45] (PDB: 6KR6) (Figure 1A). The quality of the predicted model was assessed by its C-score = −1.50. C-scores fall between −5 and 2, and more than 90% of the quality predictions are correct for models that have C-scores of −1.5 or higher [46]. Furthermore, the average template modeling (TM) score (0.53 ± 0.15) was >0.5, which indicates a model of correct topology [46]. We then superimposed the AePiwi4 model against crystalized human (Hiwi1; PDB: 3O7V) [44] and mouse (Miwi; PDB: 2XFM) [47] PAZ to determine the amino acid residues of AePiwi4 PAZ to be M270 –T380 (Figure 1B). A summary of predicted biophysical properties of AePiwi4 PAZ is provided in Supplemental Table S1. An electrostatic density map of Piwi4 (Figure 1C) revealed an inner pocket that was highly positively charged, analogous to the *Drosophila* Piwi linker regions that bind RNA nucleotides. AePiwi4 PAZ also displayed long stretches of flexibility with neighboring hydrophobic regions (Supplemental Figure S1). The AePiwi4 PAZ model suggests that the protein contains hydrophobic regions buried within a flexible protein structure, allowing AePiwi4 to bind long 3′m piRNAs.

To determine putative AePiwi4 PAZ amino acids involved in RNA binding, we aligned Piwi PAZ sequences derived from proteins whose structures had been crystalized, including *Drosophila* Piwi [45] (PDB: 6KR6), silkworm Piwi (Siwi [48]) (PDB: 5GUH), mouse Piwi (Miwi [47]) (PDB: 2XFM), and human Hili (PDB: 3O7X) and Hiwi1 (PDB: 3O6E) [44] (Figure 1D). We also included human Hiwi2 PAZ, as its binding properties to the 3′ ends of piRNAs were characterized by isothermal calorimetry in Tian et al., 2011 [44]. As a comparison, we included the outgroup Argonaute protein human Ago1 (PDB: 4KXT), whose more restricted PAZ domain dictates its RNA-binding preference for 3′nm microRNAs [44,49]. We then compared known RNA-binding residues from the crystal structures with the residues of *A. aegypti* PAZ domains (black arrows, Figure 1D; black arrows, Supplemental Figure S1). We found that the residues involved in RNA binding tended to be highly conserved across the different organisms, and many were tyrosine and phenylalanine aromatic residues, whose hydroxyl groups form hydrogen bonds with phosphate oxygens of RNA nucleotides [44]. We also noted that all Piwi PAZ domains analyzed herein displayed the Piwi PAZ specific insertion element (black box, Figure 1D) that provides the flexibility necessary for accommodating 3′m piRNA ends [44]. This insertion site lies between two beta barrels, which, when absent (as in Ago1), results in a sharp turn between the barrels and a narrower binding pocket [44]. Although the amino acids within the Piwi PAZ-specific insertion site are not conserved across organisms, or even across subfamily Piwi proteins in the same organism [44], we observed that the first five amino acids of the Piwi PAZ-specific insertion sites were highly conserved across the *A. aegypti* Piwis. The only exception to this observation was *A. aegypti* Ago3, whose residues shared no similarity to its AePiwi PAZ counterparts. Further, Ago3 displayed two amino acids within the insertion site not seen in the other PAZ sites, perhaps suggesting a more flexible binding pocket than the other *A. aegypti* Piwis.

We also generated a phylogenetic tree to compare evolutionary relatedness between Piwi PAZ sequences from the various organisms (Figure 1E). We included human Ago1 again as an outgroup. We also added the two additional *Drosophila* Piwis, Aub, and Ago3. We found that all *A. aegypti* Piwi PAZ, except for Ago3, clustered with *Drosophila* Aub. On the other hand, *A. aegypti* Ago3 was most closely related to *Drosophila* Ago3. We also observed AePiwi4 clustered with the germline AePiwis Piwi2-3 as opposed to the somatic AePiwis Piwi5, Piwi6, and Ago3.

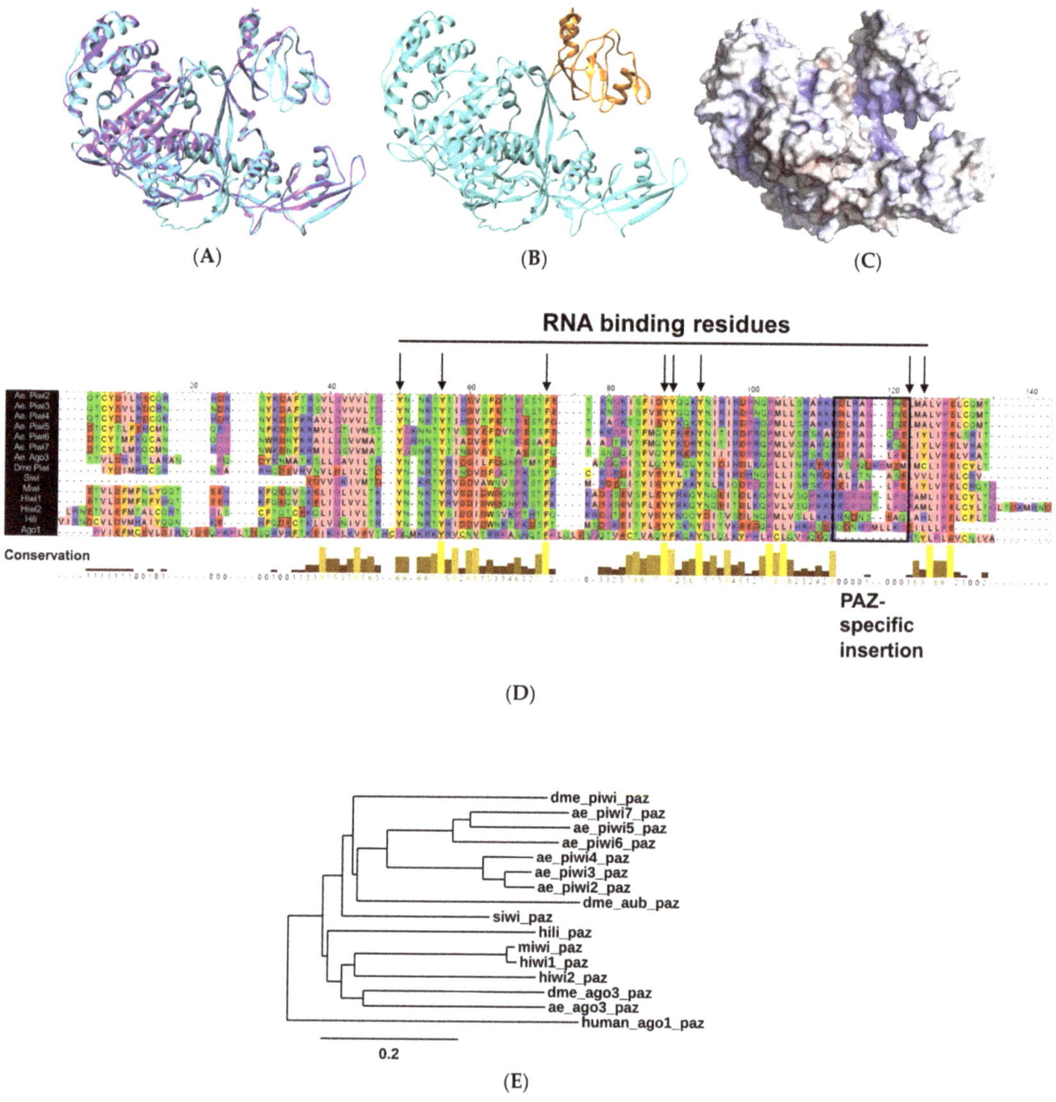

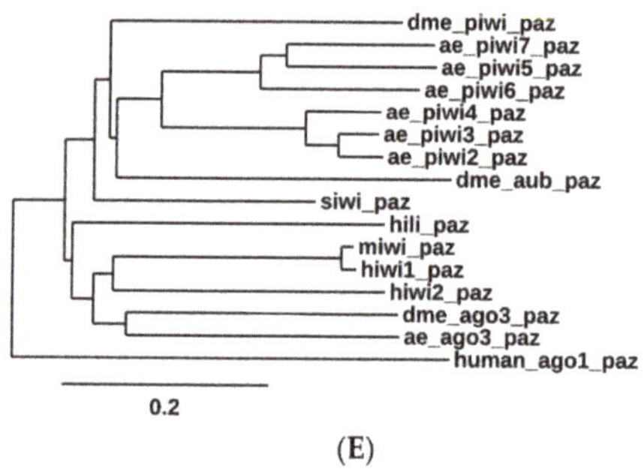

Figure 1. Predicted *A. aegypti* Piwi4 RNA-binding properties. (**A**) Predicted *A. aegypti* Piwi4 model (blue) superimposed with the crystalized *Drosophila* Piwi structure (purple); (**B**) predicted *A. aegypti* Piwi4 structure (blue) with PAZ domain highlighted in orange; (**C**) electrostatic density of *A. aegypti* Piwi4 where red = negatively charged and blue = positively charged. Structure is rotated on the right to reveal inner positively charged pocket; (**D**) alignment of Piwi PAZ domains, including all *A. aegypti* Piwis (Ae. Piwi2-7 and Ae. Ago3) and crystalized or characterized Piwi PAZ (*Drosophila* (Dme) Piwi, silkworm Piwi (Siwi), mouse Piwi (Miwi), and human Piwis Hiwi1, Hiwi2, and Hili). The human Argonaute protein Ago1 was also included as an outgroup. Black arrows indicate known RNA-binding residues by crystal structures, which in this alignment include amino acid numbers 50, 56, 71, 88, 89, 93, 123, and 125. Black box indicates the Piwi PAZ-specific insertion site. Colors (Zappo color scheme) indicate biochemical properties where peach = aliphatic/hydrophobic, aromatic = orange, blue = positively charged, red = negatively charged, green = hydrophilic, pink = conformationally special, and yellow = cysteine; (**E**) phylogenetic tree of all Piwi PAZ included in the alignment shown in 1D, with the addition of *Drosophila* Aubergine and Ago3 PAZs. Scale bar indicates number of substitutions per site.

2.2. A. aegypti Piwi4 PAZ Binds 3′ 2′ O-Methylated and Non-Methylated piRNAs in a Sequence-Independent Manner

To determine whether *A. aegypti* Piwi4 PAZ bound to both 3′ 2′ O-methylated and 3′ 2′ OH piRNAs, we cloned *AePiwi4 PAZ* with a histidine 6xHis-tag into pET-17b for bacterial expression (Supplemental Figure S2A,B), purified it, and characterized RNA-binding dynamics by SPR. We confirmed AePiwi4 PAZ expressed in BL21(DE3) pLysS *E. coli* by Western blot using antibodies against its 6xHis-tag (Supplemental Figure S2C). We purified AePiwi4 PAZ from the soluble fraction by nickel chromatography followed by size exclusion chromatography. Soluble AePiwi4 PAZ was stable in 20 mM Tris-HCl pH 7.4 and 150 mM NaCl and ran at the expected size (14 kDa) by SDS/PAGE gel electrophoresis (Supplemental Figure S2D). Protein identity was confirmed by Edman degradation.

To test RNA-binding dynamics, we performed SPR with AePiwi4 PAZ and three different RNAs: a 3′m 28 nt Phasi Charoen-like virus (PCLV)-specific piRNA, a 3′nm vpiRNA of the same sequence, and a 3′nm 28 nt scrambled sequence. We chose this piRNA sequence because a recent publication noted that PCLV piRNAs are broadly distributed across culicine mosquito cell lines, perhaps due to a PCLV-specific endogenous viral element in the genome [50]. Using synthetic RNA sequences biotinylated on the 5′ end, we immobilized the RNA on a CM5 Biacore surface that had been pre-coated with neutravidin. Immobilizing RNA by the 5′ end allowed us to test binding affinities to moieties at the 3′ end. Using increasing concentrations of AePiwi4 PAZ analytes flowed over the surface of the chip with immobilized ligand, we were able to determine the dissociation constants (K_D) from steady-state binding levels (R_{eq}) against the analyte concentration (C, in molar concentration) once binding reached equilibrium. Experiments were performed in four replicates.

We found that AePiwi4 PAZ bound the 3′m 28 nt piRNA with a K_D of 1.7 ± 0.8 µM (Figure 2A), the 3′nm 28 nt piRNA with a K_D of 5.0 ± 2.2 µM (Figure 2B), and the scrambled 28 nt 3′nm piRNA with a K_D of 2.5 ± 0.1 µM (Figure 2C). AePiwi4 PAZ bound to 3′ 2′ O-methylated piRNAs with marginally greater affinity than it did to 3′ 2′ unmethylated piRNAs ($p = 0.05$), and there was no significant difference in binding affinities for known or scrambled RNA sequences ($p = 0.25$).

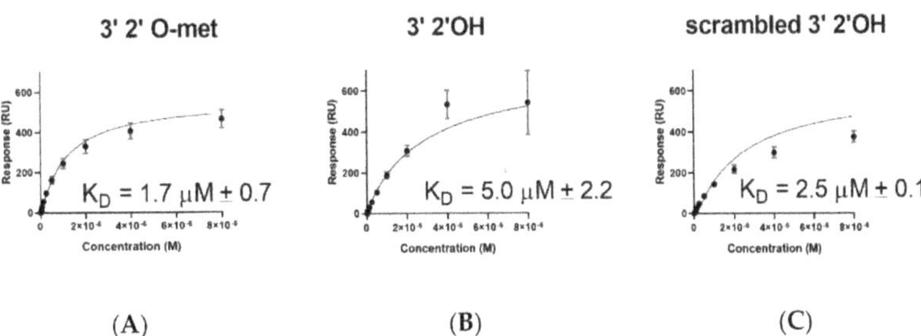

(**A**)   (**B**)   (**C**)

Figure 2. Affinity-binding equilibrium curves for AePiwi4 PAZ to piRNAs. Fitted affinity-binding equilibrium curves for AePiwi4 PAZ analyte to a 28 nt (**A**) 3′ 2′ O-methylated, (**B**) non-methylated, or (**C**) scrambled 28 nt non-methylated piRNA. Equilibrium K_D was calculated from steady-state binding levels $R_{eq} = (CR_{max})/(K_D + C)$ + offset, where C = concentration, R_{max} = analyte-binding capacity of the surface in response units (RU), K_D = dissociation constant, and offset = response at zero analyte concentration. M = molar concentration. Red bars indicate mean and standard deviation for R_{max} values.

2.3. A. aegypti Piwi4 PAZ Mutants Reveal the Amino Acids Necessary for piRNA Binding

For further insights into how the AePiwi4 PAZ structure dictates its RNA-binding preferences, we generated AePiwi4 PAZ mutants that displayed amino acid changes within predicted RNA-binding pockets. We focused our efforts around two highly conserved residues shown to form hydrogen bonds with RNA—Y40 and F55—as well as two residues that flank Y40 and appeared to be moderately conserved across the *A. aegypti* Piwi PAZ—T39 and T41. Through site-directed mutagenesis, we generated five mutants that we then expressed in bacteria and purified: T39A, Y40A, T41A, F55A, and T41R (Supplemental Figures S3 and S4). Y40A and F55A displayed alanine substitutions for the highly conserved tyrosine or phenylalanine amino acids, respectively, while T39A and T41A displayed alanine substitutions for the threonines that flank Y40. The T41R mutation reflected the arginine present in only one *A. aegypti* Piwi, Ago3, but also in most other organisms' Piwi PAZ domains analyzed herein (Figure 1D). Inserts were confirmed by Sanger sequencing, and mutations in protein sequence were confirmed by mass spectrometry. Binding behaviors were assessed by SPR, as described previously (Figure 3).

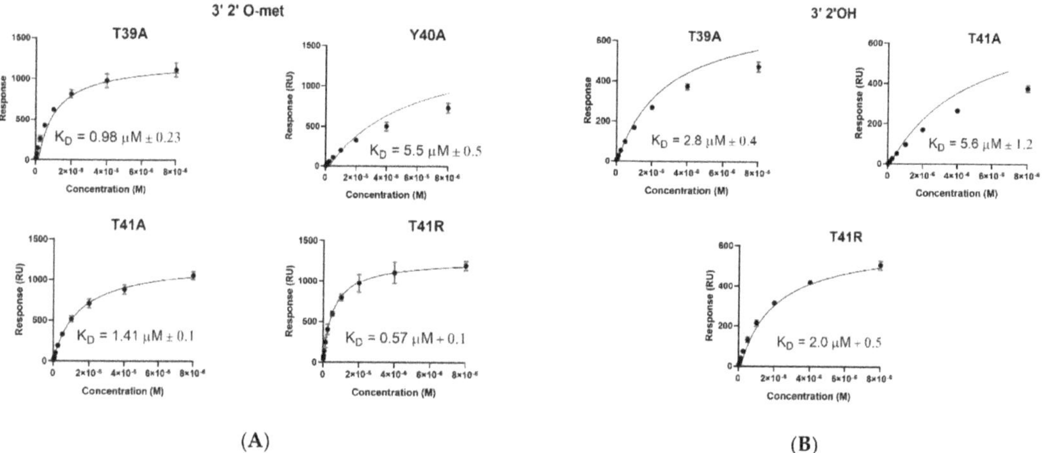

Figure 3. Affinity-binding equilibrium curves for AePiwi4 PAZ mutants to piRNAs. Fitted affinity-binding equilibrium curves for AePiwi4 PAZ mutant analytes to a 28 nt (**A**) $3'$ $2'$ O-methylated or (**B**) non-methylated piRNA. Equilibrium K_D was calculated from steady-state binding levels $R_{eq} = (CR_{max})/(K_D + C) + \text{offset}$, where C = concentration, R_{max} = analyte-binding capacity of the surface in response units (RU), K_D = dissociation constant, and offset = response at zero analyte concentration. M = molar concentration. Red bars indicate mean and standard deviation for R_{max} values.

Dissociation constants for all mutant proteins binding both the $3'$m and $3'$nm 28 nt piRNA are displayed in Figure 3 and summarized in Table 1. We found that F55 was essential for both $3'$m and $3'$nm binding because when mutated, no binding occurred for either ligand. Y40A also depleted $3'$nm binding and significantly inhibited $3'$m piRNA binding ($K_D = 5.5 \pm 0.5$ μM; $p = 0.04$). We found that disrupting the amino acids flanking Y40 with alanine mutations had no significant impact on binding $3'$m piRNAs as compared to wild-type PAZ binding to this RNA (T39A: $p = 0.2$; T41A: $p = 0.3$). However, we did observe a significantly increased affinity of T39A for the $3'$nm piRNA ($K_D = 2.8 \pm 0.4$ μM; $p = 0.02$), suggesting that this residue does have an impact on $3'$nm binding. Furthermore, we observed that mutating T41 to match the amino acid present in *A. aegypti* Ago3 PAZ tended to improve $3'$m binding ($K_D = 0.57 \pm 0.1$ μM) and significantly improved $3'$nm binding ($K_D = 2.0 \pm 0.5$ μM; $p = 0.02$).

Table 1. Summary of disassociation constants for AePiwi4 PAZ mutants binding 3′ 2′ O-methylated (met) or non-methylated (nmet) piRNA. Equilibrium K_D was calculated from steady-state binding levels $R_{eq} = (CR_{max})/(K_D + C)$ + offset, where C = concentration, R_{max} = analyte-binding capacity of the surface in response units (RU), K_D = dissociation constant, and offset = response at zero analyte concentration. * = $p \leq 0.05$ by unpaired t-test with WT as comparison group.

Immobilized Ligand	Binding AePiwi4 PAZ	K_D Values (μM)	R_{max} (RU)
3′ 2′met piRNA	WT	1.7 ± 0.7	1364 ± 3
	T39A	0.98 ± 0.2	1343 ± 2
	Y40A	5.5 ± 0.5 *	1103 ± 62
	T41A	1.4 ± 0.1	1272 ± 14
	T41R	0.57 ± 0.1	1388 ± 18
	F55A	No binding	
3′ 2′nmet piRNA	WT	5 ± 2.2	585 ± 3
	T39A	2.8 ± 0.4 *	596 ± 5
	Y40A	No binding	
	T41A	5.6 ± 1.2	580 ± 8
	T41R	2.0 ± 0.5 *	590 ± 3
	F55A	No binding	

In Hiwi1 PAZ, preferential binding of 3′m RNA over 3′nm RNA is mostly dictated by backbone confirmation of the protein rather than the amino acid composition of the binding pocket [44]. To investigate whether the mutations that impacted the RNA binding impacted the AePiwi4 PAZ secondary structure, we performed circular dichroism (CD) spectroscopy analysis with the T39A, Y40A, T41A, and T41R mutants and compared the CD curves to that of the WT AePiwi4 PAZ (Figure 4). We analyzed the data using CAPITO [51]. WT AePiwi4 PAZ displayed a CD curve most similar to proteins that had a mostly irregular structure but also that had between 30% and 49% beta strands and 6–16% alpha helices. All mutants maintained a mostly irregular secondary structure; however, they displayed different CD curves and percentages of alpha helices and beta-sheets compared to the WT protein (Figure 4 insets). T41R was most similar to proteins that were made up of between 16% and 26% beta strands and 28–46% alpha helices. Y40A also displayed a spectrum that aligned more with proteins that had a greater abundance of alpha helices—26–40% alpha helices but only 14–22% beta-strands. The T39A CD curve clustered with proteins that were made up of 9–25% alpha helices and 30–41% beta-strands, while the T41A curve clustered with proteins that were made up of 31–50% alpha helices and 4–21% beta-strands. Taken together, these data indicate that single amino acid changes in the AePiwi4 PAZ backbone can alter secondary structure, which likely impacted RNA-binding behaviors.

2.4. AePiwi4 Co-Localizes in the Cytoplasm and Nucleus in A. aegypti Tissues

For further insights into the function of *A. aegypti* Piwi4, we characterized the sub-cellular localization of the native protein in *A. aegypti* mosquitoes. While this manuscript was under preparation, Joosten and colleagues (2021) reported that Piwi4, Piwi5, and Piwi6 were in both the nucleus and the cytoplasm in an *A. aegypti* embryonic cell line infected or uninfected with Sindbis virus [26]. To determine the sub-cellular localization of native AePiwi4 in both somatic and germline tissues in the mosquito, we generated polyclonal antibodies against AePiwi4 by immunizing mice with the AePiwi4 PAZ recombinant protein. We confirmed that these antibodies recognized both recombinant AePiwi4 PAZ and full-length proteins by Western blot (Supplemental Figure S5). To confirm that the antibodies recognized AePiwi4 from mosquito tissues, we prepared whole mosquito lysates from three *A. aegypti* females 48 h post-bloodmeal (time of peak AePiwi4 expression [32]) for Western blot. Anti-AePiwi4 mouse serum reacted to whole mosquito lysate at the expected size of AePiwi4 (100 kDa) (Supplemental Figure S5). Mass spectrometry analyses further confirmed that AePiwi4-specific peptides were present at the same location on a corresponding SDS/PAGE gel slice (Supplemental Dataset 1).

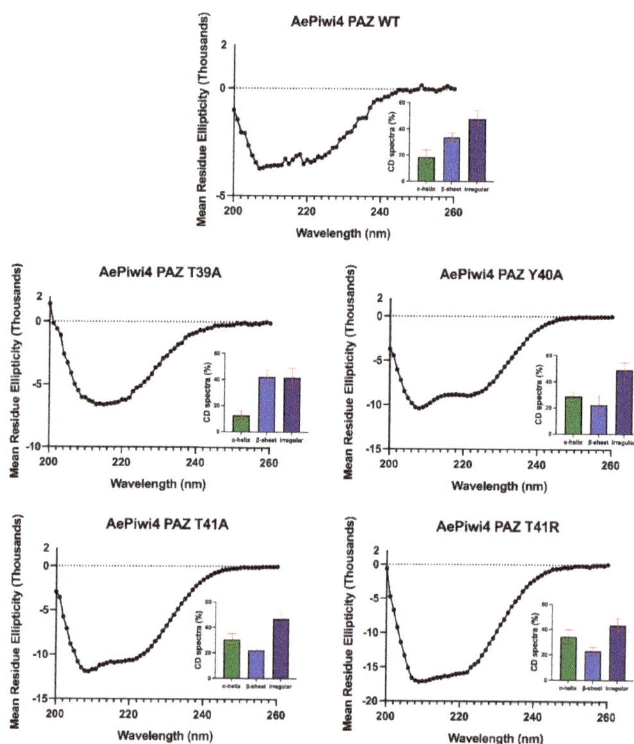

Figure 4. Circular dichroism (CD) spectra analyses of AePiwi4 PAZ mutants. CD spectra curves, by mean residue ellipticity for AePiwi4 PAZ WT and mutant proteins, recorded over 200–260 nm. Insets show the calculated percentages of secondary structures determined by CD analysis using CAPITO. Red bars indicate mean and standard deviation of three similarity hits based on lowest area differences under the curve.

To determine the sub-cellular localization of AePiwi4, we next performed immunofluorescence assays and Western blots using both somatic and germline-derived mosquito tissues. AePiwi4 tended to stain cytoplasmically in both midguts (Figure 5A; Supplemental Figure S6A) and unfertilized embryos from ovary tissues (Figure 5B, Supplemental Figure S6B). However, when we fractionated ovaries from *A. aegypti* mosquitoes 48 h post-bloodmeal into cytoplasmic and nuclear fractions for Western blot, we found that AePiwi4 was present in both fractions (Figure 5C). Antibodies targeting H3 histone were used as a marker for the nuclear fraction (Supplemental Figure S7). These results suggested AePiwi4 may be trafficked in and out of the nucleus in mosquito tissues.

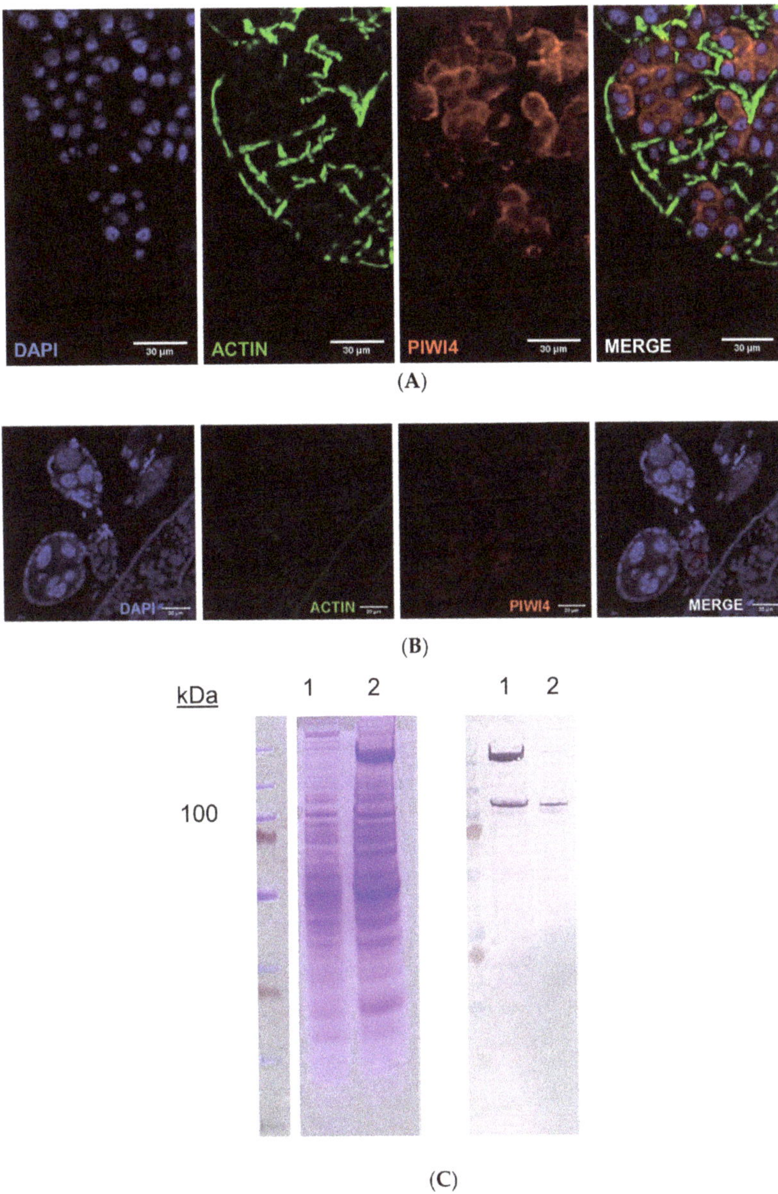

Figure 5. AePiwi4 expression in mosquito tissues. IFA of *A. aegypti* midgut (**A**) or ovaries with unfertilized embryos (**B**) stained with anti-AePiwi4 (red), phalloidin (green), or DAPI (blue). Scale bars are 30 µM and 20 µM for midguts and embryos, respectively; (**C**) Coomassie-stained SDS/PAGE gel (left) and corresponding Western blot of cytoplasmic (lane 1) or nuclear (lane 2) fractions of *A. aegypti* mosquito ovary tissue.

2.5. A. aegypti Piwi4 Expresses a Nuclear Localization Signal in the N-Terminal Region of the Protein

To further explore AePiwi4 nuclear localization, we identified a putative NLS in the N-terminal region of the protein (Supplemental Figure S8A). In *Drosophila melanogaster* Piwi, the NLS is expressed in the intrinsically disordered domain in a similar region of the N-terminal. We therefore generated a phylogenetic tree of the intrinsically disordered regions of the *A. aegypti* and *D. melanogaster* Piwi proteins to see if Piwis with known or putative NLS signals would cluster together (Supplemental Figure S8B). We found that the intrinsically disordered domains of *A. aegypti* Ago3 and *D. melanogaster* Ago3 clustered together, that *Drosophila* Aub and *A. aegypti* Piwi7 clustered together, and that *A. aegypti* Piwis2-6 clustered with *Drosophila* Piwi, the only *Drosophila* Piwi with an NLS. These results suggested that *A. aegypti* Piwi2-6 may also harbor nuclear localization signals in their intrinsically disordered domains in the N-terminal regions.

To confirm that the putative AePiwi4 NLS was responsible for protein nuclear localization, we cloned the putative AePiwi4 NLS (amino acid residues 42–83, Supplemental Figure S8A), as well as the entire N-terminal region containing the NLS (amino acid residues 1–83), fused to an eGFP; we henceforth named these constructs AePiwi4NLS-eGFP and AePiwi4Nterminal-eGFP, respectively. We used the same backbone containing either a known SV40 NLS fused to the eGFP [52] as a positive control or an eGFP alone as a negative control; we henceforth named these constructs SV40NLS-eGFP and eGFP, respectively. We then transfected HEK293 cells with these constructs and visualized eGFP and DAPI colocalization 24 h post-transfection. As expected, we found that the known SV40NLS-eGFP localized in the nucleus while the eGFP alone appeared diffused throughout the cells (Figure 6A,B). Plasmids harboring either AePiwi4NLS-eGFP or AePiwi4Nterminal-eGFP migrated into the nucleus, as evidenced by eGFP expression colocalized with DAPI staining (Figure 6C,D). The nuclear staining appeared punctated, perhaps indicative of nucleolar staining. We also observed that both AePiwi4NLS-eGFP and AePiwi4Nterminal-eGFP displayed cytoplasmic eGFP expression as well, which was not observed in cells transfected with the SV40 NLS plasmid (Figure 6C,D).

To quantitively compare eGFP fluorescent intensities across sample types, we subtracted total eGFP fluorescent intensity sums from eGFP fluorescent intensity sums in nuclear surfaces, normalized by number of cells, in three independent views across slides (Figure 6E). We found that the resulting eGFP nuclear intensity sums outside of the nuclear surfaces was significantly higher for cells transfected with the eGFP construct as compared to cells transfected with SV40NLS-eGFP ($p = 0.006$), AePiwi4NLS-EGFP ($p = 0.01$), or AePiwi4Nterminal-eGFP ($p = 0.05$). There were no significant differences in eGFP fluorescent intensity sums outside of nuclear surfaces between cells transfected with AePiwi4NLS-EGFP ($p = 0.34$) or AePiwi4Nterminal-eGFP ($p = 0.23$) compared to those transfected with the SV40NLS-eGFP positive control. Taken together, these results suggested that *A. aegypti* Piwi4 expresses an NLS in the intrinsically disordered domain in the N-terminal region of the protein.

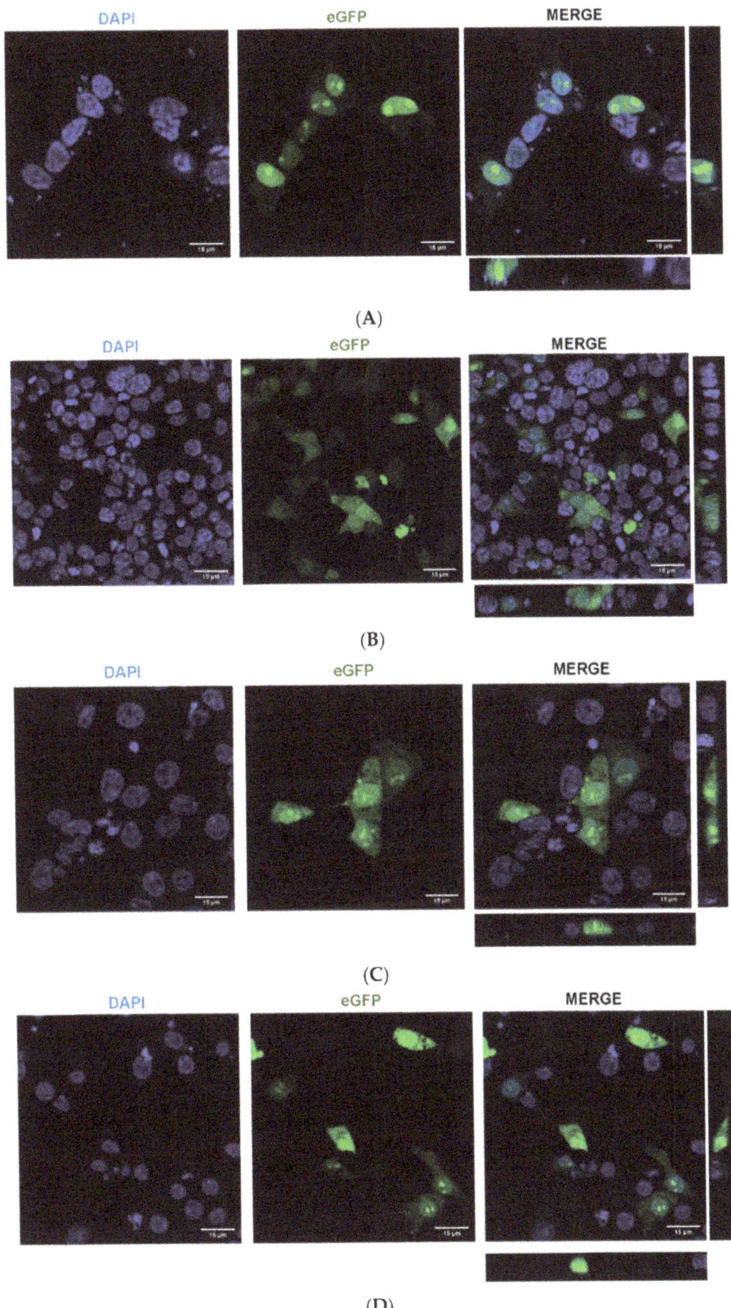

Figure 6. *Cont.*

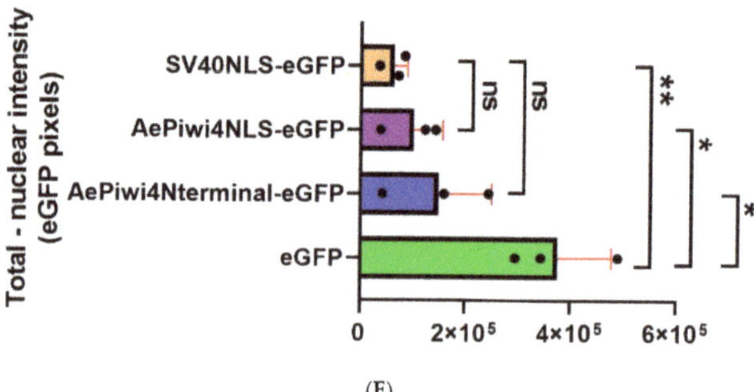

(E)

Figure 6. AePiwi4 harbors an NLS in the N-terminal region of the protein. Representative slices from image stacks of HEK293 cells transfected with (**A**) SV40NLS-eGFP, (**B**) eGFP, (**C**) AePiwi4NLS-eGFP, or (**D**) AePiwi4Nterminal-eGFP. Cells were stained with DAPI (blue) 24 h post-transfection. DAPI (left), eGFP (middle), and merged (right) channels are shown separately. Orthogonal views presented in merged channel. Scale bar = 15 µM. (**E**) Quantification of total eGFP fluorescence intensity sums subtracted from eGFP intensity sums in nuclear surfaces for each sample type, normalized by number of cells. Each black dot represents an individual picture. Red bars indicate SEM. ** = $p \leq 0.01$, * = $p \leq 0.05$, ns = non-significant.

3. Discussion

In this study, we characterized *A. aegypti* Piwi4 structural features involved in RNA binding and nuclear localization to gain insights into the protein's function. AePiwi4 had previously been associated with various 28–30 nt 3′ 2′ O-methylated piRNAs, so we focused our efforts on the PAZ domain that binds the 3′ ends of piRNAs. We assessed AePiwi4 PAZ RNA-binding dynamics by SPR and found that AePiwi4 PAZ bound to both mature and unmethylated piRNAs with micromolar affinities in a sequence independent manner. We identified key residues in AePiwi4 PAZ involved in RNA binding and found that they were highly conserved across organisms. We also highlighted a unique arginine amino acid flanking a tyrosine residue necessary for 3′nm RNA binding that was present in most other organisms' Piwi PAZ but was only present in a single *A. aegypti* Piwi PAZ (Ago3). Mutating this residue in AePiwi4 PAZ to match that of Ago3 improved both 3′m and 3′nm RNA binding. Through circular dichroism, we showed that single amino acid changes in Piwi PAZ changes the secondary structure of the protein. Finally, we found that AePiwi4 was both cytoplasmic and nuclear in mosquito tissues, and that signals in the intrinsically disordered region drove nuclear localization.

We report herein Piwi-RNA-binding affinities for a Piwi protein of an arthropod vector, which complements studies performed with human and *Drosophila melanogaster* Piwi PAZ. We found that *A. aegypti* Piwi4 PAZ bound 3′ 2′ O-methylated and non-methylated piRNAs with K_{DS} of 1.7 ± 0.8 µM and 5.0 ± 2.2 µM, respectively. The preference of AePiwi4 PAZ for 3′m piRNAs over 3′nm piRNAs was less pronounced than what has been reported for other Piwi PAZ (Supplemental Table S2). For example, Hiwi1, Hiwi2, and Hili bound 3′m piRNAs with K_{DS} of 6.5 µM, 2 µM, or 10 µM, respectively, but they bound non-methylated piRNAs with weaker affinities—K_{DS} of 16 µM, 12 µM, or 34 µM, respectively [44]. Immunoprecipitations of AePiwi4 from uninfected or infected Aag2 cells followed by sRNA sequencing of associated RNAs have revealed the protein associates with bona fide piRNAs resistant to beta-elimination, a method that selects for 3′m piRNAs and depletes 3′nm miRNAs [32,39]. Our results suggest, however, that AePiwi4 is able to bind to both 3′m and 3′nm sRNAs with only a marginally higher affinity for the former over the latter. Further investigations on AePiwi4-associated sRNAs from different cellular compartments may provide new insights on protein-RNA trafficking and the range of sRNAs with which AePiwi4 interacts. For example, the role(s) of *A. aegypti* Piwis may function with both pre-processed non-methylated RNAs and mature piRNAs across cellular compartments or with sRNA populations outside the piRNA pathway. Halbach et al. [39] compared AePiwi4-mediated silencing of a satellite repeat-derived target by way of a piRNA to that of miRNA silencing [39]. In that study, the authors found that the 3′ end of a satellite repeat-derived piRNA (tapR1) was not absolutely required for silencing, while the seed region was not sufficient for silencing, a pattern they compared to miRNA-mediated silencing [39]. In another study, Tassetto and colleagues found that silencing *AePiwi4* impacted both 3′m piRNA and siRNA production and argued that AePiwi4 links the siRNA and piRNA pathways [32]. Our AePiwi4 RNA-binding studies indicate that the PAZ domain of AePiwi4 is indeed able to interact with diverse populations of sRNAs with similar affinities, perhaps suggesting AePiwi4 has broad functions or unique roles in RNA binding that may differ from model Piwis. Future studies comparing RNA-binding dynamics across the *A. aegypti* Piwis will elucidate the roles they play in RNAi.

In this study, we studied 3′m and 3′nm RNA binding with protein partners by SPR. Other studies have characterized PAZ RNA binding by isothermal calorimetry (ITC) using small eight nucleotide RNAs, and we note that caution should be taken when comparing hard dissociation constant values across these different techniques. While ITC provides valuable information on number of binding sites and heat released from a binding reaction, we found that immobilizing the 5′ end of longer, more physiologically relevant RNAs by SPR enabled us to efficiently calculate dissociation constants for many protein-binding partners against stabilized ligands in a single experiment. This method may be useful for other studies aimed at understanding protein-RNA binding at specific motifs.

Our data suggest that small differences in Piwi PAZ amino acid composition across Piwi proteins alter protein secondary structure, which thereby impact the protein's affinity for certain RNAs. Given that preferential binding of 3′m RNA over 3′nm RNA was mostly dictated by backbone confirmation of the protein in Hiwi1, it is likely that subtle differences in Piwi protein structure may have profound impacts on preferential RNA-binding behaviors that are important for defining the functions of different Piwis. We found that although the amino acids that directly form hydrogen bonds with RNA were highly conserved, residues that flank these sites tended to be more variable. Perhaps those residues that impact the stability of the PAZ structure, rather than those that directly bind the RNA itself, drive Piwi functional divergence.

The number of Piwi proteins has expanded in culicine mosquitoes as compared to anophelines and drosophilids [53], and understanding their evolutionary relationships with other Piwis, their shared or unique structural features, and their interactions with diverse RNA populations may provide insights into how their functions have diverged. We found that the PAZ domains of *A. aegypti* Piwis 2-7 are more evolutionarily related to that of *D. melanogaster* Aubergine than to *D. melanogaster* Piwi or Ago3 PAZ. Aubergine is a cytoplasmic, germline-specific protein that participates in ping-pong amplification by binding antisense primary piRNAs and producing secondary sense piRNAs that fuel the cycle [54], a role similar to that of *A. aegypti* Piwi5 [25]. A recent study showed that piRNA binding to Aub PAZ induces a protein confirmational change that triggers symmetric dimethylarginine (sDMA) methylation of the Aub intrinsically disordered domain in the N-terminal region [55]. The sDMA modification then serves as a binding site for Krimper, which simultaneously binds unmethylated Ago3 to bring the proteins in close proximity for RNA transfer during ping-pong amplification [55]. Indeed, Joosten and colleagues recently characterized a Krimper ortholog in *Aedes*, Atari, which bound Ago3 without sDMA modifications [26]. Perhaps a similar mode of RNA binding-dependent autoregulation and sDMA signaling also govern the *Aedes aegypti* Piwis Piwi2-6.

Drosophila Piwi, the only nuclear Piwi in the fly, expresses a bipartite NLS in its intrinsically disordered domain [56]. *Drosophila* Piwi nuclear localization is autoregulated by confirmational changes that occur once the protein binds piRNAs; the NLS remains buried within the protein structure until RNA binding triggers a conformational change and exposes the NLS [56]. Once the protein is imported into the nucleus and releases the piRNA, the protein is trafficked back into the cytoplasm. We found that *A. aegypti* Piwi4 also expresses signals in the intrinsically disordered region that drive proteins to the nucleus (Figure 6C,D), which, if similar to *Drosophila* Piwi, could autoregulate protein trafficking based on protein confirmational changes that dictate signal exposure. We observed that the AePiwi4 NLS did not drive complete expression of eGFP into the nucleus, as evidenced by diffused cytoplasmic fluorescence in addition to punctated nuclear staining (Figure 6C,D) It is possible that the intrinsically disordered region of *AePiwi4* contains both an NLS and nuclear export signals (NES) that drive protein trafficking in and out of the nucleus. Investigations on how AePiwi4 regulates its trafficking into different cellular compartments require future studies and could be useful in understanding its role in different RNAi-mediated processes. It is also possible that several *A. aegypti* Piwis autoregulate their subcellular localization in similar manners as AePiwi4. Our phylogenetic analyses revealed that the regions of the *AePiwi4* and *Drosophila melanogaster* Piwi proteins that harbored nuclear localization signals, the intrinsically disordered domains, also clustered with *A. aegypti* Piwi2, Piwi3, Piwi5, and Piwi6 (Supplemental Figure S8B). Different Piwis likely have sophisticated and diverse regulation mechanisms that control their expression patterns in different compartments of the cell.

Growing evidence reveals that the piRNA pathway is involved in gene regulation in somatic tissues and contributes to diverse human diseases including cancer [23,57] and neurodegenerative disorders [58]. Because somatic piRNAs and Piwi expression are common in arthropods [24], they could be valuable models for understanding the molecular mechanisms underlying the lesser understood Piwi or piRNA functions. Future studies aimed at understanding how Piwi-RNA binding impacts protein structure and function will be useful for learning more about how this pathway is involved in immunity, gene regulation, and disease in arthropod vectors as well as in other organisms.

4. Materials and Methods

4.1. A. aegypti Piwi4 Structure Model Prediction

A model of the *A. aegypti* Piwi4 structure was generated using the I-TASSER software (version 5.1 Zhang Lab, University of Michigan, Ann Arbor, MI, USA [46,59,60]) and visualized using Chimera (University of California, San Francisco, CA, USA). This software predicts secondary and tertiary structures based on the similarity of other proteins whose structures have been solved. The AePiwi4 amino acid sequence was queried against the *Drosophila* Piwi structure that was crystalized in Yamaguchi et al., 2020 [45], which allowed us to determine the predicted *A. aegypti* Piwi4 PAZ domain. AePiwi4 PAZ was then superimposed to other crystalized PAZ proteins, including Hili PAZ [44], Hiwi1 PAZ [44], Miwi PAZ [47], and Siwi PAZ [48].

4.2. Cloning

Aedes aegypti Piwi4 (*AAEL007698*), including a 6xHis-tag, was synthesized by BioBasic Inc. (Markham, ON, Canada). Both *AePiwi4* full length (FL) and PAZ domain (residues 270–380) nucleotide sequences were sub-cloned into pCR-Blunt II-TOPO vector using the Zero Blunt TOPO PCR Cloning Kit (Thermo Fisher Scientific, Waltham, MA, USA) following the manufacturer's instructions. pCR-Blunt II-TOPO vectors containing either *AePiwi4* FL or *AePiwi4* PAZ were transformed in OneShot Top10 chemically competent *E. coli* (Invitrogen, Waltham, MA, USA). Using standard restriction enzyme-mediated cloning and the pCR-Blunt II-TOPO vectors described above as PCR templates, *AePiwi4* FL or *AePiwi4* PAZ were then cloned into pET-17b vectors. pET-17b vectors containing either *AePiwi4* FL or *AePiwi4* PAZ were transformed in OneShot Top10 chemically competent *E. coli* (Invitrogen, Waltham, MA, USA). Inserts were confirmed by Sanger sequencing. Primers used in this study are displayed in Supplemental Table S3.

The putative AePiwi4 NLS, as well as the entire N-terminal region of AePiwi4 containing the putative NLS, were cloned from the *AePiwi4* FL-containing pCR-Blunt II-TOPO vector into a backbone containing an eGFP by In-Fusion cloning (TakaRa, San Jose, CA, USA) following the manufacturer's instructions. The parent SV40NLS-eGFP backbone was a gift from Rob Parton (Addgene plasmid # 67652; http://n2t.net/addgene:67652 (accessed on 28 April 2021); RRID: Addgene_67652) [52]. Briefly, an eGFP-alone plasmid was generated by NcoI digestion and religation with the T4 DNA ligation Mighty Mix (TakaRa, San Jose, CA, USA) of the SV40NLS-eGFP plasmid. AePiwi4NLS-eGFP or AePiwi4Nterminal-eGFP were then cloned into the eGFP alone plasmid using In-Fusion primers listed in Supplemental Table S3. All constructs were transformed into Stellar Competent Cells (TakaRa, San Jose, CA, USA), and inserts were confirmed by Sanger sequencing.

4.3. Recombinant Protein Expression

pET-17b vectors containing either *AePiwi4* FL or PAZ were transformed into BL21(DE3) pLysS chemically competent *E. coli* cells (Thermo Fisher Scientific, Waltham, MA, USA). Transformed *E. coli* were plated onto Luria–Bertani (LB) agar plates with 100 µg/mL ampicillin and 34 µg/mL chloramphenicol that were left O/N at 37 °C. Individual colonies were picked into starter cultures of 4 mL LB broth (supplemented with 100 µg/mL ampicillin and 34 µg/mL chloramphenicol) that were left shaking at 220 RPM O/N at 37 °C. Starter cultures were then added to 150 mL LB broth (supplemented with 100 µg/mL ampicillin and 34 µg/mL chloramphenicol) that were left shaking at 220 RPM 37 °C until OD_{600} = 0.4 (~1 h). Protein expression was then induced with 0.1 mM isopropyl β- d-1-thiogalactopyranoside (IPTG) for 4 h shaking at 160 RPM at 25 °C. Bacteria was then pelleted and stored at −30 °C until protein purification.

For larger scale expression, 150 mL LB broth (supplemented with 100 µg/mL ampicillin and 34 µg/mL chloramphenicol) starter cultures that had been inoculated with glycerol scrapings of BL21(DE3) pLysS *E. coli* containing either pET-17b-*AePiwi4* FL or pET-17b-*AePiwi4* PAZ were left shaking at 220 RPM O/N at 37 °C. Starter cultures were then added to 1 L LB broth (supplemented with 100 µg/mL ampicillin and 34 µg/mL chloramphenicol) and expression was induced following the above protocol.

Expression was confirmed by SDS-PAGE separation and anti-6xHis-tag Western blot in both the soluble and inclusion body fractions for both proteins.

4.4. Recombinant Protein Purification

The soluble AePiwi4 PAZ protein was purified by affinity chromatography followed by size-exclusion chromatography using Nickel-charged HiTrap Chelating HP (GE Healthcare, Chicago, IL, USA) and Superdex 200 10/300 GL columns (GE Healthcare, Chicago, IL, USA), respectively. Frozen *E. coli* pellets were resuspended with Buffer A (10 mM Tris, 500 mM NaCl, 5 mM imidazole, pH 8), left on ice for 10–15 min, and pulse sonicated 4× for 30 s—2 min. The lysates were then spun at $15,000 \times g$ for 30 min at 4 °C. The resulting supernatants were filtered with a 0.8 µM filter (MilliporeSigma, Burlington, MA, USA) and loaded onto a pre-equilibrated Nickel-charged HiTrap Chelating HP column using a peristaltic pump. The column was then pre-washed with 3 column volumes (CV) of Buffer A, followed by a 3 CV wash with Wash Buffer 1 (10 mM Tris, 500 mM NaCl, 20 mM imidazole, pH 8) and 3 CV wash with Wash Buffer 2 (10 mM Tris, 500 mM NaCl, 100 mM imidazole, pH 8). The protein was eluted from the column using 3 CV of Elution Buffer (10 mM Tris, 500 mM NaCl, 300 mM imidazole, and pH 8) and visualized by SDS-PAGE gel electrophoresis.

Eluted protein was concentrated down to ~500 µL using an Amicon stirred cell with a cellulose membrane of 3 kDa nominal molecular weight (MilliporeSigma, Burlington, MA, USA). The resulting concentrated protein was spun down at $4000 \times g$ for 10 min to remove large debris and loaded onto a Superdex 200 10/300 GL column that had been equilibrated with 20 mM Tris-HCl, 150 mM NaCl, pH 7.4. Peak elutions that corresponded to the correct size of Piwi4 PAZ (14 kDa) were confirmed by SDS-PAGE gel electrophoresis. N-terminal protein sequence was also confirmed by Edman degradation.

4.5. SDS-PAGE

All proteins were heated to 95 °C for 5 min under reducing conditions in 1X LDS (Thermo Fisher Scientific, Waltham, MA, USA) and were separated using 4–12% Bis Tris protein gels (Thermo Fisher Scientific, Waltham, MA, USA). Gels were stained with Coomassie Brilliant Blue (GenScript, Piscataway, NJ, USA). Protein concentrations were determined using the Nanodrop ND-1000 spectrophotometer adjusted by the molar extinction coefficient.

4.6. Western Blot

Aedes aegypti mosquito midguts and ovaries, as well as recombinant proteins, were separated by SDS-PAGE gel electrophoresis for Western blots. *A. aegypti* that had been fed defibrinated sheep blood (Denver Serum Company, Denver, CO, USA) were collected 48 h post-bloodmeal, and their midguts (cleaned of blood in $1 \times$ PBS) and ovaries were dissected and flash frozen on dry ice. 15 midguts or ovaries/tube were resuspended in 100 µL cold hypotonic lysis buffer (10 mM Hepes pH 7.9, 1.5 mM $MgCl_2$, 10 mM KCl, 0.2 mM PMSF) and left on ice for 15 min. Samples were vortexed vigorously for 30 s and then pelleted at $1000 \times g$ for 15 min. The supernatant was collected as the cytoplasmic fraction. The remaining pellets were then resuspended in 100 µL solubilization buffer (15 mM Tris, 150 mM NaCl, 5 mM EDTA, 0.5% Triton X-100, 10% glycerol, 0.2 mM PMSF) and spun down at $100,000 \times g$. The supernatant was collected as the nuclear fraction.

30 µg of protein was processed for SDS/PAGE separation, as described previously, and run alongside 10 µM Piwi4 PAZ or Piwi4 FL inclusion bodies. Proteins were transferred to a PVDF membrane (iBlot, Invitrogen, Waltham, MA, USA) that was blocked for 2 h at RT in blocking buffer (5% powdered milk (Carnation), 50 mM Tris-HCl pH 7.4, 150 mM NaCl, 1% Tween 20 (TBST)). Membranes were incubated O/N at 4 °C with anti-Piwi4 PAZ mouse serum (1:500 in blocking buffer), a 6xHis-tag monoclonal antibody (ThermoFisher Scientific, Waltham, MA, USA, diluted 1:5000 in blocking buffer), or anti-Histone H3 as a nuclear marker (Novus Biologicals, Littleton, CO, USA; generated in rabbit, diluted 1:1000 in blocking buffer).

Membranes were washed with TBST ($3 \times$ for 10 min) and with TBS ($1 \times$ 10 min) and incubated at RT for 1–2 h with goat anti-mouse or anti-rabbit antibodies conjugated to alkaline phosphatase (1 mg/mL, diluted 1:10,000). Membranes were again washed with TBST and TBS, and proteins were detected for 5–10 min using Western Blue Stabilized alkaline phosphatase substrate (Promega, Madison, WI, USA).

4.7. Mosquito Rearing

A. aegypti mosquitoes (Liverpool (LVP) strain) were reared in standard insectary conditions at the Laboratory of Malaria and Vector Research, NIAID, NIH (28 °C, 60–70% humidity, 14:10 h light/dark cycle) under the expert supervision of Karina Sewell, Andre Laughinghouse, Kevin Lee, and Yonas Gebremicale. Mosquitoes had a solution of 10% sucrose ad libitum and were offered defibrinated sheep blood (Denver Serum Company, Denver, CO, USA) in an artificial feeding system. Larva were fed Tetramin.

4.8. Sequence Alignment

Nucleotide and amino acid sequences were retrieved from the NCBI databases. Multiple alignments and phylogenetic trees were obtained by Clustal Omega [61] and visualized on Jalview [62].

4.9. Site Directed Mutagenesis

AePiwi4 PAZ protein mutants were generated using the QuikChange II Site-Directed mutagenesis kit (Agilent, Santa, Clara, CA, USA) following the manufacturer's instructions.

Primers were designed using PrimerX (https://www.bioinformatics.org/primerx/cgi-bin/DNA_1.cgi, (accessed on 14 March 2021)) and are displayed in Supplemental Table S3. The pET-17b vector containing *AePiwi4* FL was used as template for the reactions, and XL1-Blue supercompetent cells (Agilent, Santa, Clara, CA, USA) were transformed with the mutant plasmids. Mutation nucleotide sequences were confirmed by Sanger sequencing, and protein mutant sequences were confirmed by mass spectrometry.

The pET-17b vectors containing the AePiwi4 PAZ mutations were transformed into BL21(DE3) pLysS chemically competent *E. coli* cells (Thermo Fisher Scientific, Waltham, MA, USA), and all proteins were expressed and purified, as described previously.

4.10. Surface Plasmon Resonance (SPR)

All SPR experiments were carried out in a T100 instrument (GE Healthcare, Chicago, IL, USA) following the manufacturer's instructions. Sensor CM5, amine coupling reagents, and HBS-P buffers were also purchased from GE Healthcare (Chicago, IL, USA). HBS-P was supplemented with EDTA (HBS-PE, 10 mM Hepes pH 7.4, 150 mM NaCl, 3 mM EDTA, and 0.005% (v/v) P20 surfactant) and was used as the running buffer while Conditioning Solution 2 (50 mM NaOH, 1 M NaCl) was used as the regeneration and conditioning solution for all experiments. Briefly, the CM5 sensor was coated 40 μg/mL neutravidin and pre-conditioned with 3 × 60 s injections of Conditioning Solution 2. ~500–1000 RUs of biotinylated RNAs were then captured to flow cells 2 or 4, which were then conditioned with 3 × 60 s injections of Conditioning Solution 2. Protein analyte was introduced unto the surface with 180 s injections (30 μL/s). Results were analyzed using the Biacore T200 Evaluation software v2.0.3 provided by GE Healthcare (Chicago, IL, USA). Equilibrium dissociation constants were calculated from steady-state binding levels (R_{eq}) against molar concentration of the analyte (C). The fitted equation was $R_{eq} = ((CR_{max})/(K_D + C)) + \text{offset}$, where R_{max} = analyte-binding capacity of the surface in response units (RU) and offset = response at zero analyte concentration, which accounts for buffer-mediated effects on the refractive index. SPR experiments were carried out 2–4×.

4.11. Circular Dichroism

0.1 mg/mL of purified AePiwi4 PAZ WT, T39A, Y40A, T41A, or T41R in 20 mM Tris 75 mM, NaCl pH 7.4 were used for CD analyses. Continuous measurements with a pitch of 0.2 nm were recorded from 200–260 nm wavelengths with a bandwidth of 1 nm. Mean residue ellipticity was calculated with the following equation: (molecular weight of each protein in daltons/((number of amino acids − 1) × θ_λ))/(10 × pathlength in cm × protein concentration in g/mL). All readings were normalized by subtracting with blank (buffer) mean residue ellipticity. Data were analyzed using CAPITO [51].

4.12. RNA Synthesis

The 3′ 2′ O-CH$_3$ and 3′ 2′ OH 28 nt RNAs were synthesized by Eurofins Genomics (Louisville, KY, USA). Sequences are listed in Supplemental Table S4. RNA was resuspended at 1–2 mM in DEPC-treated water and stored at −80 °C.

4.13. Mouse Polyclonal Antibody Production

Polyclonal antibodies against *A. aegypti* Piwi4 PAZ were raised in mice. Mice (Balb/c; Charles River, Frederick, MD, USA) were IM immunized with 10 μg of AePiwi4 PAZ in combination with Magic Mouse Adjuvant (CD Creative Diagnostics, Shirley, NY, USA). Negative control mice were immunized with Magic Mouse Adjuvant alone. At 21 d post-immunization, mice received a 2nd booster immunization with 10 μg of AePiwi4 PAZ in combination with Magic Mouse Adjuvant (or adjuvant alone for negative control group). Blood was collected 35 d post-immunization. The antibody levels were confirmed by ELISA.

4.14. Mass Spectrometry

Mosquito tissue samples and recombinant proteins were prepared and separated by an SDS-PAGE gel as previously described, which was then stained with Coomassie blue. Bands of interest were excised from the gel and submitted for liquid chromatography coupled with mass spectrometry at the Research and Technology Branch (NIAID, NIH Rockville, MD, USA). Briefly, the gel slices from the SDS-PAGE gel were cut into small pieces and subjected to in-gel trypsin digestion. The gel slices were destained to remove Coomassie blue staining and were then reduced and alkylated. After dehydration with acetonitrile and air-drying, a sequencing grade trypsin (Promega, Madison, WI, USA) solution was added onto the gel slices and was allowed to be absorbed into the gel slice. The gel slices were then incubated overnight at 30 °C for in-gel digestion. The peptides released from in-gel digestion were extracted by acetonitrile and then applied for LC-MS/MS analysis. Proteomic analyses were performed, as previously described [63].

4.15. IFA

Mosquito midguts or ovaries were dissected 48 h post-bloodmeal, flash fixed for 30 s in cold 4% paraformaldehyde (PFA) in PBS and cleaned of blood in cold PBS. The midguts or ovaries were then left shaking in 4% PFA in PBS O/N at 4 °C. The next day, midguts or ovaries were washed 3X in PBS and blocked O/N in blocking buffer (2% BSA, 0.5% Triton-X-100, PBS). The midguts or ovaries were then incubated with either serum from mice immunized with AePiwi4 PAZ or with Magic Mouse adjuvant alone (1:500, diluted in blocking buffer) O/N at 4 °C. The midguts or ovaries were washed with blocking buffer a minimum of 3× for 30 min and were then incubated with secondary goat anti-mouse antibodies conjugated to Alexa Fluor 594 (Thermo Fisher Scientific, Waltham, MA, USA; diluted 1:1000 in blocking buffer) for 1 h at RT. The midguts or ovaries were again washed with blocking buffer a minimum of 3× for 30 min, followed by incubation with 1 μg/mL DAPI (diluted 1:1000 in blocking buffer) and phalloidin conjugated to Alexa Fluor 488 (Thermo Fisher Scientific, Waltham, MA, USA; diluted 1:250 in blocking buffer) for 20 min at RT. The midguts or ovaries were washed 2× for 20 min with blocking buffer and 1× with 0.5% Triton-X-100 in PBS and were then mounted onto slides with ProLong Gold antifade mountant with DAPI (Thermo Fisher Scientific, Waltham, MA, USA).

4.16. HEK293 Cell Culture and Transfection

HEK293 cells were cultured in 35 mm dishes with a No. 15 coverslip pre-coated with Poly-D-Lysine (MatTek Life Sciences, Ashland, MA, USA). Briefly 300,000 cells were plated on individual dishes in DMEM media. The next day, the cells were transfected with transfection complex containing 500 ng of (1) SV40NLS-eGFP, (2) eGFP alone, (3) AePiwi4NLS-eGFP, or (4) AePiwi4Nterminal-eGFP in 0.5 μL Lipofectamine 3000 (Invitrogen, Waltham, MA, USA) in serum free Opti-MEM media, according to the manufacturer's protocol. Twenty-four hours post-transfection, the cells were washed 3X with PBS and fixed with 4% PFA in PBS for 30 min at RT. The cells were then washed 3X with PBS and permeabilized with 0.5% Triton-X-100 in PBS for 30 min at RT. The cells were then stained with DAPI (1 μg/mL in 2% BSA, 0.5% Triton-X-100, PBS). The cells were visualized using a Leica Confocal SP8 microscope. Images were processed with Imaris software version 9.2.1 and post-processing was carried out in Fiji ImageJ version 1.52n for representative purposes.

4.17. Statistics

Surface Plasmon Resonance equilibrium curves were fitted with a non-linear regression generated by the Biacore Evaluation software v2.0.3 provided by GE Healthcare (described in "Surface Plasmon Resonance" Methods section), which were then visualized with GraphPad Prism. The equilibrium dissociation constants, calculated based on steady state, were generated by that same software. Differences between dissociation constants were compared using an unpaired two-tailed t-test with GraphPad Prism.

Quantifications of eGFP fluorescent intensities were calculated by subtracting eGFP pixel total intensity sums by average nuclear intensity sums, normalized by the number of cells, in three independent views across a slide. Nuclear surfaces were determined by DAPI display and eGFP pixel intensity values were extracted using Imaris software version 9.2.1. Differences between eGFP intensities outside of nuclear surfaces were compared using an unpaired two-tailed T-test with GraphPad Prism.

Supplementary Materials: The following are available online at https://www.mdpi.com/article/10.3390/ijms222312733/s1.

Author Contributions: Conceptualization, A.E.W., K.E.O. and E.C.; methodology, A.E.W., G.S., A.G.G., S.G., I.M.-M., P.C.V.L., K.E.O. and E.C.; software, A.E.W., A.G.G., S.G., I.M.-M. and E.C.; validation, A.E.W., G.S., A.G.G., S.G. and E.C.; formal analysis, A.E.W., A.G.G., S.G. and E.C.; investigation, A.E.W., G.S., A.G.G., S.G., I.M.-M. and P.C.V.L.; resources, A.G.G., S.G. and E.C.; data curation, A.E.W., G.S., A.G.G., S.G., I.M.-M., P.C.V.L., K.E.O. and E.C.; writing—original draft preparation, A.E.W., K.E.O. and E.C.; writing—review and editing, A.E.W., G.S., A.G.G., S.G., I.M.-M., P.C.V.L., K.E.O. and E.C.; visualization, A.E.W., A.G.G., S.G., I.M.-M., K.E.O. and E.C.; supervision, A.E.W., A.G.G., S.G., I.M.-M., K.E.O. and E.C.; project administration, A.E.W., K.E.O. and E.C.; funding acquisition, K.E.O. and E.C. All authors have read and agreed to the published version of the manuscript.

Funding: This research was funded by the Intramural Research Program of the NIH/NIAID (AI001246). The research was also supported by NIH/NIAID, grant number R01 AI130085-02.

Institutional Review Board Statement: Public Health Service Animal Welfare Assurance #A4149-01 guidelines were followed according to the National Institute of Allergy and Infectious Diseases (NIAID) and the National Institutes of Health (NIH) Animal Office of Animal Care and Use (OACU). These studies were carried out according to the NIAID-NIH animal study protocols (ASP) approved by the NIH Office of Animal Care and Use Committee (OACUC), with the approval ID ASP-LMVR3. Mice used in this study were housed in one of the animal facilities from the NIAID/NIH and were humanly treated according to OACU regulations.

Informed Consent Statement: Not applicable.

Acknowledgments: The authors thank Karina Sewell, Kevin Lee, Andre Laughinghouse, and Yonas Gebremicale for excellent and essential mosquito rearing. The authors also thank Dave Garboczi and Jose Ribeiro for stimulating scientific discussion as well as John Andersen for technical assistance with SPR studies. The authors thank Glenn Nardone, Lisa (Renne) Olano, and Ming Zhao, Research Technology Branch, NIH, for mass spectrometry analysis, and Brian Martin, Research Technology Branch, NIH, for N-terminal sequencing. Finally, the authors thank Carol Blair for manuscript feedback and review.

Conflicts of Interest: The authors declare no conflict of interest. The funders had no role in the design of the study; in the collection, analyses, or interpretation of data; in the writing of the manuscript, or in the decision to publish the results.

References

1. Aravin, A.A.; Sachidanandam, R.; Girard, A.; Fejes-Toth, K.; Hannon, G.J. Developmentally Regulated piRNA Clusters Implicate MILI in Transposon Control. *Science* **2007**, *316*, 744–747. [CrossRef]
2. O'Donnell, K.A.; Boeke, J. Mighty Piwis Defend the Germline against Genome Intruders. *Cell* **2007**, *129*, 37–44. [CrossRef]
3. Senti, K.A.; Brennecke, J. The piRNA pathway: A fly's perspective on the guardian of the genome. *Trends Genet.* **2010**, *26*, 499–509. [CrossRef] [PubMed]
4. Tóth, K.F.; Pezic, D.; Stuwe, E.; Webster, A. The piRNA Pathway Guards the Germline Genome Against Transposable Elements. *Adv. Exp. Med. Biol.* **2016**, *886*, 51–77. [CrossRef]
5. Höck, J.; Meister, G. The Argonaute protein family. *Genome Biol.* **2008**, *9*, 210. [CrossRef] [PubMed]
6. Czech, B.; Munafo, M.; Ciabrelli, F.; Eastwood, E.L.; Fabry, M.H.; Kneuss, E.; Hannon, G.J. piRNA-Guided Genome Defense: From Biogenesis to Silencing. *Annu. Rev. Genet.* **2018**, *52*, 131–157. [CrossRef] [PubMed]
7. Ozata, D.M.; Gainetdinov, I.; Zoch, A.; O'Carroll, D.; Zamore, P.D. PIWI-interacting RNAs: Small RNAs with big functions. *Nat. Rev. Genet.* **2019**, *20*, 89–108. [CrossRef]
8. Brennecke, J.; Aravin, A.A.; Stark, A.; Dus, M.; Kellis, M.; Sachidanandam, R.; Hannon, G.J. Discrete Small RNA-Generating Loci as Master Regulators of Transposon Activity in Drosophila. *Cell* **2007**, *128*, 1089–1103. [CrossRef]

9. Malone, C.D.; Brennecke, J.; Dus, M.; Stark, A.; McCombie, W.R.; Sachidanandam, R.; Hannon, G.J. Specialized piRNA Pathways Act in Germline and Somatic Tissues of the *Drosophila* Ovary. *Cell* **2009**, *137*, 522–535. [CrossRef] [PubMed]
10. Théron, E.; Maupetit-Mehouas, S.; Pouchin, P.; Baudet, L.; Brasset, E.; Vaury, C. The interplay between the Argonaute proteins Piwi and Aub within *Drosophila* germarium is critical for oogenesis, piRNA biogenesis and TE silencing. *Nucleic Acids Res.* **2018**, *46*, 10052–10065. [CrossRef]
11. Le Thomas, A.; Rogers, A.K.; Webster, A.; Marinov, G.K.; Liao, S.E.; Perkins, E.M.; Hur, J.K.; Aravin, A.A.; Tóth, K.F. Piwi induces piRNA-guided transcriptional silencing and establishment of a repressive chromatin state. *Genes Dev.* **2013**, *27*, 390–399. [CrossRef]
12. Senti, K.-A.; Jurczak, D.; Sachidanandam, R.; Brennecke, J. piRNA-guided slicing of transposon transcripts enforces their transcriptional silencing via specifying the nuclear piRNA repertoire. *Genes Dev.* **2015**, *29*, 1747–1762. [CrossRef]
13. Fabry, M.H.; Ciabrelli, F.; Munafo, M.; Eastwood, E.L.; Kneuss, E.; Falciatori, I.; Falconio, F.A.; Hannon, G.J.; Czech, B. piRNA guided co-transcriptional silencing coopts nuclear export factors. *eLife* **2019**, *8*, e47999. [CrossRef]
14. Cox, D.N.; Chao, A.; Baker, J.; Chang, L.; Qiao, D.; Lin, H. A novel class of evolutionarily conserved genes defined by piwi are essential for stem cell self-renewal. *Genes Dev.* **1998**, *12*, 3715–3727. [CrossRef]
15. Aravin, A.A.; Naumova, N.M.; Vagin, V.V.; Rozovsky, Y.M.; Gvozdev, V.A. Double-stranded RNA-mediated silencing of genomic tandem repeats and transposable elements in the *D. melanogaster* germline. *Curr. Biol.* **2001**, *11*, 1017–1027. [CrossRef]
16. Klattenhoff, C.; Bratu, D.P.; McGinnis-Schultz, N.; Koppetsch, B.S.; Cook, H.A.; Theurkauf, W.E. *Drosophila* rasiRNA Pathway Mutations Disrupt Embryonic Axis Specification through Activation of an ATR/Chk2 DNA Damage Response. *Dev. Cell* **2007**, *12*, 45–55. [CrossRef] [PubMed]
17. Li, C.; Vagin, V.V.; Lee, S.; Xu, J.; Ma, S.; Xi, H.; Seitz, H.; Horwich, M.D.; Syrzycka, M.; Honda, B.M.; et al. Collapse of germline piRNAs in the absence of Argonaute3 reveals somatic piRNAs in flies. *Cell* **2009**, *137*, 509–521. [CrossRef] [PubMed]
18. Muerdter, F.; Guzzardo, P.M.; Gillis, J.; Luo, Y.; Yu, Y.; Chen, C.; Fekete, R.; Hannon, G.J. A Genome-wide RNAi Screen Draws a Genetic Framework for Transposon Control and Primary piRNA Biogenesis in Drosophila. *Mol. Cell* **2013**, *50*, 736–748. [CrossRef]
19. Handler, D.; Meixner, K.; Pizka, M.; Lauss, K.; Schmied, C.; Gruber, F.S.; Brennecke, J. The Genetic Makeup of the *Drosophila* piRNA Pathway. *Mol. Cell* **2013**, *50*, 762–777. [CrossRef] [PubMed]
20. Theron, E.; Dennis, C.; Brasset, E.; Vaury, C. Distinct features of the piRNA pathway in somatic and germ cells: From piRNA cluster transcription to piRNA processing and amplification. *Mob. DNA* **2014**, *5*, 28. [CrossRef] [PubMed]
21. Mani, S.R.; Juliano, C.E. Untangling the web: The diverse functions of the PIWI/piRNA pathway. *Mol. Reprod. Dev.* **2013**, *80*, 632–664. [CrossRef] [PubMed]
22. Weick, E.-M.; Miska, E.A. piRNAs: From biogenesis to function. *Development* **2014**, *141*, 3458–3471. [CrossRef] [PubMed]
23. Han, Y.-N.; Li, Y.; Xia, S.-Q.; Zhang, Y.-Y.; Zheng, J.-H.; Li, W. PIWI Proteins and PIWI-Interacting RNA: Emerging Roles in Cancer. *Cell. Physiol. Biochem.* **2017**, *44*, 1–20. [CrossRef]
24. Lewis, S.; Quarles, K.; Yang, Y.; Tanguy, M.; Frezal, L.; Smith, S.; Sharma, P.; Cordaux, R.; Gilbert, C.; Giraud, I.; et al. Pan-arthropod analysis reveals somatic piRNAs as an ancestral defence against transposable elements. *Nat. Ecol. Evol.* **2018**, *2*, 174–181. [CrossRef]
25. Miesen, P.; Girardi, E.; van Rij, R.P. Distinct sets of PIWI proteins produce arbovirus and transposon-derived piRNAs in *Aedes aegypti* mosquito cells. *Nucleic Acids Res.* **2015**, *43*, 6545–6556. [CrossRef]
26. Joosten, J.; Taşköprü, E.; Jansen, P.W.; Pennings, B.; Vermeulen, M.; Van Rij, R.P. PIWI proteomics identifies Atari and Pasilla as piRNA biogenesis factors in Aedes mosquitoes. *Cell Rep.* **2021**, *35*, 109073. [CrossRef]
27. Schnettler, E.; Donald, C.; Human, S.; Watson, M.; Siu, R.W.C.; McFarlane, M.; Fazakerley, J.; Kohl, A.; Fragkoudis, R. Knockdown of piRNA pathway proteins results in enhanced Semliki Forest virus production in mosquito cells. *J. Gen. Virol.* **2013**, *94*, 1680–1689. [CrossRef] [PubMed]
28. Miesen, P.; Joosten, J.; Van Rij, R.P. PIWIs Go Viral: Arbovirus-Derived piRNAs in Vector Mosquitoes. *PLoS Pathog.* **2016**, *12*, e1006017. [CrossRef] [PubMed]
29. Varjak, M.; Dietrich, I.; Sreenu, V.B.; Till, B.E.; Merits, A.; Kohl, A.; Schnettler, E. Spindle-E Acts Antivirally Against Alphaviruses in Mosquito Cells. *Viruses* **2018**, *10*, 88. [CrossRef]
30. Varjak, M.; Leggewie, M.; Schnettler, E. The antiviral piRNA response in mosquitoes? *J. Gen. Virol.* **2018**, *99*, 1551–1562. [CrossRef]
31. Lee, W.-S.; Webster, J.A.; Madzokere, E.T.; Stephenson, E.B.; Herrero, L.J. Mosquito antiviral defense mechanisms: A delicate balance between innate immunity and persistent viral infection. *Parasites Vectors* **2019**, *12*, 165. [CrossRef] [PubMed]
32. Tassetto, M.; Kunitomi, M.; Whitfield, Z.J.; Dolan, P.T.; Sánchez-Vargas, I.; Knight, M.A.G.; Ribiero, I.; Chen, T.E.; Olson, K.; Andino, R. Control of RNA viruses in mosquito cells through the acquisition of vDNA and endogenous viral elements. *eLife* **2019**, *8*, 8. [CrossRef]
33. Blair, C.D. Deducing the Role of Virus Genome-Derived PIWI-Associated RNAs in the Mosquito–Arbovirus Arms Race. *Front. Genet.* **2019**, *10*, 1114. [CrossRef]
34. Joosten, J.; Miesen, P.; Taşköprü, E.; Pennings, B.; Jansen, P.W.T.C.A.; Huynen, M.; Vermeulen, M.; Van Rij, R.P. The Tudor protein Veneno assembles the ping-pong amplification complex that produces viral piRNAs in Aedes mosquitoes. *Nucleic Acids Res.* **2019**, *47*, 2546–2559. [CrossRef] [PubMed]

35. Suzuki, Y.; Baidaliuk, A.; Miesen, P.; Frangeul, L.; Crist, A.B.; Merkling, S.H.; Fontaine, A.; Lequime, S.; Moltini-Conclois, I.; Blanc, H.; et al. Non-retroviral Endogenous Viral Element Limits Cognate Virus Replication in *Aedes aegypti* Ovaries. *Curr. Biol.* **2020**, *30*, 3495–3506. [CrossRef] [PubMed]

36. Petit, M.; Mongelli, V.; Frangeul, L.; Blanc, H.; Jiggins, F.; Saleh, M.-C. piRNA pathway is not required for antiviral defense in Drosophila melanogaster. *Proc. Natl. Acad. Sci. USA* **2016**, *113*, E4218–E4227. [CrossRef] [PubMed]

37. Vodovar, N.; Bronkhorst, A.W.; Van Cleef, K.W.R.; Miesen, P.; Blanc, H.; Van Rij, R.P.; Saleh, M.-C. Arbovirus-Derived piRNAs Exhibit a Ping-Pong Signature in Mosquito Cells. *PLoS ONE* **2012**, *7*, e30861. [CrossRef]

38. Morazzani, E.M.; Wiley, M.R.; Murreddu, M.G.; Adelman, Z.N.; Myles, K.M. Production of virus-derived ping-pong-dependent piRNA-like small RNAs in the mosquito soma. *PLoS Pathog.* **2012**, *8*, e1002470. [CrossRef]

39. Halbach, R.; Miesen, P.; Joosten, J.; Taskopru, E.; Rondeel, I.; Pennings, B.; Vogels, C.B.F.; Merkling, S.H.; Koenraadt, C.J.; Lambrechts, L.; et al. A satellite repeat-derived piRNA controls embryonic development of Aedes. *Nature* **2020**, *580*, 274–277. [CrossRef]

40. Betting, V.; Joosten, J.; Halbach, R.; Thaler, M.; Miesen, P.; Van Rij, R.P. A piRNA-lncRNA regulatory network initiates responder and trailer piRNA formation during mosquito embryonic development. *RNA* **2021**, *27*, 1155–1172. [CrossRef]

41. Arensburger, P.; Hice, R.H.A.; Wright, J.; Craig, N.L.; Atkinson, P.W. The mosquito *Aedes aegypti* has a large genome size and high transposable element load but contains a low proportion of transposon-specific piRNAs. *BMC Genom.* **2011**, *12*, 606. [CrossRef]

42. Girardi, E.; Miesen, P.; Pennings, B.; Frangeul, L.; Saleh, M.-C.; Van Rij, R.P. Histone-derived piRNA biogenesis depends on the ping-pong partners Piwi5 and Ago3 in *Aedes aegypti*. *Nucleic Acids Res.* **2017**, *45*, 4881–4892. [CrossRef] [PubMed]

43. Varjak, M.; Maringer, K.; Watson, M.; Sreenu, V.B.; Fredericks, A.C.; Pondeville, E.; Donald, C.; Sterk, J.; Kean, J.; Vazeille, M.; et al. *Aedes aegypti* Piwi4 Is a Noncanonical PIWI Protein Involved in Antiviral Responses. *mSphere* **2017**, *2*, e00144-17. [CrossRef] [PubMed]

44. Tian, Y.; Simanshu, D.K.; Ma, J.-B.; Patel, D.J. Structural basis for piRNA 2′-O-methylated 3′-end recognition by Piwi PAZ (Piwi/Argonaute/Zwille) domains. *Proc. Natl. Acad. Sci. USA* **2011**, *108*, 903–910. [CrossRef] [PubMed]

45. Yamaguchi, S.; Oe, A.; Nishida, K.M.; Yamashita, K.; Kajiya, A.; Hirano, S.; Matsumoto, N.; Dohmae, N.; Ishitani, R.; Saito, K.; et al. Crystal structure of Drosophila Piwi. *Nat. Commun.* **2020**, *11*, 858. [CrossRef] [PubMed]

46. Roy, A.; Kucukural, A.; Zhang, Y. I-TASSER: A unified platform for automated protein structure and function prediction. *Nat. Protoc.* **2010**, *5*, 725–738. [CrossRef] [PubMed]

47. Simon, B.; Kirkpatrick, J.P.; Eckhardt, S.; Reuter, M.; Rocha, E.A.; Andrade, M.; Sehr, P.; Pillai, R.S.; Carlomagno, T. Recognition of 2′-O-Methylated 3′-End of piRNA by the PAZ Domain of a Piwi Protein. *Structure* **2011**, *19*, 172–180. [CrossRef]

48. Matsumoto, N.; Nishimasu, H.; Sakakibara, K.; Nishida, K.M.; Hirano, T.; Ishitani, R.; Siomi, H.; Siomi, M.C.; Nureki, O. Crystal Structure of Silkworm PIWI-Clade Argonaute Siwi Bound to piRNA. *Cell* **2016**, *167*, 484–497. [CrossRef]

49. Ma, J.-B.; Ye, K.; Patel, D.J. Structural basis for overhang-specific small interfering RNA recognition by the PAZ domain. *Nat. Cell Biol.* **2004**, *429*, 318–322. [CrossRef] [PubMed]

50. Ma, Q.; Srivastav, S.P.; Gamez, S.; Dayama, G.; Feitosa-Suntheimer, F.; Patterson, E.I.; Johnson, R.M.; Matson, E.M.; Gold, A.S.; Brackney, D.E.; et al. A mosquito small RNA genomics resource reveals dynamic evolution and host responses to viruses and transposons. *Genome Res.* **2021**, *31*, 512–528. [CrossRef]

51. Wiedemann, C.; Bellstedt, P.; Görlach, M. CAPITO—A web server-based analysis and plotting tool for circular dichroism data. *Bioinformatics* **2013**, *29*, 1750–1757. [CrossRef] [PubMed]

52. Ariotti, N.; Hall, T.; Rae, J.; Ferguson, C.; McMahon, K.-A.; Martel, N.; Webb, R.E.; Webb, R.I.; Teasdale, R.; Parton, R.G. Modular Detection of GFP-Labeled Proteins for Rapid Screening by Electron Microscopy in Cells and Organisms. *Dev. Cell* **2015**, *35*, 513–525. [CrossRef]

53. Gamez, S.; Srivastav, S.; Akbari, O.S.; Lau, N.C. Diverse Defenses: A Perspective Comparing Dipteran Piwi-piRNA Pathways. *Cells* **2020**, *9*, 2180. [CrossRef] [PubMed]

54. Wang, W.; Han, B.; Tipping, C.; Ge, D.T.; Zhang, Z.; Weng, Z.; Zamore, P.D. Slicing and Binding by Ago3 or Aub Trigger Piwi-Bound piRNA Production by Distinct Mechanisms. *Mol. Cell* **2015**, *59*, 819–830. [CrossRef] [PubMed]

55. Huang, X.; Hu, H.; Webster, A.; Zou, F.; Du, J.; Patel, D.J.; Sachidanandam, R.; Toth, K.F.; Aravin, A.A.; Li, S. Binding of guide piRNA triggers methylation of the unstructured N-terminal region of Aub leading to assembly of the piRNA amplification complex. *Nat. Commun.* **2021**, *12*, 4061. [CrossRef]

56. Yashiro, R.; Murota, Y.; Nishida, K.M.; Yamashiro, H.; Fujii, K.; Ogai, A.; Yamanaka, S.; Negishi, L.; Siomi, H.; Siomi, M.C. Piwi Nuclear Localization and Its Regulatory Mechanism in *Drosophila* Ovarian Somatic Cells. *Cell Rep.* **2018**, *23*, 3647–3657. [CrossRef] [PubMed]

57. Liu, Y.; Dou, M.; Song, X.; Dong, Y.; Liu, S.; Liu, H.; Tao, J.; Li, W.; Yin, X.; Xu, W. The emerging role of the piRNA/piwi complex in cancer. *Mol. Cancer* **2019**, *18*, 123. [CrossRef]

58. Wu, X.; Pan, Y.; Fang, Y.; Zhang, J.; Xie, M.; Yang, F.; Yu, T.; Ma, P.; Li, W.; Shu, Y. The Biogenesis and Functions of piRNAs in Human Diseases. *Mol. Ther.-Nucleic Acids* **2020**, *21*, 108–120. [CrossRef] [PubMed]

59. Yang, J.; Yan, R.; Roy, A.; Xu, D.; Poisson, J.; Zhang, Y. The I-TASSER Suite: Protein structure and function prediction. *Nat. Methods* **2015**, *12*, 7–8. [CrossRef]

60. Zhang, Y. I-TASSER server for protein 3D structure prediction. *BMC Bioinform.* **2008**, *9*, 40. [CrossRef]

61. Sievers, F.; Wilm, A.; Dineen, D.; Gibson, T.J.; Karplus, K.; Li, W.; López, R.; McWilliam, H.; Remmert, M.; Söding, J.; et al. Fast scalable generation of high-quality protein multiple sequence alignments using Clustal Omega. *Mol. Syst. Biol.* **2011**, *7*, 539. [CrossRef] [PubMed]
62. Waterhouse, A.M.; Procter, J.B.; Martin, D.M.A.; Clamp, M.; Barton, G.J. Jalview Version 2—A multiple sequence alignment editor and analysis workbench. *Bioinformatics* **2009**, *25*, 1189–1191. [CrossRef] [PubMed]
63. Martin-Martin, I.; Paige, A.; Leon, P.C.V.; Gittis, A.G.; Kern, O.; Bonilla, B.; Chagas, A.C.; Ganesan, S.; Smith, L.B.; Garboczi, D.N.; et al. ADP binding by the Culex quinquefasciatus mosquito D7 salivary protein enhances blood feeding on mammals. *Nat. Commun.* **2020**, *11*, 2911. [CrossRef] [PubMed]

International Journal of
Molecular Sciences

MDPI

Article

MAEBL Contributes to *Plasmodium* Sporozoite Adhesiveness

Mónica Sá [1,2,3], David Mendes Costa [1,2], Ana Rafaela Teixeira [1,2,3], Begoña Pérez-Cabezas [1,2,†],
Pauline Formaglio [4], Sylvain Golba [5], Hélèna Sefiane-Djemaoune [5], Rogerio Amino [4] and Joana Tavares [1,2,3,*]

[1] Host-Parasite Interactions Group, Instituto de Investigação e Inovação em Saúde, Universidade do Porto, 4200-135 Porto, Portugal; monica.sa@ibmc.up.pt (M.S.); david.costa@ibmc.up.pt (D.M.C.); ana.teixeira@ibmc.up.pt (A.R.T.); bperezcabezas@gmail.com (B.P.-C.)

[2] Instituto de Biologia Molecular e Celular, Universidade do Porto, 4200-135 Porto, Portugal

[3] Departamento de Ciências Biológicas, Faculdade de Farmácia, Universidade do Porto, 4050-313 Porto, Portugal

[4] Unit of Malaria Infection and Immunity, Institut Pasteur, 75015 Paris, France; pauline.formaglio@pasteur.fr (P.F.); rogerio.amino@pasteur.fr (R.A.)

[5] Center for Production and Infection of *Anopheles*, Institut Pasteur, 75015 Paris, France; sylvain.golba@pasteur.fr (S.G.); helena.sefiane-djemaoune@pasteur.fr (H.S.-D.)

* Correspondence: jtavares@ibmc.up.pt; Tel.: +351-226074919

† Current address: Instituto de Ciências Biomédicas Abel Salazar, Universidade do Porto, 4050-313 Porto, Portugal.

Citation: Sá, M.; Costa, D.M.; Teixeira, A.R.; Pérez-Cabezas, B.; Formaglio, P.; Golba, S.; Sefiane-Djemaoune, H.; Amino, R.; Tavares, J. MAEBL Contributes to *Plasmodium* Sporozoite Adhesiveness. *Int. J. Mol. Sci.* **2022**, 23, 5711. https://doi.org/10.3390/ijms23105711

Academic Editor: Michail Kotsyfakis

Received: 15 April 2022
Accepted: 13 May 2022
Published: 20 May 2022

Publisher's Note: MDPI stays neutral with regard to jurisdictional claims in published maps and institutional affiliations.

Abstract: The sole currently approved malaria vaccine targets the circumsporozoite protein—the protein that densely coats the surface of sporozoites, the parasite stage deposited in the skin of the mammalian host by infected mosquitoes. However, this vaccine only confers moderate protection against clinical diseases in children, impelling a continuous search for novel candidates. In this work, we studied the importance of the membrane-associated erythrocyte binding-like protein (MAEBL) for infection by *Plasmodium* sporozoites. Using transgenic parasites and live imaging in mice, we show that the absence of MAEBL reduces *Plasmodium berghei* hemolymph sporozoite infectivity to mice. Moreover, we found that *maebl* knockout (*maebl-*) sporozoites display reduced adhesion, including to cultured hepatocytes, which could contribute to the defects in multiple biological processes, such as in gliding motility, hepatocyte wounding, and invasion. The *maebl-* defective phenotypes in mosquito salivary gland and liver infection were reverted by genetic complementation. Using a parasite line expressing a C-terminal myc-tagged MAEBL, we found that MAEBL levels peak in midgut and hemolymph parasites but drop after sporozoite entry into the salivary glands, where the labeling was found to be heterogeneous among sporozoites. MAEBL was found associated, not only with micronemes, but also with the surface of mature sporozoites. Overall, our data provide further insight into the role of MAEBL in sporozoite infectivity and may contribute to the design of future immune interventions.

Keywords: *Plasmodium*; sporozoite; MAEBL; liver; adhesion; genetic complementation; in vivo bioluminescence imaging

1. Introduction

In 2020 alone, more than 240 million cases of malaria were reported leading to 627,000 deaths. These values represent a substantial increase in the number of malaria case incidence and deaths estimated globally, fueled by the disruptions caused by the COVID-19 pandemic [1]. In 2021, the World Health Organization recommended for the first time a malaria vaccine, RTS,S/AS01, for use in children living in endemic areas with moderate to high transmission [1]. However, this vaccine only confers moderate protection against clinical disease by *Plasmodium falciparum*, the most dangerous human malaria parasite [2]. RTS,S/AS01 targets the circumsporozoite protein (CSP), the protein that densely coats the surface of sporozoites, the parasite stage deposited in the skin of the mammalian host by infected mosquitoes. Sporozoites actively migrate in the skin and invade blood vessels to complete their development in the liver. Inside hepatocytes, a single sporozoite will transform and multiply into thousands of merozoites, the red blood cells infective forms. Sporozoites and ensuing liver stages, called the pre-erythrocytic phase, represent an attractive target for immune interventions [3].

Sera from individuals immunized with radiation-attenuated *P. falciparum* sporozoites, the gold standard malaria vaccine, contain antibodies against multiple pre-erythrocytic antigens highly associated with sporozoite-induced protection [4]. In an attempt to find novel pre-erythrocytic antigens, Peng and colleagues screened a library of *P. falciparum* antigens with sera from volunteers immunized by mosquito bite under chemoprophylaxis with chloroquine [5]. One of the antigens recognized by the sera from most of the individuals was the membrane-associated erythrocyte binding-like protein (MAEBL) [5].

MAEBL is a large type I transmembrane protein composed of two N-terminal cysteine-rich adhesion domains homologous to the apical membrane antigen 1 (AMA-1), named M1 and M2, and a C-terminal cysteine-rich region (C-cys) structurally related with *Plasmodium* Duffy binding-like family of erythrocyte binding proteins [6]. Conserved among *Plasmodium* species [7], MAEBL was initially reported as an erythrocytic-binding protein present in blood-stage parasites [6,8], but was later found to be expressed in sporozoites and late liver stages [9–12]. Although dispensable for asexual blood-stage growth [13–16], immunization with MAEBL M2 domain protects animals from dying of a challenge with the lethal *Plasmodium yoelii* YM strain infected red blood cells [17].

MAEBL is required for the colonization of the mosquito salivary glands by sporozoites [13,14,16]. Two main *maebl* transcripts are expressed in sporozoites as a result of the alternative splicing in 3′ exons, encoding a canonical transmembrane and a putative soluble MAEBL isoform [12]. However, only the transmembrane isoform is essential for *P. falciparum* sporozoite infection of salivary glands [16].

In sporozoites, MAEBL is found associated with the micronemes [13,14]. However, immunolabelling studies indicate that its subcellular localization is developmentally regulated during parasite maturation, as it changes from being restricted to the apical pole in immature sporozoites, to covering the surface of mature parasites colocalizing with CSP [11]. In salivary gland sporozoites, the protein was detected both internally and on the parasites surface [11,18]. Nevertheless, antibodies generated against MAEBL domains often recognize multiple bands on western blot analysis of parasite extracts that might hinder conclusions on the localization, particularly when sera reactivity is not evaluated also in a knockout line [18].

While MAEBL was suggested to be dispensable for liver infection by *P. berghei* sporozoites collected from the midgut of mosquitoes [13], MAEBL-deficient *P. falciparum* sporozoites from the hemolymph have been shown to exhibit impaired hepatocyte wounding and invasion capacities along with reduced liver infection of humanized chimeric mice [14]. Indeed, antibodies against MAEBL partially inhibit hepatocyte invasion by sporozoites and/or liver-stage development [5,18], supporting a role for MAEBL in sporozoite infectivity in the mammalian host. In this study, and using the rodent malaria model *P. berghei*, we aimed at understanding the contribution of this protein in the sequence of events that lead to a successful establishment of liver infection by sporozoites.

2. Results

2.1. Genetic Complementation Reverts the Phenotype of maebl- Parasites in the Mosquito

A *maebl* knockout (*maebl-*) line was generated in a bioluminescent background of *P. berghei*, by replacing the open reading frame (ORF) of *maebl* with the selectable marker *Toxoplasma gondii* dihydrofolate reductase-thymidylate synthase by double-crossover homologous recombination (Figure S1A). Three *maebl-* isogenic lines (*maebl-* B2, B3, and G3) were generated and their genotype was verified by PCR and Southern blot analysis (Figure S1B,C). The absence of *maebl* transcripts was confirmed for the *maebl-* lines by RT-PCR (Figure S1D) and the data presented throughout this work refers to *maebl-* G3 clone.

A genetic complementation approach was simultaneously adopted to directly link the defective phenotypes of *maebl-* parasites to the absence of MAEBL. As the *P. falciparum maebl* is transcribed along with the upstream gene as a bicistronic transcript [19], the full-length gene was re-introduced into the original locus of *maebl* together with the human dihydrofolate reductase selectable marker cassette, by a single-crossover homologous recombination event (Figure S1E). A *maebl* complemented isogenic line (*maebl_comp* V3) with the expected genotype was isolated and used in further studies (Figure S1F).

To investigate the development of *maebl-* and *maebl_comp* mutant lines in the vector, mosquitoes were fed on mice infected with control, *maebl-*, or *maebl_comp* parasites. Between days 18 and 26 post-infection, mosquitoes were dissected and the numbers of sporozoites collected from their midguts, hemolymph, and salivary glands were determined. While there were no significant differences between the numbers of midgut sporozoites among all lines, we frequently found higher numbers of *maebl-* sporozoites in the hemolymph, an observation consistent with the inability of these parasites to colonize the salivary glands (Figure 1A) [13]. Transmission electron microscopy (TEM) analysis showed no *maebl-* sporozoites inside salivary glands even when these are collected at a late time point post-infection such as day 27 (Figure 1B), suggesting that the few sporozoites recovered most likely result from contamination with hemolymph. Importantly, we found no differences between the number of control and *maebl_comp* salivary gland sporozoites (Figure 1A), which indicates the genetic complementation rescued sporozoite infectivity to the vector.

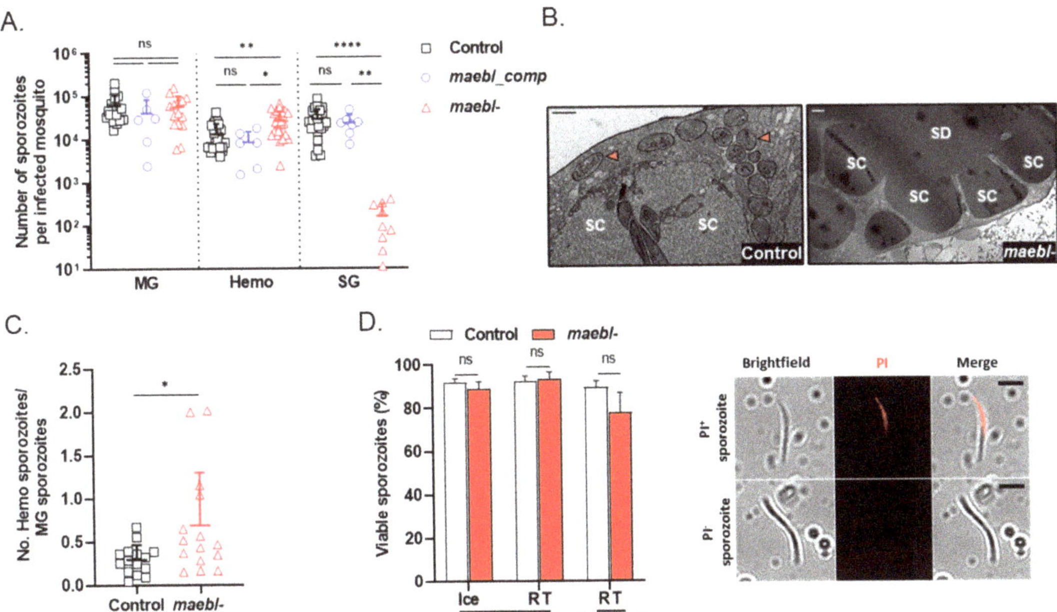

Figure 1. Development of *maebl-* and *maebl_comp* parasites in the mosquito. (**A**) Number of sporozoites in the midgut (MG), hemolymph (Hemo), and salivary glands (SG) of mosquitoes infected with Control, *maebl_comp*, or *maebl-* parasites, on days 18 to 26 post-infection. Symbols represent the counts in independent experiments and bars indicate the mean + SD. Statistical significance was determined using one-way ANOVA (Kruskal–Wallis test with Dunn's multiple comparisons test). (**B**) Transmission electron micrographs of salivary glands of Control- or *maebl-* -infected mosquitoes, dissected on day 27 post-infection. SC, secretory cavity; SD, salivary duct; red arrows, sporozoites. Scale bar, 1 μm (**left** panel) or 2 μm (**right** panel). (**C**) Ratio of hemolymph (Hemo) to midgut (MG) sporozoites in Control- or *maebl-* infected mosquitoes, between days 18 and 21 post-infection. Symbols

represent values of independent experiments and bars indicate the mean + SD. Statistical significance was determined using the Mann–Whitney test. (**D**) Viability of sporozoites. Hemolymph sporozoites were collected from Control- or *maebl-* -infected mosquitoes, on days 18 and 19 post-infection. Sporozoites were activated with DMEM supplemented with 5% FBS at room temperature (activated RT) or incubated with saline phosphate buffer on ice or at room temperature (non-activated Ice and RT, respectively). Propidium iodide (PI) was then added to the parasite suspensions and sporozoites were immediately imaged. **Left**: sporozoites were manually classified as PI+ or PI− sporozoites (dead or viable, respectively). The graphic shows the mean of two independent experiments performed in duplicated + SD. At least 150 sporozoites were analyzed per replicate. Statistical analysis was performed using the unpaired *t*-test. Right: representative images of PI+ or PI− sporozoites. Scale bar, 5 μm. ns, non-significant. * $p < 0.05$; ** $p < 0.01$; **** $p < 0.0001$.

In agreement with what was previously reported [13], our data show that *maebl-* sporozoites accumulate in the vector circulatory system (Figure 1A,C). To test whether the accumulation of sporozoites in the hemolymph of infected mosquitoes led to a reduction in parasite viability, we performed a standard propidium iodide (PI) exclusion assay. The percentage of viable *maebl-* hemolymph sporozoites, collected from mosquitoes at day 18/19 post-infection, was not significantly different from that of the control even following activation in the presence of serum (Figure 1D). These results validate the use of *maebl-* hemolymph sporozoites in subsequent experiments.

2.2. maebl- Sporozoites Exhibit Decreased Infectivity to Mice

It has been previously suggested that MAEBL is dispensable for the infection of rat livers by *P. berghei* midgut sporozoites [13]. In contrast, *maebl- P. falciparum* sporozoites collected from the mosquito hemolymph showed reduced infectivity to chimeric mice with humanized livers [14]. Therefore, to evaluate whether this phenotype is species-specific [13,14], we assessed the infectivity of *maebl-* and *maebl_comp* hemolymph sporozoites to C57BL/6 mice using in vivo bioluminescence imaging. Mice were inoculated intravenously (i.v.) with control, *maebl-* or *maebl_comp* hemolymph sporozoites, and the bioluminescent signal in the liver was quantified 1- and 2-days post-infection (D1 and D2, respectively). Animals infected with *maebl-* sporozoites showed a reduced liver burden compared to both control and *maebl_comp* at D1 and D2 (Figure 2A). *maebl-* parasites only emerged in the blood of 3 out of 4 mice and after a delay of 2 days comparing with the other lines (Figure 2B). These observations were consistently reproducible, as we frequently observed 1 to 2 days of delay in the prepatent period of mice inoculated i.v. with *maebl-* sporozoites, in several independent experiments (data not shown). Once in the blood, *maebl-* parasites exhibited normal asexual growth kinetics as determined by counting the percentage of infected red blood cells (Figure 2B). Although mice inoculated with *maebl_comp* hemolymph sporozoites displayed lower parasite loads in the liver at D1 compared to control-infected animals (~3.0-fold reduction), the reduction was no longer observed at a later time-point (Figure 2A). In agreement with this observation, no differences were seen in the prepatent periods or in the blood-stage growth of *maebl_comp* and control parasites (Figure 2B).

Genetically complemented *maebl-* sporozoites successfully enter the mosquito salivary glands (Figure 1A). To assess whether *maebl_comp* sporozoites have completed their maturation in the vector and efficiently infect the mammalian host, mice were inoculated i.v. with *maebl_comp* or control sporozoites collected from the mosquito salivary glands. As expected, no differences were observed in the bioluminescent signal between experimental groups at D1 and D2 (Figure 2C), as well as in the parasitemia of animals (Figure 2D). The *maebl-* line was not used in these experiments due to its reduced number of salivary glands-associated sporozoites. Therefore, mice were inoculated i.v. with the few parasites we could collect but no animal ever became blood-stage positive, contrarily to mice infected with similar numbers of control salivary gland sporozoites (Table S1). Altogether, our data demonstrate that in the absence of MAEBL, *P. berghei* hemolymph sporozoites exhibit an impaired ability to infect the liver of mice.

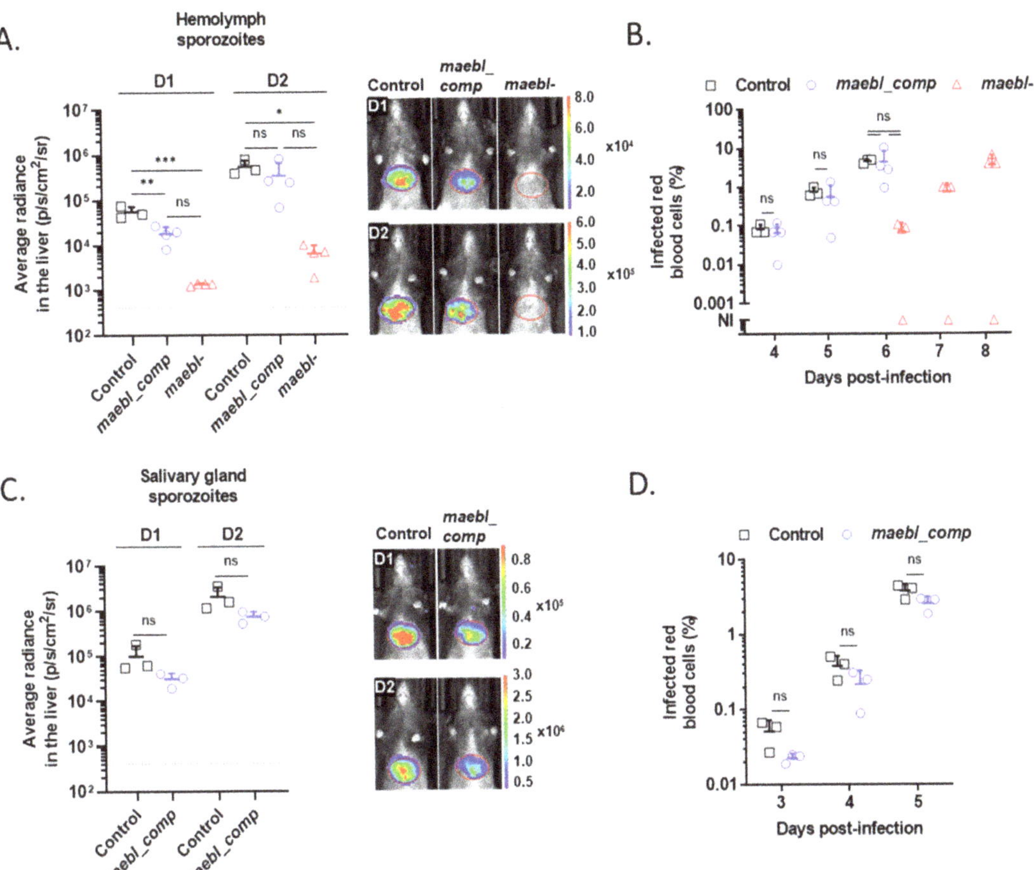

Figure 2. Infectivity of *maebl-* and *maebl_comp* sporozoites to mice. (**A–D**) Infectivity of *maebl-* and *maebl_comp* sporozoites to C57BL/6 mice. Mice were injected intravenously with 3.5×10^4 Control, *maebl-* and *maebl_comp* hemolymph sporozoites (panels **A,B**) or with 2.5×10^4 Control and *maebl_comp* salivary gland sporozoites (panels **C,D**), collected from mosquitoes on days 20 or 21 post-infection. (**A,C**) **Left**: parasite burdens in the liver were quantified as average radiance (photons/s/cm^2/steradian) one (D1) and two (D2) days post-infection. Symbols represent values for individual animals and bars indicate the mean + SD (*n* = 3–4). Dotted line: background level, calculated using non-infected mice. Statistical analysis was performed using one-way ANOVA (Tukey's multiple comparisons test; panel (**A**) or unpaired *t*-test (panel **C**). **Right**: representative images of infected mice, on D1 and D2 post-infection. (**B,D**) Parasitemia of infected mice, determined daily by a Giemsa-stained blood smear. Symbols represent values for individual animals and bars indicate the mean + SD. Statistical analysis was performed using unpaired *t*-test (**B,D**) or one-way ANOVA (Tukey's multiple comparisons test) (**B**). ns, non-significant; * $p < 0.05$; ** $p < 0.01$; *** $p < 0.001$; NI, non-infected.

2.3. maebl- Hemolymph Sporozoites Present Hampered Invasion and Wounding of Host Cells In Vitro

Next, we conducted several in vitro experiments to further explore the infectivity o maebl- sporozoites to the mammalian host. We started by evaluating sporozoite invasion and liver stage development inside the hepatoma cell line HepG2, using immunofluorescence microscopy. To that end, sporozoites were collected from the hemolymph or salivary glands and incubated with cells for 2 h to evaluate the invasion of host cells. Parasite development was analyzed at 48 h after infection. The percentage of cells with intracellular sporozoites was significantly reduced for the maebl- line compared to the control and maebl_comp line (Figure 3A), leading to the formation of a lower number of exoerythrocytic forms (EEFs) (Figure 3B). No differences in the size of EEFs were observed among all lines (Figure 3C), suggesting that MAEBL is not required for liver stage development. As expected, maebl_comp sporozoites did not show impaired hepatocyte invasion nor exoerythrocytic development (Figure 3A–C).

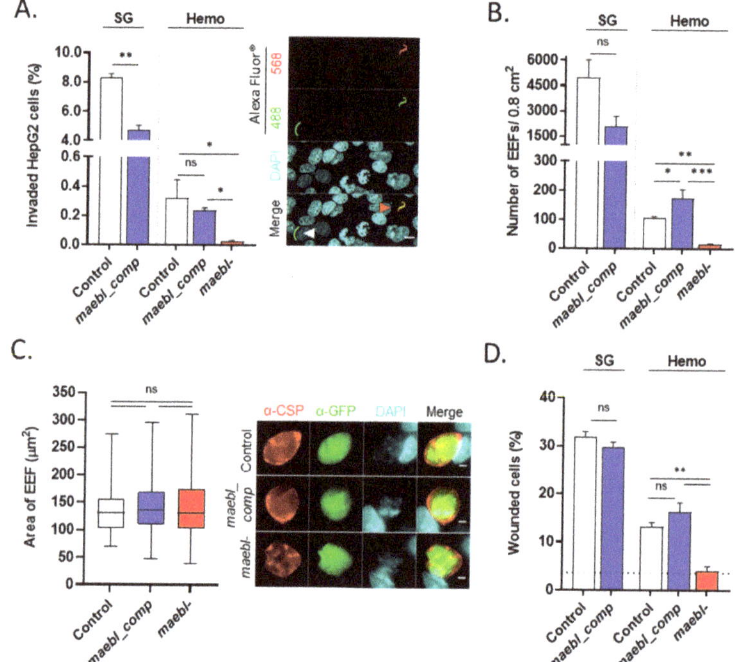

Figure 3. Evaluation of HepG2 cells invasion and cell wounding activity of *maebl*- sporozoites. (**A–C**) Invasion of cells by *maebl*- and *maebl_comp* sporozoites and exoerythrocytic forms (EEFs) development. Control, *maebl*- or *maebl_comp* hemolymph (Hemo) or salivary glands (SG) sporozoites, collected from mosquitoes on days 19 to 21 post-infection, were incubated with cells for 2 h (panel **A**) or 48 h (panel **B,C**). (**A**) **Left**: percentage of cells containing intracellular sporozoites. Bars represent the mean + SD of experimental replicates. Statistical analysis was performed using unpaired *t*-test (SG) or one-way ANOVA (Tukey's multiple comparisons test; Hemo). **Right**: representative immunofluorescence images of intracellular (white arrow) and extracellular sporozoites (red arrow). DAPI-stained cell nuclei, cyan. Scale bar, 10 μm. (**B**) Number of EEFs in cells per well. Bars represent the mean + SD of experimental replicates. Values are representative of at least two independent experiments (panel **A,B**). Statistical analysis was performed using unpaired *t*-test (SG) or one-way ANOVA (Tukey's multiple comparisons test; Hemo). (**C**) **Left**: area of EEFs. Box plots showing the median, maximum, minimum, and the 25th and 75th percentiles of the area of individual EEFs. At least 45 EEFs were analyzed per

condition. Statistical significance was determined using one-way ANOVA (Kruskal–Wallis test with Dunn's multiple comparisons test). **Right**: representative immunofluorescence images of EEFs. CSP, red; GFP, green; DAPI-stained nuclei, cyan. Scale bar, 3 μm. (**D**) Cell wounding capacity of Control, *maebl-* and *maebl_comp* Hemo or SG sporozoites collected from mosquitoes on day 19 post-infection. Sporozoites were incubated with cells for 60 min in the presence of propidium iodide (PI). The graph shows the percentage of wounded cells (PI+) assessed by flow cytometry analysis. Bars represent the mean + SD of experimental replicates; values are representative of 3 independent experiments. Horizontal dotted line: percentage of PI+ cells following incubation with medium only. Statistical analysis was performed using unpaired *t*-test (SG) or one-way ANOVA (Tukey's multiple comparisons test; Hemo). ns, non-significant. * $p < 0.05$; ** $p < 0.01$; *** $p < 0.001$.

It has also been demonstrated that cell traversal activity is disrupted in the MAEBL-deficient *P. falciparum* sporozoites [14]. To assess whether this process is also affected in the *P. berghei* knockout line, we performed a standard in vitro cell wounding assay using PI [20]. During traversal, the plasma membrane of host cells is breached, allowing the incorporation of cell-impermeant dyes, such as PI. Thus, sporozoites were allowed to traverse HepG2 cells in the presence of PI for 1 h before quantification of the percentage of wounded cells by flow cytometry analysis. Whereas the percentage of PI+ cells obtained upon incubation with control and *maebl_comp* hemolymph sporozoites was $13.3 \pm 0.8\%$ and $16.4 \pm 1.8\%$, respectively, *maebl-* sporozoites induced PI-incorporation levels on host cells close to those of cells incubated with medium alone ($4.1 \pm 0.9\%$, Figure 3D and Figure S2). No differences were observed in the percentage of wounded cells by control and *maebl_comp* sporozoites collected from either the hemolymph or the salivary glands (Figure 3D). Altogether, our data indicate that the absence of MAEBL results in a decrease of host cell invasion and wounding by *P. berghei* sporozoites in vitro.

2.4. maebl- Hemolymph Sporozoites Glide at Lower Average Speed and Exhibit Defective Attachment

Since hepatocyte invasion and traversal are two processes dependent on the parasite actin-myosin molecular motor, as well as gliding motility, we next evaluated the motile behaviors of *maebl-* sporozoites in vitro by time-lapse microscopy. Sporozoites were allowed to glide in a polystyrene plate and classified as attached, waving, floating, or motile. As expected, most control sporozoites collected from salivary glands showed vigorous circular gliding contrarily to less mature parasites from the hemolymph of mosquitoes (Figure 4A). Instead, the latter mostly displayed a waving behavior, characterized by the attachment of sporozoites to the surface only by one pole or part of the body (Figure 4A). Although no statistically significant differences were observed between the percentage of control hemolymph and *maebl-* sporozoites that glide at least one complete circle (Figure 4A), mutant parasites showed a significant reduction in their average speed when compared to control sporozoites (Figure 4B).

Strikingly, the *maebl-* line exhibited a significantly higher percentage of floating sporozoites compared to the control (Figure 4A). Based on these observations, we hypothesized that MAEBL could be involved in sporozoite adhesion to the substrate. Indeed, a flawed adhesion could explain the multitude of defects associated with *maebl-* sporozoites (Figures 2A,B, 3 and 4A). To test this hypothesis, we assessed the capacity of *maebl-* sporozoites to attach to HepG2 cells under static conditions. HepG2 cells were incubated with sporozoites for 30 min at 37 °C in the presence or absence of cytochalasin D, an actin polymerization inhibitor, known to impair sporozoite invasion but not adhesion [21], and the numbers of attached and non-attached sporozoites were counted using flow cytometry. Interestingly, we found that mutant sporozoites consistently adhered less than control sporozoites, in the presence of cytochalasin D or the DMSO control only (Figure 4C). Sporozoites collected from the salivary glands tend to adhere more to HepG2 cells than their hemolymph counterparts. However, this trend failed to reach statistical significance (Figure 4C).

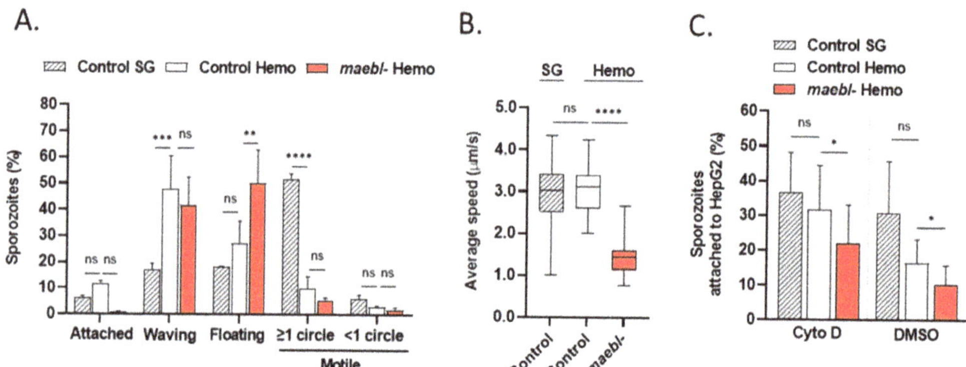

Figure 4. Analysis of *maebl-* sporozoites gliding motility and adhesive properties. (**A,B**) Gliding motility of Control salivary gland (SG) and hemolymph (Hemo) sporozoites, and *maebl-* Hemo sporozoites. Sporozoites were allowed to glide in medium supplemented with FBS at 37 °C. Bright-field images were acquired every second for 1 min using an inverted epifluorescence microscope. (**A**) Gliding behaviors of *maebl-* sporozoites. Sporozoites (Control SG $n = 316$, Control Hemo $n = 296$, and *maebl-* Hemo $n = 839$) were classified according to their motility behaviors. Bars indicate the mean of two independent experiments + SD. Statistical analysis was performed using two-way ANOVA (Tukey's multiple comparisons test). (**B**) Gliding speed of *maebl-* sporozoites. Box plots showing the median, maximum, minimum, and the 25th and 75th percentiles of the average speed of individual sporozoites (Control SG $n = 20$, Control Hemo $n = 12$, and *maebl-* Hemo $n = 22$). Data were pooled from two independent experiments. Statistical analysis was performed using one-way ANOVA (Tukey's multiple comparisons test). (**C**) Adhesion of Control SG and Hemo sporozoites, and *maebl-* Hemo sporozoites to HepG2 cells. Sporozoites were added on top of cells, in the presence of cytochalasin D (Cyto D) or DMSO. After 30 min incubation at 37 °C, the supernatant was removed to quantify the number of unattached sporozoites. After trypsinization, the number of extracellular parasites (attached sporozoites) was quantified. Quantification of sporozoites was performed by flow cytometry. The high levels of autofluorescence of HepG2 cells precluded the quantification of intracellular sporozoites. The graph shows the mean of three independent experiments performed at least in duplicate + SD. Statistical analysis was performed with repeated measures one-way ANOVA (Tukey's multiple comparisons test). ns, non-significant; * $p < 0.05$; ** $p < 0.01$; *** $p < 0.001$; **** $p < 0.0001$.

2.5. Carboxy-Terminal Myc Tagging of MAEBL Does Not Affect Protein Function

We next aimed to investigate in detail the expression and localization of the full-length MAEBL in mature and immature sporozoites. To this end, we engineered a parasite line that expresses a version of MAEBL tagged at the C-terminus end using a functional complementation approach. Briefly, the transfection vector used to generate the *maebl_comp* line was modified as to insert a sequence encoding two myc tag epitopes before the stop codon of *maebl*, and then used to complement the *maebl-* G3 line (Figure S3A). PCR analysis confirmed the correct integration of the construct and the presence of the tagged *maebl* ORF in the genome of three clonal populations (Figure S3B). The clonal population R2 (henceforth named *maebl:myc*) was used throughout this study.

To assess if the presence of the tag impaired the function of MAEBL, mosquitoes were fed with *maebl:myc* parasites, and the number of sporozoites in the salivary glands was quantified. No differences were observed between the *maebl:myc* and the control line (Figure 5A), indicating that complementation with a C-terminus myc-tagged MAEBL successfully reverted the *maebl-* phenotype in the mosquito. Subsequently, the infectivity of *maebl:myc* sporozoites to the mammalian host was evaluated using in vivo bioluminescence imaging. The parasite load in the liver of C57BL/6 mice inoculated i.v. with *maebl:myc* sporozoites was comparable to control, both at D1 and D2 post-infection, and no delay in the emergence and growth of blood-stage parasites was seen (Figure 5B,C).

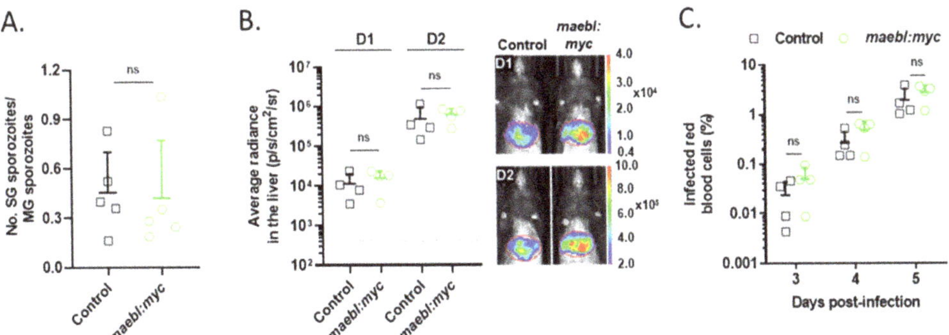

Figure 5. *maebl:myc* sporozoite infectivity to the mosquito and mammalian hosts. (**A**) Ratio of salivary gland (SG) sporozoites to midgut (MG) sporozoites in the Control and *maebl:myc* lines. Sporozoites were collected from mosquitoes on days 17 to 24 post-infection. Symbols represent the counts of independent experiments and bars indicate the mean + SD. Statistical significance was determined using unpaired *t*-test. (**B**) Infectivity of *maebl:myc* sporozoites to C57BL/6 mice by intravenous inoculation with 2.0×10^4 Control or *maebl:myc* salivary gland sporozoites. **Left**: parasite burden in the liver quantified as average radiance (photons/s/cm^2/steradian) one (D1) and two (D2) days post-infection. Symbols represent values for individual animals and bars indicate the mean + SD ($n = 4$). Dotted line: background level, calculated using non-infected mice. Statistical analysis was performed using unpaired *t*-test. **Right**: Representative images of the infected mice, on D1 and D2 post-infection. (**C**) Parasitemia of the infected mice, determined daily by a Giemsa-stained blood smear. Symbols represent values for individual animals and bars indicate the mean + SD. Statistical analysis was performed using unpaired *t*-test. ns, non-significant.

2.6. Characterization of C-Terminal Myc-Tagged MAEBL Expression and Localization in Sporozoites

A high molecular weight band corresponding to the expected size of the full-length myc-tagged MAEBL (~224 kDa) was detected by Western blot using a monoclonal anti-myc antibody in extracts of *maebl:myc* sporozoites collected from midguts (Figure 6A). Using the same antibody, we next evaluated MAEBL expression in sporozoites by immunofluorescence microscopy. The fluorescence intensity in sporozoites was quantified, and the percentage of positively stained parasites was calculated for parasites presenting an integrated fluorescence value higher than the highest value obtained in control sporozoites. To have a complete overview of the full-length MAEBL expression profile during sporozoite maturation in the vector, sporozoites were collected from different anatomical compartments of the mosquito and at several days post-blood meal. Specifically, *maebl:myc* and control sporozoites were collected as follows: (i) on days 17/18 post-infection, for midgut and hemolymph sporozoites and (ii) on day 21, for hemolymph and salivary gland sporozoites (Figure 6B–D). On days 17/18, all midgut sporozoites analyzed were positive for myc-tagged MAEBL, while only a minor sporozoite population collected from the hemolymph were negative (15%; Figure 6D). However, no significant differences were observed in the signal intensity between both conditions (Figure 6C). When we compare hemolymph parasites collected from mosquitoes a few days later (D21), the percentage of myc-tagged MAEBL-expressing sporozoites reached up to 91% (Figure 6D). On the other hand, salivary gland sporozoites collected at the same day showed not only a lower percentage for the positive population (76%) but also lower signal intensity values compared to hemolymph parasites (Figure 6C,D). This is in agreement with proteomic studies showing that MAEBL is less abundant in salivary gland sporozoites [22,23]. Together, our data indicate that full-length MAEBL is expressed by the large majority of sporozoites during their journey in the vector. Although being abundantly expressed in oocysts, we also found high levels in hemolymph sporozoites, an expected observation considering the role of this protein in sporozoite colonization of the mosquito salivary glands (Figure 1A,B). Importantly, our data clearly shows heterogeneity in full-length MAEBL expression by salivary gland sporozoites.

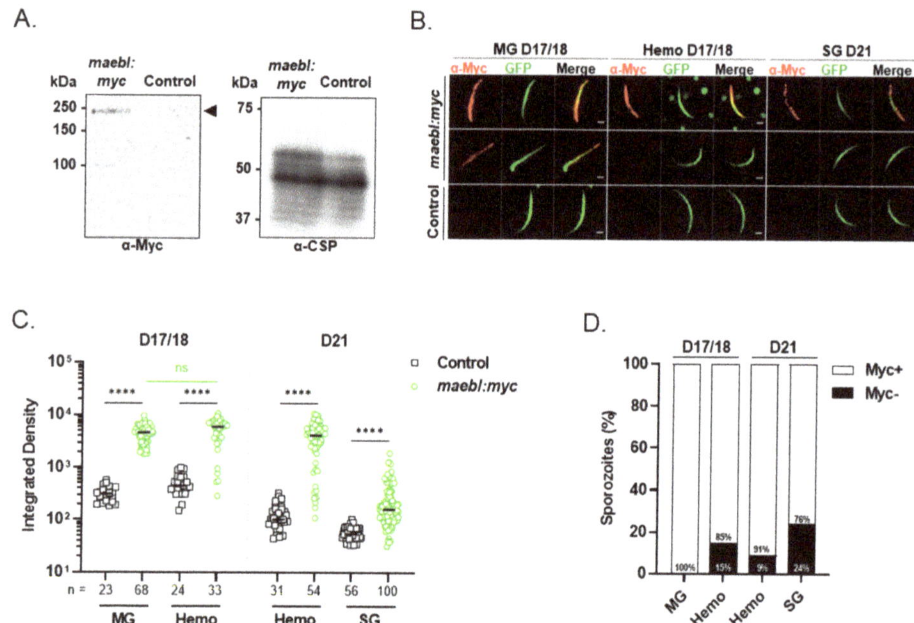

Figure 6. Quantification of myc-tagged MAEBL expression in sporozoites. (**A**) Western blot analysis of myc-tagged MAEBL expression in *maebl:myc* and Control midgut (MG) sporozoite extracts. Denatured lysates of 8.0×10^4 sporozoites were separated on 8% SDS gel and probed with a mouse monoclonal anti-myc tag antibody (clone 4A6). CSP was used as loading control.(**B–D**) Quantification of myc-tagged MAEBL expression in sporozoites by immunofluorescence. Sporozoites were collected from the midgut (MG), hemolymph (Hemo) and salivary glands (SG) of mosquitoes infected with Control or *maebl:myc* parasites, on the indicated days. Sporozoites were fixed, permeabilized and labeled anti-myc tag antibody (clone 4A6). (**B**) Representative immunofluorescence images of *maebl:myc* and Control sporozoites stained with anti-myc antibodies visualized in red and the GFP reporter in green. For the representation purpose only, the Smooth filter was applied to the GFP channel. Scale bar, 2 μm. (**C**) Fluorescence intensity of *maebl:myc* sporozoites, quantified as integrated density. Symbols represent individual sporozoites and bars indicate the median value of each population. Statistical significance was determined using the one-way ANOVA (Kruskal–Wallis test with Dunn's multiple comparisons test). (**D**) Percentage of myc-positive (Myc+) and negative (Myc−) sporozoites. *maebl:myc* sporozoites with a fluorescence intensity superior to the highest value obtained for Control sporozoites were considered positive. ns, non-significant; **** $p < 0.0001$.

Finally, we evaluated the subcellular localization of myc-tagged MAEBL in sporozoites by confocal microscopy. The staining pattern observed in midgut sporozoites was heterogeneous: whereas some parasites showed strong staining restricted to the apical (and sometimes posterior) pole, in others the signal was found to be more disperse (Figure 7A). In hemolymph sporozoites, myc-tagged MAEBL was frequently distributed throughout the body, and occasionally concentrated towards one (Figure 7A) or both poles (data not shown). Importantly, similar patterns were observed in salivary glands sporozoites (Figure 7A). Partial co-localization with thrombospondin related anonymous protein (TRAP) was frequently observed in midgut and hemolymph sporozoites, confirming the micronemal localization of MAEBL [13,14]. However, colocalization with TRAP in salivary gland sporozoites was not as evident in some sporozoites, as TRAP was frequently found uniformly spread over the sporozoite surface, unlike myc-tagged MAEBL (Figure 7A).

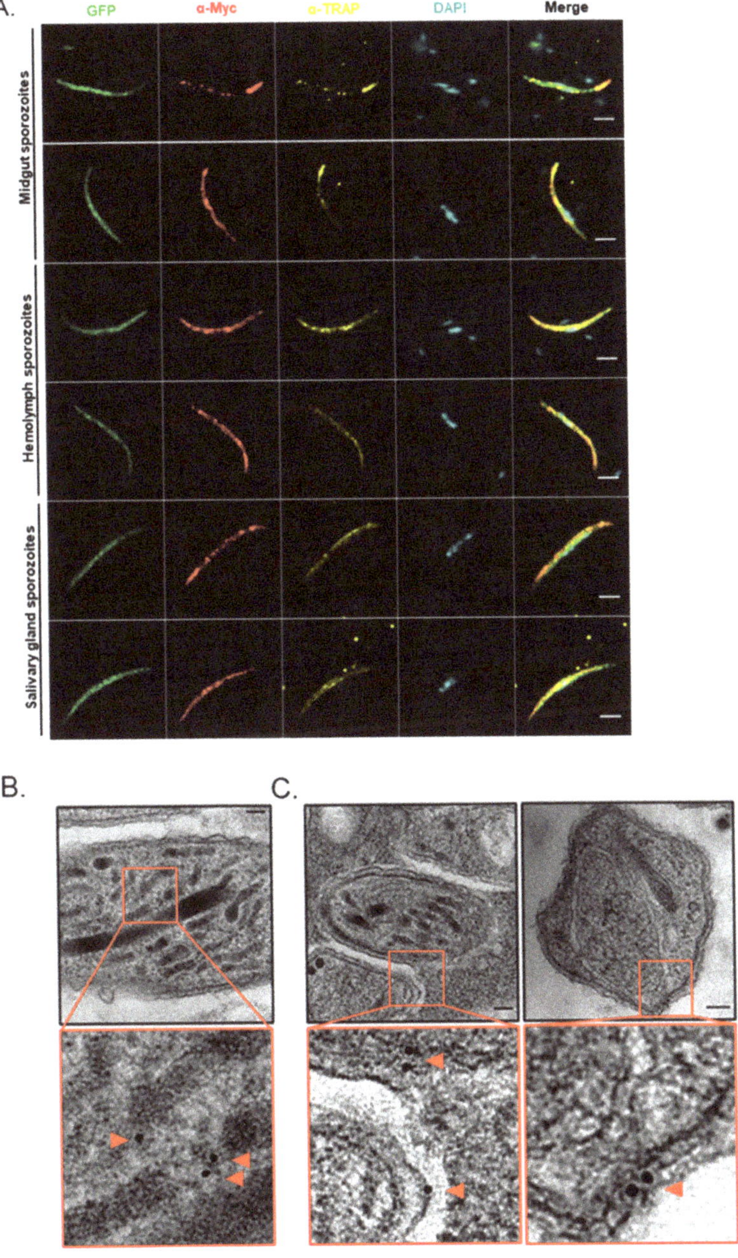

Figure 7. Localization of myc-tagged MAEBL in sporozoites. (**A**) Immunofluorescence analysis of

myc-tagged MAEBL distribution by confocal microscopy. *maebl:myc* sporozoites were collected on day 18 (for midgut and hemolymph sporozoites) and day 20 (for salivary glands sporozoites) post-infection, and labeled with a mouse monoclonal anti-myc tag antibody (clone 4A6) and a rabbit polyclonal anti-TRAP repeats antibody. GFP is visualized in green, myc-tagged MAEBL in red, TRAP in yellow, and DAPI-stained nuclei in cyan. For representation purpose only, the Smooth filter was applied in the GFP channel. Images are maximal Z-projections of 7 to 12 contiguous stacks separated by 0.17 to 0.25 µm. Scale bar, 3 µm. (**B,C**) Immunoelectron microscopy analysis of salivary glands of *maebl:myc*-infected mosquitoes on day 18 post-infection. Samples were stained with the antibody used in panel (**A**) and secondary antibodies conjugated with 6 nm gold particles. Transmission electron micrographs showing myc-tagged MAEBL in parasite micronemes (panel **B**) and associated with the sporozoite surface (panel **C**). Red arrows, gold particles. Scale bar, 100 nm.

A previous study reported that sera from *P. falciparum* sporozoite-immunized individuals under chloroquine cover recognized MAEBL and antibodies against two MAEBL isoforms blocked the liver stage in vitro, suggesting the protein can reach the surface of the sporozoite and become accessible to antibodies [5]. As our fluorescence microscopy approach required permeabilization of sporozoites before staining as the myc tag is placed at the intracellular portion of the protein, we resorted to immunoelectron microscopy to confirm the subcellular localization of MAEBL in salivary glands sporozoites. Indeed, myc-tagged MAEBL was detected not only associated with micronemes but also to the surface of sporozoites, i.e., near the plasma membrane and/or inner membrane complex (Figure 7B,C).

3. Discussion

While MAEBL is dispensable for the asexual growth of parasites in the blood [13–16], MAEBL-deficient *P. falciparum* sporozoites show impairment in hepatocyte wounding and invasion in vitro, as well as decreased infectivity to humanized chimeric mice [14]. Since previous studies indicate that antibodies against MAEBL can inhibit sporozoites invasion of hepatocytes and/or liver stage development [5,18], its exact contribution to the sequence of events that precedes sporozoite hepatocyte infection is worth exploring. To that end, GFP:luciferase-expressing *P. berghei* parasites were genetically modified to generate several MAEBL mutant lines (Figures S1 and S3) and their phenotype in both the vertebrate and invertebrate hosts analyzed.

Our results indicate that in the absence of MAEBL, *P. berghei* sporozoites show reduced infectivity to the mammalian host (Figure 2A,B), in contrast to previous work [13]. These contradictory findings probably result from the distinct experimental approaches used in both studies, as our experiments were performed with sporozoites collected from the mosquito's hemolymph instead of parasites collected from the oocysts. Since *maebl*-sporozoites can successfully egress from oocysts but fail to colonize the salivary glands (Figure 1A), we used parasites collected from the mosquito hemocoel, as this transient sporozoite population shows intermediate infectivity to the mammalian host [24]. Noteworthy, we failed every attempt to infect animals with *maebl*- salivary gland-associated sporozoites (Table S1). Considering that we do not find mutant parasites inside salivary glands of mosquitoes (Figure 1B), it is possible that we inoculated mice with hemolymph sporozoites that were collected with the salivary glands. Such low numbers of sporozoites are most likely insufficient to reliably yield productive infections. In addition to the inherently reduced infectivity of hemolymph parasites, *maebl*- sporozoites are 10- to 100-fold less infectious than control parasites, as delays of 1 to 2 days in the prepatent period are frequently observed in mice. Yet, we cannot exclude the possibility of other proteins being up or downregulated in *maebl*- sporozoites, thus, also contributing to the observed phenotype of this line. However, genetic complementation of *maebl*- parasites, rescued the defective phenotype of sporozoites both in the mosquito and the mammalian host (Figures 1A and 2). This confirms that the major impairments associated with *maebl*- sporozoites result from the loss of MAEBL and not from the disarrangement of the *maebl* locus.

On the other hand, the defective phenotype of the *maebl-* line in this study, i.e., loss of cell wounding activity (Figure 3D) and decreased infectivity in vitro (Figure 3A–C) and in vivo (Figure 2A,B), resembles that of MAEBL-deficient *P. falciparum* sporozoites [14], suggesting that the function of this protein is preserved in rodent and human *Plasmodium* infecting species. MAEBL is a conserved protein that predates *Plasmodium* speciation and contains two extracellular N-terminal cysteine-rich domains, named M1 and M2. Each of these domains contains two APPLE domains that are found in bacterial and eukaryotic adhesion molecules [6,25]. Sequence analysis shows that both the number and location of all cysteine residues present in the M1 and M2 domains are evolutionary conserved, suggesting that both regions have significant and similar functions across different *Plasmodium* species [7,10,26]. In blood-stages, it was suggested that MAEBL localizes to the rhoptries and surface of merozoites [7,8] and was shown to possess erythrocyte-binding capacity mainly through the M2 domain [6]. Moreover, it was suggested that MAEBL also participates in the binding of sporozoites to the vector salivary glands [13]. Based on the nature of the M1 and M2 domains and in the multiple defects of *maebl-* sporozoites exhibit in vitro (Figures 3 and 4A,B), we hypothesized that MAEBL contributes to sporozoite adhesion. To test whether MAEBL could be involved in adhesion to host cells, we evaluated the binding of mutant sporozoites to HepG2 using a flow cytometry-based assay. Our data shows that in the absence of MAEBL, sporozoites adhere less to HepG2 cells, both in the presence and absence of cytochalasin D (Figure 4C). These observations, together with the increased percentage of floating *maebl-* sporozoites observed in the gliding assays (Figure 4A), suggest that MAEBL may contribute to the overall adhesion of sporozoites.

As reported previously we did not observe a change in the proportion of motile hemolymph sporozoites in the absence of MAEBL (Figure 4A). Nonetheless, *maebl-* sporozoites glided at a lower average speed compared to controls (Figure 4B), challenging previous conclusions [13]. This could indicate that, for example, the loss of MAEBL may disturb the normal dynamics of discrete adhesion sites formed by sporozoites, as it could participate directly in their formation and/or the turnover, or indirectly, by interfering with the function of other adhesins, such as TRAP or TRAP-related proteins present in such sites [27].

It is likely that the decreased *maebl-* sporozoite infectivity, observed in vitro (Figure 3) and in vivo (Figure 2A,B), results from multiple adhesion-dependent defects. Nevertheless, we cannot exclude a possible role for MAEBL in other steps of hepatocyte invasion, through the interaction with members of the rhoptry neck protein (RON) complex, for example. However, to our knowledge, unlike the structurally related AMA-1, MAEBL was not found associated with RONs [28], suggesting that it acts independently of these proteins during host cell invasion.

We also generated a complemented parasite line expressing a myc-tagged MAEBL (Figure S3). As complementation rescued the phenotype of *maebl-* sporozoites, both in the vector and in the mammalian host (Figure 5), proving it is fully functional, the *maebl:myc* line was used for protein quantification and immunolocalization studies.

In this study, the myc-tag epitope coding sequences were inserted at the C-terminus of *maebl* ORF right before the stop codon. This means that, theoretically, our tagging strategy only allows the detection of the transmembrane isoform (Figure S3C). However, we cannot exclude that due to the alternative splicing other myc-tagged putative soluble and processed forms are also being detected [12]. Nevertheless, our data indicate that MAEBL levels peak in midgut and hemolymph sporozoites (Figure 6C). These observations are in agreement with the crucial function of MAEBL in sporozoite colonization of the vector salivary glands as well as with previous transcriptomic and proteomic studies [22,23,29]. Furthermore, not all salivary gland sporozoites are positive for myc-tagged MAEBL immunolabeling and importantly, protein expression varies within the positive population. Whether MAEBL is expressed de novo in the salivary glands or is carried over from hemolymph parasites remains unknown. Interestingly, recent data from single-cell RNA sequencing reveals extensive transcription heterogeneity among the sporozoite from the same anatomical compartment [29,30]. This could conceivably be an explanation for the fact that we were unable to detect myc-tagged MAEBL in some sporozoites residing in the salivary glands (Figure 6D).

In terms of protein localization, myc-tagged MAEBL frequently colocalized with TRAP (Figure 7A), in agreement with the literature that MAEBL is associated with the micronemes of sporozoites [13,14]. Notably, our data unequivocally indicate that MAEBL was found not only in the micronemes but also associated to the surface of salivary gland sporozoites (Figure 7C), a finding that might be relevant for the design of future immune interventions against *Plasmodium* sporozoites.

4. Materials and Methods

4.1. Mice, Mosquitoes and Parasites

The *Plasmodium berghei* ANKA strain clone 676cl1 expressing a GFP-Luciferase fusion gene via the *pbef1α* promoter [31], henceforth referred to as control line, was used to generate all mutant lines. C57BL/6, NMRI, and CD1 mice were purchased from Charles River (France) or obtained from the IBMC/i3S animal facility. *Anopheles stephensi* mosquitoes (Sda500 strain) were reared in the Centre for Production and Infection of *Anopheles* (CEPIA) at the Pasteur Institute using standard procedures.

4.2. Generation of Transfection Vectors

PCR reactions were performed using a high-fidelity *Taq* DNA polymerase with proofreading activity (Takara Bio, Otsu, Japan) and genomic DNA of control parasites as the template. Primers used for the generation and genotyping of all mutant lines are shown in Table S2. All PCR products were cloned into the pGEM-T Easy vector (Promega, Madison, WI, USA; unless stated otherwise), sequenced (LightRun, Eurofins Genomics, Ebersberg, Germany), and verified against the *P. berghei* genome database (PlasmoDB, http://plasmodb.org/plasmo/, accessed on 22 September 2015) using the Basic Local Alignment Search Tool (BLAST). As a matter of convenience, the intergenic regions upstream and downstream of the *maebl* gene (PBANKA_0901300.2) will be referred to as 5′ and 3′ UTR, respectively.

For the generation of the *maebl* knockout line (*maebl-*), the *maebl* open reading frame (ORF), along with the last 462 bp of the *maebl* 5′UTR, were replaced by the *Toxoplasma gondii* dihydrofolate reductase–thymidylate synthase gene (*TgDHFR/ts*) selectable marker, by a double cross-over homologous recombination event. Part of the *maebl* 5′ (499 bp) and 3′ (493 bp) UTRs were used as homology regions and were amplified using the primer pairs P1/P2 and P3/P4, respectively. The PCR products were subcloned into the plasmid pL0001 (MRA-770; MR4), on each side of the selectable marker, using the restriction sites KpnI/ClaI or EcoRI/BamHI. The final vector was digested with KpnI and BamHI before transfection.

Genetic complementation of *maebl-* clone G3 parasites was achieved by reinserting the wild type coding sequence of *maebl* (*maebl_comp*) or the same gene fused with a sequence encoding two tandem myc tag epitopes (*maebl:myc*), along with the last 462 bp of the *maebl* 5′UTR, into the recombinant locus by a single cross-over homologous recombination event.

The transfection vector containing the untagged *maebl* ORF was obtained as follows. The *maebl* 3′UTR (510 bp) and a 1907 bp DNA fragment corresponding to the first 87 bp of *maebl* ORF, the complete *maebl* 5′UTR and the last 621 nucleotides of the gene upstream of *maebl* (PBANKA_0901400), henceforth referred to as 5′ fragment, were amplified using the primer pairs P5/P6 and P7/P8, respectively. The 3′UTR and the 5′ fragment were inserted into the pGEM-T Easy vector (Promega) or the pCR®-TOPO-XL® vector (Invitrogen, Thermo Fisher Scientific, Waltham, MA, USA), respectively, and subcloned into the XhoI/NheI or XmaI/EcoRI sites of a pL0007 vector (MRA-776; MR4). The resulting plasmid was digested with HincII/BsgI to allow the insertion of a 6664 bp fragment that includes the complete coding sequence of *maebl*, flanked by the last 140 bp of the *maebl* 5′UTR and the first 387 bp of the 3′UTR, obtained through the digestion of the *P. berghei* artificial chromosome PbAC02-99h11 (PlasmoGem, Wellcome Sanger Institute, Hinxton, Cambridge,

UK) [32,33] with the same restriction enzymes. Finally, the entire DNA sequence ranging from the beginning of the 5' fragment to the end of the *maebl* 3'UTR (8446 bp) was inserted into a new pL0007 vector digested with HindIII, originating to the final transfection vector pL0007_MAEBLcomp. The correct orientation of the insert was confirmed by EcoRI/HincII digestion. pL0007_MAEBLcomp was linearized with PmeI before transfection.

The transfection vector containing the tagged *maebl* ORF was obtained through modification of the pL0007_MAEBLcomp plasmid by inserting a sequence encoding 2 copies of the myc tag epitope (2× EQKLISEEDL) right before the stop codon. The *maebl* 3'UTR (497 bp) and the last 669 bp of the *maebl* ORF (excluding the stop codon) were amplified using the primer pairs P5/P9 and P10/P11, respectively; the latter primer including the coding sequence of the tag. Both fragments were subcloned into the XhoI/EcoRV and EcoRV/EcoRI sites of a pL0007 vector, originating the plasmid pL0007_MAEBLmyc_3'UTR. The pL0007_MAEBLcomp plasmid was digested with the BstBI restriction enzyme to remove a 6689 bp fragment corresponding to the last 13 bp of the PBANKA_0901400 gene sequence, the entire *maebl* 5'UTR and the first 5475 bp of the *maebl* ORF, henceforth named 5'UTR_ORF fragment. After religation, the resultant plasmid was digested with BstBI/HincII to replace the *maebl* 3'UTR and the final portion of the *maebl* ORF by a myc-tagged version, obtained through the digestion of pL0007_MAEBLmyc_3'UTR with the same restriction enzymes. Finally, the plasmid was digested with BstBI to allow the insertion of the 5'UTR_ORF fragment, originating the final transfection vector. The correct orientation of the insert and the presence of the myc tag epitope coding sequence in the transfection vector were confirmed by DNA sequencing. The plasmid was linearized with PmeI before transfection.

4.3. Transfection and Cloning of Mutant Lines

Transfection of schizonts was performed as previously reported [34]. Immediately after electroporation, parasites were injected intravenously into 2 mice (parental populations) and selected with the appropriate drug, starting the day after parasite inoculation. Pyrimethamine (0.07 mg/mL) was given in drinking water, to select *maebl-* parasites. Once parasitemia was above 1%, blood from each animal was transferred into 2 naïve mice (transfer populations) for another round of selection. Selection of *maebl_comp* and *maebl:myc* parasites was performed with WR99210 (Jacobus Pharmaceutical Company, Inc., Princeton, NJ, USA). WR99210 (3.2 mg/mL) was dissolved in dH$_2$O 40% (v/v) ethanol, 3% (v/v) benzyl alcohol [34], and administrated subcutaneously (16 mg/Kg) for 3 successive days. Once parasitemia was above 1%, blood from each animal was transferred into a naïve mouse (transfer population). The treatment was repeated, starting from the day of infection. Cloning populations were obtained by limiting dilution [35].

4.4. Genotypic Analysis of Mutant Parasites by PCR and Southern Blot

Blood from infected mice was collected, filtered through a Plasmodipur filter (Euro-Proxima, Arnhem, The Netherlands), and lysed with 0.15% (v/v) saponin. Genomic DNA extraction and purification were done using the QIAamp DNA Blood Mini kit (Qiagen, Hilden, Germany). The integration of the constructs in the expected loci (primers P14/P15 and P7/P18, for the *maebl-* and *maebl_myc* genotyping strategies, respectively), the presence or absence of the *maebl* ORF in the genome of parasites (primers P12/P13), and the presence of the myc tag epitope in the final portion of *maebl* coding sequence (primers P19/P20), were evaluated by PCR using a high-fidelity DNA polymerase (Phusion®, New England Biolabs, Ipswich, MA, USA).

For Southern blot analysis, 2.3 to 10 µg of genomic DNA were digested with HindIII/NruI, separated by 0.8% (w/v) agarose gel electrophoresis, and transferred to a Nytran-N membrane (Amersham Hybond N+, Cytiva, Marlborough, MA, USA). The hybridization probe was obtained by PCR amplification of control DNA, using the primers P1/P2. Labelling of the probe and signal generation were performed using the AlkPhos Direct™ Labeling and Detection System with CDP-Star chemiluminescent detection reagent (Cytiva), respectively.

4.5. Evaluation of Gene Expression by Reverse Transcription PCR (RT-PCR)

Total RNA was isolated from midgut sporozoites using the NucleoSpin RNA II kit (Macherey-Nagel, Düren, Germany) and converted into cDNA using the NZY First-Strand cDNA Synthesis kit (NZYTech, Lisbon, Portugal). Detection of the *maebl* cDNA by PCR was done using a high-fidelity DNA polymerase (Phusion®, New England Biolabs) and the primers P12/P13. A region of the tubulin beta chain (PBANKA_1206900) was amplified using the primers P16/P17 and used as an internal control. Primer sequences are given in Table S2.

4.6. Mosquito Infections and Isolation of Sporozoites

Female mosquitoes were fed on infected NMRI or CD1 mice as described elsewhere [36]. Sporozoites were isolated from mosquitoes 17 to 27 days after the infectious bloodmeal. Midguts and salivary glands were collected into cold Dulbecco's Phosphate Buffered Saline (DPBS; Gibco, Thermo Fisher Scientific) and disrupted with a pestle immediately before use. Hemolymph sporozoites were isolated by flushing the mosquitos with 10 to 15 μL DPBS and left on ice until use. The total number of sporozoites obtained was determined using a plastic slide with a grid (KOVA® Glasstic® Slides, Kova International, Inc., Garden Grove, CA, USA) and a light microscope.

4.7. Sporozoite In Vitro Assays

4.7.1. Viability Assay

Hemolymph sporozoites, collected from mosquitoes on days 18 and 19 post-infection, were incubated for 15 min in DPBS on ice or at room temperature (RT), or at RT after dilution with an equal volume of Dulbecco's Modification of Eagle's Medium (DMEM;Lonza, Basel, Switzerland) supplemented with 10% (v/v) fetal bovine serum (FBS; Biowest, Nuaillé, France). Propidium iodide (PI; 5 μg/mL; Sigma, Merck, Darmstadt, Germany) was added to the suspensions, which were loaded into Ibidi 18-well μ-Slides (Ibidi GmbH, Gräfelfing, Germany), at a density of 5×10^3 to 1×10^4 parasites per well, and immediately imaged using IN Cell Analyzer 2000 (Cytiva). Based on PI incorporation, sporozoites were manually classified as dead or viable (PI+ or PI− sporozoites, respectively), using ImageJ/Fiji analysis software version 1.53f51 (ImageJ, National Institutes of Health, Bethesda, MD, USA). At least 150 sporozoites were analyzed per well and the percentage of viable sporozoites was calculated by dividing the number of PI− sporozoites by the total number of analyzed sporozoites.

4.7.2. Invasion and Liver Stage Development Assays

Host cell invasion and development assays were performed with sporozoites collected from mosquitoes on days 19 to 21 post-infection. The 8-well Lab-Tek chamber slides (Thermo Fisher Scientific) were precoated with 10 μg/cm^2 of collagen type I from rat tail (Sigma), overnight at 4 °C, if required. HepG2 cells (ATCC) were seeded at 1×10^5 cells per well in DMEM high glucose supplemented with 10% FBS (v/v) and cultured at 37 °C, 5% CO$_2$, for 24 h. Infections of hepatoma cells were performed with 1.4×10^4 to 2.0×10^4 sporozoites in DMEM supplemented with 5% FBS (v/v), penicillin–streptomycin (100 U/mL; Lonza), for 2 or 48 h at 37 °C, 5% CO$_2$, to assess sporozoite invasion and liver stage development, respectively. Preparations were fixed with 4% paraformaldehyde (PFA) (w/v) in DPBS, for 30 min, and stored at 4 °C until use.

Processing of samples was performed at RT, unless stated otherwise, and the incubation time of all antibodies was 1 h. The percentage of invaded cells was accessed using a double staining strategy [37]. Briefly, samples were blocked with 5% FBS (v/v) in DPBS, for 30 min, and extracellular sporozoites were labeled with the anti-CSP 3D11 mouse monoclonal antibody (2 μg/mL; MR4) and a goat anti-mouse Alexa Fluor 568 antibody (4 μg/mL; Invitrogen). Following cell permeabilization with 1% (v/v) Triton X-100 (Sigma), for 4 min, sporozoites were labeled with the same primary antibody in combination with goat anti-mouse Alexa Fluor 488 antibodies (4 μg/mL; Invitrogen™). Cell nuclei were stained with DAPI. Antifade mounting medium [90% (v/v) glycerol (Alfa Aesar, Thermo

Fisher Scientific), 0.5% (*w/v*) n-propyl gallate (Sigma), 20 mM Tris-HCl (Sigma), pH 8.0] was added to the preparations and slides were stored at 4 °C until use. Image acquisition was performed using IN Cell Analyzer 2000 (Cytiva). The numbers of sporozoites and HepG2 cell nuclei were determined using the ImageJ/Fiji analysis software (ImageJ, National Institutes of Health) or using an automated counting system, as previously described [38]. The percentage of infected cells was calculated by dividing the total number of intracellular sporozoites by the total number of HepG2 cell nuclei.

To evaluate the development of parasites, slides were blocked with 5% FBS (*v/v*) in DPBS, for 30 min, permeabilized with 1% (*v/v*) Triton X-100, for 4 min, and labeled with an anti-CSP 3D11 mouse monoclonal antibody (2 µg/mL; MR4) and an anti-GFP rabbit antibody (1:250; MBL, Tokyo, Japan) in combination with secondary antibodies goat anti-mouse Alexa Fluor 568 (4 µg/mL; Invitrogen™) and goat anti-rabbit Alexa Fluor 488 (4 µg/mL; Invitrogen). Nuclei were stained with DAPI. Antifade mounting medium were added to the preparations and slides were stored at 4 °C until use. EEFs were counted by microscopic visualization using a Zeiss Axio Imager Z1 microscope (Carl Zeiss, Oberkochen, Germany) and AxioVision software version 4.9 (Carl Zeiss, Germany), or using an automated counting system, as described previously [38]. To quantify the size of EEFs, images were taken of random EEFs using a Zeiss Axio Imager Z1 microscope (Carl Zeiss, Germany) and the area was manually determined based on the CSP staining, using the ImageJ/Fiji analysis software (ImageJ, National Institutes of Health).

4.7.3. Cell Wounding Assay

The capacity of sporozoites to wound cells was addressed using a standard flow cytometry-based cell-wounding assay [20]. In brief, HepG2 cells were seeded on a 96-well plate at a density of 8×10^4 cells per well in DMEM supplemented with 10% FBS (*v/v*) and cultured at for 24 h at 37 °C, 5% CO_2. The cells were then incubated with $\sim 3 \times 10^4$ sporozoites, isolated from mosquitoes on day 19 post-infection, in the presence of 5 µg/mL of PI for 60 min at 37 °C, 5% CO_2. Uninfected cells, incubated with or without PI, were used as controls. Cells were washed twice with warm DPBS and trypsinized. At least 7.8×10^3 events were analyzed with a FACS Canto II flow cytometer (BD Biosciences, Franklin Lakes, NJ, USA). Data analysis was performed using the FlowJo software version 10.7.1 (FlowJo LLC, Ashland, OR, USA).

4.7.4. Motility Assay

Sporozoites collected from mosquitoes on day 24 post-infection into DPBS were mixed with an equal volume of DMEM supplemented with 10% FBS (*v/v*) and transferred into a 384-microwell plate with an optical bottom (Greiner AG, Kremsmünster, Austria). After centrifugation for 5 min at 500× *g*, the plate was placed into the temperature-controlled microscope chamber held at 37 °C. Bright-field images were acquired every second for 1 min, using a widefield inverted Leica DMI6000 (Leica Microsystems GmbH, Wetzlar, Germany) microscope and LAS X software version 3.7.4.23463 (Leica Microsystems GmbH, Germany). Image analysis was performed using the ImageJ/Fiji analysis software (ImageJ, National Institutes of Health). Sporozoites were classified as follows: (i) attached, defined as sporozoites that were completely immobilized at the bottom of the well during the entire video; (ii) waving, defined as sporozoites that were attached only by a portion of the body; (iii) floating or (iv) motile. Motile sporozoites were further subclassified based on the completion or not of a full circle. The average speed was calculated by manually tracking at the apical end on sporozoites that glide at least one complete circle.

4.7.5. Adhesion Assay

Cell adhesion assays were performed with sporozoites collected from mosquitoes on days 19 to 21 post-infection. HepG2 cells were seeded in 96-well plates at a density of 5.0×10^4 to 1.50×10^5 cells per well, in DMEM supplemented with 10% FBS (*v/v*) and 1× MEM non-essential amino acids (Sigma), and cultured at 37 °C, 5% CO_2 until reach confluency. Sporozoites (1.25×10^4), diluted in an equal volume of DMEM supplemented with 10% FBS (v/v), $1\times$ MEM non-essential amino acids solution (v/v) and penicillin-streptomycin (200 U/mL; Sigma), were incubated with cells for 30 min at 37 °C, 5% CO$_2$ under static conditions, in the presence of 1 µM cytochalasin D (Sigma) or DMSO (Sigma). Following incubation, the supernatant was removed, and cells were washed twice with warm DPBS. Unattached sporozoites, defined as the number of GFP$^+$-parasites present in the supernatants, were quantified by flow cytometry. Subsequently, cells were trypsinized and analyzed by flow cytometry, to determine the number of extracellular GFP$^+$-sporozoites that were attached to cells. The high levels of autofluorescence of HepG2 cells precluded the quantification of intracellular sporozoites. Data were acquired using a CytoFLEX S flow cytometer (Beckman Coulter, Inc., Brea, CA, USA) and analyzed with the CytExpert version 2.0 (Beckman Coulter, Inc.). The numbers of attached and unattached sporozoites were determined based on sample volume and cell concentration. The percentage of the attached sporozoites was calculated by dividing the number of attached parasites by the total number of sporozoites recovered (attached and unattached).

4.7.6. Quantification and Subcellular Localization of Myc-Tagged MAEBL by Immunofluorescence

Sporozoites were collected from mosquitoes and transferred to an Ibidi 18-well µ-Slides (Ibidi GmbH). Sample processing was performed at RT, unless stated otherwise. Preparations were fixed with 4% PFA (w/v) in DPBS, for 30 min, permeabilized with 1% (v/v) Triton X-100, for 4 min, and blocked with 5% FBS (v/v) in DPBS, for 30 min. Sporozoites were stained with a mouse monoclonal anti-myc tag antibody clone 4A6 (5 µg/mL, Merck), overnight at 4 °C, and with goat anti-mouse Alexa Fluor 568 antibodies (2 µg/mL; Invitrogen™), for 30 min. Between each step (except after blocking), wells were washed five times with DPBS. Antifade mounting medium was added to the preparations and slides were immediately imaged using an inverted epifluorescence Leica DMI6000 microscope (Leica Microsystems) and LAS X software version 3.7.4.23463 (Leica Microsystems). Sporozoite signal intensity was quantified as integrated density (the product of the sporozoite area and the mean grey value), using ImageJ/Fiji software (ImageJ, National Institutes of Health). The background fluorescence was subtracted from the integrated density value for every sporozoite. For each day and condition, control sporozoites were used as negative control. The percentage of myc-positive sporozoites was calculated by dividing the number of *maebl:myc* sporozoites with a fluorescence intensity superior to highest value detected for control sporozoites. For illustrative purpose only, the Smooth filter of the ImageJ/Fiji software was applied to the GFP channel in the representative sporozoite images.

To study of the subcellular localization of myc-tagged MAEBL, *maebl:myc* sporozoites were stained with anti-myc tag antibodies as described above. Additionally, sporozoites were probed with a rabbit polyclonal anti-TRAP repeats antibody (1:10,000), overnight at 4 °C, stained with goat Alexa Fluor® 647 anti-rabbit IgG antibodies (2 µg/mL; Life Technologies, Thermo Fisher Scientific), for 30 min at RT, and incubated with DAPI (1:5000), for 10 min at RT. The wells were washed with DPBS between each step. Finally, preparations were mounted with antifade solution and immediately imaged. Images were acquired using an inverted microscope Leica TCS SP5 II (Leica Microsystems) and LAS AF software version 2.6.3.8173 (Leica Microsystems), and processed using ImageJ/Fiji software (ImageJ, National Institutes of Health) by projecting the maximum intensity of 7 to 12 contiguous z-stacks, separated by 0.17 to 0.25 µm. For illustrative purpose only, the Smooth filter of the ImageJ/Fiji software was applied to the GFP channel in the representative images of sporozoites.

4.7.7. Transmission Electron Microscopy

Infected salivary glands were collected 27 days post-infections and fixed in 2.5% (w/v) glutaraldehyde (Electron Microscopy Sciences, Hatfield, PA, USA) and 2% (w/v) formaldehyde (Electron Microscopy Sciences) in 0.1 M sodium cacodylate buffer (pH 7.4) for 24 h.

Samples were washed in buffer, postfixed with 2% (*w/v*) osmium tetroxide (Electron Microscopy Sciences) in 0.1 M sodium cacodylate buffer (pH 7.4) for 2 h, washed with water and incubated with 1% (*w/v*) uranyl acetate (Electron Microscopy Sciences) overnight. Subsequently, samples were dehydrated with ethanol and embedded in EPON resin (Electron Microscopy sciences). Ultrathin sections of 50 nm thickness were cut using an ultramicrotome (RMC PowerTome XL, Boeckeler Instruments, Inc., Tucson, AZ, USA), mounted on mesh copper grids, and stained with uranyl acetate substitute (Electron Microscopy sciences) and lead citrate (Electron Microscopy sciences) for 5 min each. Samples were visualized using a JEOL JEM 1400 transmission electron microscope (JEOL Ltd., Tokyo, Japan). Images were digitally recorded using a CCD digital camera Orius 1100 W (Japan) and analyzed using ImageJ/Fiji software (ImageJ, National Institutes of Health).

For the detection of myc-tagged MAEBL in salivary gland sporozoites by immunoelectron microscopy, salivary glands of mosquitoes were collected on day 18 post-infection, fixed in 0.05% (*w/v*) glutaraldehyde, 2% (*w/v*) PFA, 4% (*w/v*) sucrose in 0.1 M PBS, for 1 h, and washed with PBS. Samples were sequentially postfixed with 2% (*w/v*) osmium tetroxide in 0.1 M sodium cacodylate buffer (pH 7.4) for 1 h, washed with water, incubated with 1% (*w/v*) uranyl acetate for 45 min, dehydrated with ethanol and embedded in EPON resin. Ultrathin sections of 60 nm thickness were mounted on mesh nickel grids and processed as follows. Sections washed with Tris-buffered saline (TBS), incubated with 14.4% (*w/v*) sodium metaperiodate (Merck) for 1 h, washed with TBS, immersed with 10–20 mM glycine (±0.15%; *w/v*) for 5 min, blocked with 2% (*w/v*) BSA (AURION BSA-c™, Wageningen, The Netherlands) in TBS for 30 min and incubated with a mouse monoclonal anti-myc tag antibody clone 4A6 (100 μg/mL, Merck) in 2% (*w/v*) BSA 3% (*w/v*) NaCl in TBS, overnight at 4 °C. Sections were washed with 0.1% (*w/v*) BSA in TBS, incubated with 1% (*w/v*) BSA in TBS for 20 min, and then with goat anti-mouse secondary antibodies conjugated to 6 nm gold particles (1:20, Abcam, Cambridge, UK) diluted in 1% (*w/v*) BSA in TBS, for 1 h. Finally, sections were washed with TBS, post-fixed in 1% (*w/v*) glutaraldehyde in TBS for 5 min, washed with water and stained with uranyl acetate substitute and lead citrate for 1 min each. Samples were viewed using a JEOL JEM 1400 transmission electron microscope (JEOL Ltd.). Images were digitally recorded using a CCD digital camera Orius 1100 W (Japan) and analyzed using ImageJ/Fiji software (ImageJ, National Institutes of Health).

4.7.8. Western Blot Analysis

Sporozoites were collected from the midgut of mosquitos on days 17 and 18, mechanically liberated from oocysts, filtered using a 35 μm cell strainer cap (Falcon), and stored at −80 °C until use. Lysates of 8.0×10^4 sporozoites supplemented with cOmplete™, EDTA-free Protease Inhibitor Cocktail (Roche, Basel, Switzerland) were denatured in 2× Laemmli buffer (0.25 M Tris-HCl, pH 6.8, 5% SDS, 20% glycerol 0.02% bromophenol blue, 2.5% β-Mercaptoethanol), for 10 min at 95 °C. Samples were diluted to 1× Laemmli buffer and separated on an 8% (*w/v*) acrylamide gel by SDS-PAGE. Proteins were allowed to transfer to a PVDF membrane using a wet transfer system, for 16 h at 20 V, in 1× Towbin buffer [25 mM Tris, 192 mM Glycine, 20% (*v/v*) methanol] with 0.025% (*w/v*) SDS. After transfer, the membrane was rinsed with PBS and blocked with 5% (*w/v*) skim milk, 0.1% Tween 20 (Sigma) in PBS, for at least 1 h at RT. Incubations with a mouse monoclonal anti-myc tag antibody clone 4A6 (5 μg/mL, Merck) or with anti-CSP 3D11 mouse monoclonal antibody (0.3 μg/mL; MR4), diluted in blocking solution, were performed overnight at 4 °C or for 1 h at RT, respectively. The membranes were washed, probed with horseradish peroxidase-conjugated goat anti-mouse secondary antibodies (1:5000; SouthernBiotech, Birmingham, AL, USA) diluted in blocking buffer, for 1 h at RT, and washed again. Signal detection was performed using SuperSignal West Pico Chemiluminescent Substrate (Thermo Scientific, Thermo Fisher Scientific) and Amersham Hyperfilm ECL (Cytiva). Films were revealed using the Fujifilm FPM-100A film processor (Fujifilm, Tokyo, Japan).

4.8. Sporozoite In Vivo Assays

To assess mutant sporozoite infectivity and liver-stage development in vivo, C57BL/6 mice were injected i.v. with hemolymph or salivary gland sporozoites, isolated from mosquitoes at day 20 to 25 post-infection.

4.8.1. Bioluminescence Imaging

Bioluminescence imaging was performed as previously described [39], using the IVIS Lumina LT System (PerkinElmer, Inc., Waltham, MA, USA). Mice were imaged 1- and 2-days post-infection to quantify parasite loads in the liver. Before image acquisition, the ventral fur of mice was depilated with an appropriate clipper. Animals were anesthetized with isoflurane and injected subcutaneously with 2.4 mg of D-luciferin potassium salt (PerkinElmer, Inc.) dissolved in DPBS, 5 min before image acquisition. A non-infected mouse was routinely imaged in parallel to evaluate background noise signal. Quantitative analysis in the anatomical region of interest (ROI) encompassing the liver was performed using the Living Image software version 4.4 (PerkinElmer, Inc.), as previously described [39].

4.8.2. Parasitemia

Parasitemia was assessed daily by analysis of Giemsa-stained thin blood smears, starting on day 3 or 4 post-inoculation. The prepatent period was defined as the number of days until mice reached 0.1% parasitemia.

4.9. Statistical Analysis

Statistical analyses were performed using the GraphPad Prism Software (version 9.3.0). Statistical significance: $p < 0.05$ (*), $p < 0.01$ (**), $p < 0.001$ (***), $p < 0.0001$ (****).

5. Conclusions

In conclusion, we show that MAEBL is required for the optimal adhesiveness of sporozoites. Indeed, we propose that a flawed adhesion is likely to impair subsequent processes such as gliding motility, hepatocyte traversal, and invasion, and ultimately lead to decreased infectivity in vivo. Our work contributes to a better understanding of the role MAEBL plays in sporozoite infectivity to the mammalian host.

Supplementary Materials: The following supporting information can be downloaded at: https://www.mdpi.com/article/10.3390/ijms23105711/s1.

Author Contributions: Conceptualization, J.T. and M.S.; methodology, M.S., D.M.C. and J.T.; validation, M.S., R.A. and J.T.; formal analysis, M.S., R.A. and J.T.; investigation, M.S., D.M.C., A.R.T., B.P.-C., P.F. and J.T.; resources, S.G. and H.S.-D.; writing—original draft preparation, M.S. and J.T.; writing—review and editing, M.S., D.M.C., A.R.T., R.A. and J.T.; visualization, M.S.; supervision, J.T. and R.A.; project administration, J.T.; funding acquisition, J.T. and R.A. All authors have read and agreed to the published version of the manuscript.

Funding: This work was financed by FEDER—Fundo Europeu de Desenvolvimento Regional funds through the COMPETE 2020—Operacional Programme for Competitiveness and Internationalisation (POCI), Portugal 2020, and by Portuguese funds through FCT—Fundação para a Ciência e a Tecnologia/Ministério da Ciência, Tecnologia e Ensino Superior in the framework of the project POCI-01-0145-FEDER-031340 (PTDC/SAU-PAR/31340/2017) and the research unit no. 4293. The work also received funds from the French Government's Investissement d'Avenir program, Laboratoire d'Excellence "Integrative Biology of Emerging Infectious Diseases" (ANR-10-LABX-62-IBEID). J.T. is an investigator funded by national funds through FCT and co-funded through the European Social Fund within the Human Potential Operating Programme (CEECIND/02362/2017). M.S. and A.R.T. are funded by FCT individual fellowships SFRH/BD/133485/2017 and SFRH/BD/133276/2017, respectively.

Institutional Review Board Statement: All experiments carried out on mice were approved by the i3S Animal Welfare and Ethics Committee and the Animal Care and Use Committee of Institut Pasteur and conducted following the statements on the directive 2010/63/EU of the European Parliament and Council.

Informed Consent Statement: Not applicable.

Data Availability Statement: The original contributions presented in the study are included in the article, further inquiries can be directed to the corresponding authors.

Acknowledgments: We would like to thank the Sanger Institute for providing the PlasmoGEM construct PbAC02-99h11 [32,33]. The following reagents were obtained through BEI Resources, NIAID, NIH: (I) Plasmid pL0001, for Transfection in *P. berghei*, MRA-770, contributed by Andrew P. Waters; (II) Plasmid pL0007, for Transfection in *P. berghei*, MRA-776, contributed by Andrew P. Waters; (III) Hybridoma 3D11 Anti-*Plasmodium berghei* 44-Kilodalton Sporozoite Surface Protein (Pb44), MRA-100, contributed by Victor Nussenzweig. The authors also acknowledge the support of the following i3S Scientific Platforms: Animal Facility, BioSciences Screening, Advanced Light Microscopy, Histology and Electron Microscopy, all members of the national infrastructure PPBI-Portuguese Platform of BioImaging (supported by POCI-01-0145-FEDER-022122).

Conflicts of Interest: The authors declare no conflict of interest. The funders had no role in the design of the study; in the collection, analyses, or interpretation of data; in the writing of the manuscript, or in the decision to publish the results.

References

1. World Health Organization. *World Malaria Report 2021*; WHO: Geneva, Switzerland, 2021.
2. RTS,S Clinical Trials Partnership. Efficacy and Safety of RTS,S/AS01 Malaria Vaccine with or without a Booster Dose in Infants and Children in Africa: Final Results of a Phase 3, Individually Randomised, Controlled Trial. *Lancet* **2015**, *386*, 31–45. [CrossRef]
3. Ménard, R.; Tavares, J.; Cockburn, I.; Markus, M.; Zavala, F.; Amino, R. Looking under the Skin: The First Steps in Malarial Infection and Immunity. *Nat. Rev. Microbiol.* **2013**, *11*, 701–712. [CrossRef] [PubMed]
4. Trieu, A.; Kayala, M.A.; Burk, C.; Molina, D.M.; Freilich, D.A.; Richie, T.L.; Baldi, P.; Felgner, P.L.; Doolan, D.L. Sterile Protective Immunity to Malaria Is Associated with a Panel of Novel P. Falciparum Antigens. *Mol. Cell. Proteom.* **2011**, *10*, M111.007948. [CrossRef]
5. Peng, K.; Goh, Y.S.; Siau, A.; Franetich, J.-F.; Chia, W.N.; Ong, A.S.M.; Malleret, B.; Wu, Y.Y.; Snounou, G.; Hermsen, C.C.; et al. Breadth of Humoral Response and Antigenic Targets of Sporozoite-Inhibitory Antibodies Associated with Sterile Protection Induced by Controlled Human Malaria Infection. *Cell. Microbiol.* **2016**, *18*, 1739–1750. [CrossRef] [PubMed]
6. Kappe, S.H.I.; Noe, A.R.; Fraser, T.S.; Blair, P.L.; Adams, J.H. A Family of Chimeric Erythrocyte Binding Proteins of Malaria Parasites. *Proc. Natl. Acad. Sci. USA* **1998**, *95*, 1230–1235. [CrossRef] [PubMed]
7. Blair, P.L.; Kappe, S.H.I.; Maciel, J.E.; Balu, B.; Adams, J.H. Plasmodium Falciparum MAEBL Is a Unique Member of the Ebl Family. *Mol. Biochem. Parasitol.* **2002**, *122*, 35–44. [CrossRef]
8. Noe, A.R.; Adams, J.H. Plasmodium Yoelii YM MAEBL Protein Is Coexpressed and Colocalizes with Rhoptry Proteins. *Mol. Biochem. Parasitol.* **1998**, *96*, 27–35. [CrossRef]
9. Kappe, S.H.I.; Gardner, M.J.; Brown, S.M.; Ross, J.; Matuschewski, K.; Ribeiro, J.M.; Adams, J.H.; Quackenbush, J.; Cho, J.; Carucci, D.J.; et al. Exploring the Transcriptome of the Malaria Sporozoite Stage. *Proc. Natl. Acad. Sci. USA* **2001**, *98*, 9895–9900. [CrossRef]
10. Ghai, M.; Dutta, S.; Hall, T.; Freilich, D.; Ockenhouse, C.F. Identification, Expression, and Functional Characterization of MAEBL, a Sporozoite and Asexual Blood Stage Chimeric Erythrocyte-Binding Protein of Plasmodium Falciparum. *Mol. Biochem. Parasitol.* **2002**, *123*, 35–45. [CrossRef]
11. Srinivasan, P.; Abraham, E.G.; Ghosh, A.K.; Valenzuela, J.; Ribeiro, J.M.C.; Dimopoulos, G.; Kafatos, F.C.; Adams, J.H.; Fujioka, H.; Jacobs-Lorena, M. Analysis of the Plasmodium and Anopheles Transcriptomes during Oocyst Differentiation. *J. Biol. Chem.* **2004**, *279*, 5581–5587. [CrossRef]
12. Singh, N.; Preiser, P.; Rénia, L.; Balu, B.; Barnwell, J.; Blair, P.; Jarra, W.; Voza, T.; Landau, I.; Adams, J.H. Conservation and Developmental Control of Alternative Splicing in Maebl Among Malaria Parasites. *J. Mol. Biol.* **2004**, *343*, 589–599. [CrossRef] [PubMed]
13. Kariu, T.; Yuda, M.; Yano, K.; Chinzei, Y. MAEBL Is Essential for Malarial Sporozoite Infection of the Mosquito Salivary Gland. *J. Exp. Med.* **2002**, *195*, 1317–1323. [CrossRef] [PubMed]
14. Yang, A.S.P.; Lopaticki, S.; O'Neill, M.T.; Erickson, S.M.; Douglas, D.N.; Kneteman, N.M.; Boddey, J.A. AMA1 and MAEBL Are Important for Plasmodium Falciparum Sporozoite Infection of the Liver. *Cell. Microbiol.* **2017**, *19*, e12745. [CrossRef] [PubMed]
15. Fu, J.; Sáenz, F.E.; Reed, M.B.; Balu, B.; Singh, N.; Blair, P.L.; Cowman, A.F.; Adams, J.H. Targeted Disruption of Maebl in Plasmodium Falciparum. *Mol. Biochem. Parasitol.* **2005**, *141*, 113–117. [CrossRef] [PubMed]
16. Saenz, F.E.; Balu, B.; Smith, J.; Mendonca, S.R.; Adams, J.H. The Transmembrane Isoform of Plasmodium Falciparum MAEBL Is Essential for the Invasion of Anopheles Salivary Glands. *PLoS ONE* **2008**, *3*, e2287. [CrossRef]
17. Leite, J.A.; Bargieri, D.Y.; Carvalho, B.O.; Albrecht, L.; Lopes, S.C.P.; Kayano, A.C.A.V.; Farias, A.S.; Chia, W.N.; Claser, C.; Malleret, B.; et al. Immunization with the MAEBL M2 Domain Protects against Lethal Plasmodium Yoelii Infection. *Infect. Immun.* **2015**, *83*, 3781–3792. [CrossRef]

18. Preiser, P.; Rénia, L.; Singh, N.; Balu, B.; Jarra, W.; Voza, T.; Kaneko, O.; Blair, P.; Torii, M.; Landau, I.; et al. Antibodies against MAEBL Ligand Domains M1 and M2 Inhibit Sporozoite Development In Vitro. *Infect. Immun.* **2004**, 72, 3604–3608. [CrossRef]
19. Balu, B.; Blair, P.L.; Adams, J.H. Identification of the Transcription Initiation Site Reveals a Novel Transcript Structure for Plasmodium Falciparum Maebl. *Exp. Parasitol.* **2009**, 121, 110–114. [CrossRef]
20. Formaglio, P.; Tavares, J.; Ménard, R.; Amino, R. Loss of Host Cell Plasma Membrane Integrity Following Cell Traversal by Plasmodium Sporozoites in the Skin. *Parasitol. Int.* **2014**, 63, 237–244. [CrossRef]
21. Pinzon-Ortiz, C.; Friedman, J.; Esko, J.; Sinnis, P. The Binding of the Circumsporozoite Protein to Cell Surface Heparan Sulfate Proteoglycans Is Required for Plasmodium Sporozoite Attachment to Target Cells. *J. Biol. Chem.* **2001**, 276, 26784–26791. [CrossRef]
22. Lindner, S.E.; Swearingen, K.E.; Shears, M.J.; Walker, M.P.; Vrana, E.N.; Hart, K.J.; Minns, A.M.; Sinnis, P.; Moritz, R.L.; Kappe, S.H.I. Transcriptomics and Proteomics Reveal Two Waves of Translational Repression during the Maturation of Malaria Parasite Sporozoites. *Nat. Commun.* **2019**, 10, 4964. [CrossRef] [PubMed]
23. Lasonder, E.; Janse, C.J.; van Gemert, G.-J.; Mair, G.R.; Vermunt, A.M.W.; Douradinha, B.G.; van Noort, V.; Huynen, M.A.; Luty, A.J.F.; Kroeze, H.; et al. Proteomic Profiling of Plasmodium Sporozoite Maturation Identifies New Proteins Essential for Parasite Development and Infectivity. *PLoS Pathog.* **2008**, 4, e1000195. [CrossRef] [PubMed]
24. Vanderberg, J.P. Development of Infectivity by the Plasmodium Berghei Sporozoite. *J. Parasitol.* **1975**, 61, 43–50. [CrossRef] [PubMed]
25. Anantharaman, V.; Iyer, L.M.; Balaji, S.; Aravind, L. Adhesion Molecules and Other Secreted Host-Interaction Determinants in Apicomplexa: Insights from Comparative Genomics. In *International Review of Cytology*; Academic Press: Cambridge, MA, USA, 2007; Volume 262, pp. 1–74.
26. Michon, P.; Stevens, J.R.; Kaneko, O.; Adams, J.H. Evolutionary Relationships of Conserved Cysteine-Rich Motifs in Adhesive Molecules of Malaria Parasites. *Mol. Biol. Evol.* **2002**, 19, 1128–1142. [CrossRef]
27. Münter, S.; Sabass, B.; Selhuber-Unkel, C.; Kudryashev, M.; Hegge, S.; Engel, U.; Spatz, J.P.; Matuschewski, K.; Schwarz, U.S.; Frischknecht, F. Plasmodium Sporozoite Motility Is Modulated by the Turnover of Discrete Adhesion Sites. *Cell Host Microbe* **2009**, 6, 551–562. [CrossRef]
28. Fernandes, P.; Loubens, M.; Le Borgne, R.; Marinach, C.; Ardin, B.; Briquet, S.; Vincensini, L.; Hamada, S.; Hoareau-Coudert, B.; Verbavatz, J.-M.; et al. The AMA1-RON Complex Drives Plasmodium Sporozoite Invasion in the Mosquito and Mammalian Hosts. *bioRXiv* **2022**, bioRXiv:474787. [CrossRef]
29. Bogale, H.N.; Pascini, T.V.; Kanatani, S.; Sá, J.M.; Wellems, T.E.; Sinnis, P.; Vega-Rodríguez, J.; Serre, D. Transcriptional Heterogeneity and Tightly Regulated Changes in Gene Expression during Plasmodium Berghei Sporozoite Development. *Proc. Natl. Acad. Sci. USA* **2021**, 118, e2023438118. [CrossRef]
30. Ruberto, A.A.; Bourke, C.; Merienne, N.; Obadia, T.; Amino, R.; Mueller, I. Single-Cell RNA Sequencing Reveals Developmental Heterogeneity among Plasmodium Berghei Sporozoites. *Sci. Rep.* **2021**, 11, 4127. [CrossRef]
31. Franke-Fayard, B.; Djokovic, D.; Dooren, M.W.; Ramesar, J.; Waters, A.P.; Falade, M.O.; Kranendonk, M.; Martinelli, A.; Cravo, P.; Janse, C.J. Simple and Sensitive Antimalarial Drug Screening in Vitro and in Vivo Using Transgenic Luciferase Expressing Plasmodium Berghei Parasites. *Int. J. Parasitol.* **2008**, 38, 1651–1662. [CrossRef]
32. Godiska, R.; Mead, D.; Dhodda, V.; Wu, C.; Hochstein, R.; Karsi, A.; Usdin, K.; Entezam, A.; Ravin, N. Linear Plasmid Vector for Cloning of Repetitive or Unstable Sequences in Escherichia Coli. *Nucleic Acids Res.* **2010**, 38, e88. [CrossRef]
33. Gomes, A.R.; Bushell, E.; Schwach, F.; Girling, G.; Anar, B.; Quail, M.A.; Herd, C.; Pfander, C.; Modrzynska, K.; Rayner, J.C.; et al. A Genome-Scale Vector Resource Enables High-Throughput Reverse Genetic Screening in a Malaria Parasite. *Cell Host Microbe* **2015**, 17, 404–413. [CrossRef] [PubMed]
34. Janse, C.J.; Ramesar, J.; Waters, A.P. High-Efficiency Transfection and Drug Selection of Genetically Transformed Blood Stages of the Rodent Malaria Parasite Plasmodium Berghei. *Nat. Protocols.* **2006**, 1, 346–356. [CrossRef] [PubMed]
35. Ménard, R.; Janse, C. Gene Targeting in Malaria Parasites. *Methods* **1997**, 13, 148–157. [CrossRef] [PubMed]
36. Amino, R.; Thiberge, S.; Martin, B.; Celli, S.; Shorte, S.; Frischknecht, F.; Ménard, R. Quantitative Imaging of Plasmodium Transmission from Mosquito to Mammal. *Nat. Med.* **2006**, 12, 220–224. [CrossRef]
37. Rénia, L.; Miltgen, F.; Charoenvit, Y.; Ponnudurai, T.; Verhave, J.P.; Collins, W.E.; Mazier, D. Malaria Sporozoite Penetration A New Approach by Double Staining. *J. Immunol. Methods* **1988**, 112, 201–205. [CrossRef]
38. Costa, D.M.; Sá, M.; Teixeira, A.R.; Loureiro, I.; Thouvenot, C.; Golba, S.; Amino, R.; Tavares, J. TRSP Is Dispensable for the Plasmodium Pre-Erythrocytic Phase. *Sci. Rep.* **2018**, 8, 15101. [CrossRef]
39. Tavares, J.; Costa, D.M.; Teixeira, A.R.; Cordeiro-da-Silva, A.; Amino, R. In Vivo Imaging of Pathogen Homing to the Host Tissues. *Methods* **2017**, 127, 37–44. [CrossRef]

MDPI
St. Alban-Anlage 66
4052 Basel
Switzerland
Tel. +41 61 683 77 34
Fax +41 61 302 89 18
www.mdpi.com

International Journal of Molecular Sciences Editorial Office
E-mail: ijms@mdpi.com
www.mdpi.com/journal/ijms

9 783036 566979